KT-447-450

Dennis Smith has been called 'the Poet Laureate of Firefighters' by the *New York Post*. His classic bestseller *Report from Engine Company 82* was published in 1972, and was about being a working fireman on the front line of a South Bronx firehouse that was the world's busiest at that time. It sold over two million copies and was translated into thirteen languages. Dennis Smith is the author of seven bestselling non-fiction books in all, three novels, and a book for children. He is the founding editor of *Firehouse Magazine*, and lives in New York City.

www.**booksattransworld**.co.uk

Report from Ground Zero

The Heroic Story of the Rescuers at
the World Trade Center

Dennis Smith

CORGI BOOKS

The author has assigned a significant percentage of his royalty from this book to the Foundation for American Firefighters (a 501(c) 3 charity registered in New York State) which conveys contributions to the New York Police & Fire Widows' & Children's Benefit Fund; the John Redmond Fund of the International Association of Fire Fighters (AFL); the Health and Safety Fund of the International Association of Fire Chiefs; and the Congressional Fire Services Institute, an adjunct of the Fire Service Caucus of the United States Congress.

REPORT FROM GROUND ZERO
A CORGI BOOK : 0 552 150061

Originally published in Great Britain by Doubleday,
a division of Transworld Publishers

PRINTING HISTORY
Doubleday edition published 2002
Corgi edition published 2003

3 5 7 9 10 8 6 4 2

Copyright © Dennis Smith 2002

Grateful acknowledgement is made for permission to reprint an excerpt from The Cure at Troy: A Version of Sophocles' Philoctetes by Seamus Heaney. Copyright © 1990 by Seamus Heaney. Reprinted by permission of Farrar, Straus and Giroux, LLC. and Faber and Faber Ltd.

Set in Bembo by Falcon Oast Graphic Art Ltd.

Corgi Books are published by Transworld Publishers,
61–63 Uxbridge Road, London W5 5SA,
a division of The Random House Group Ltd,
in Australia by Random House Australia (Pty) Ltd,
20 Alfred Street, Milsons Point, Sydney, NSW 2061, Australia,
in New Zealand by Random House New Zealand Ltd,
18 Poland Road, Glenfield, Auckland 10, New Zealand
and in South Africa by Random House (Pty) Ltd,
Endulini, 5a Jubilee Road, Parktown 2193, South Africa.

Printed and bound in Great Britain by
Cox & Wyman Ltd, Reading, Berkshire.

This Book Is Dedicated to These 403 Brave Souls Who Went in to Help Others Get Out

New York City
Fire Department

Joseph Agnello
Brian Ahearn
Eric Allen
Richard Allen
James Amato
Calixto Anaya, Jr.
Joseph Angelini
Joseph Angelini, Jr.
Faustino Apostol, Jr.
David Arce
Louis Arena
Carl Asaro
Gregg Atlas
Gerald Atwood
Gerald Baptiste
Gerard Barbara
Matthew Barnes
Arthur Barry
Steven Bates
Carl Bedigian
Stephen Belson
John Bergin
Paul Beyer
Peter Bielfeld
Brian Bilcher
Carl Bini
Christopher
   Blackwell
Michael Bocchino
Frank Bonomo
Gary Box
Michael Boyle
Kevin Bracken
Michael Brennan
Peter Brennan

Daniel Brethel
Patrick Brown
Andrew Brunn
Vincent Brunton
Ronald Bucca
Greg Buck
William Burke, Jr.
Donald Burns
John Burnside
Thomas Butler
Patrick Byrne
George Cain
Salvatore Calabro
Frank Callahan
Michael Cammarata
Brian Cannizzaro
Dennis Carey
Michael Carlo
Michael Carroll
Peter Carroll
Thomas Casoria
Michael Cawley
Vernon Cherry
Nicholas Chiofalo
John Chipura
Michael Clarke
Steven Coakley
Tarel Coleman
John Collins
Robert Cordice
Ruben Correa
James Coyle
Robert Crawford
John Crisci
Dennis Cross
Thomas Cullen III
Robert Curatolo
Edward D'Atri

Michael Dauria
Scott Davidson
Edward Day
Thomas DeAngelis
Manuel Delvalle
Martin Demeo
David DeRubbio
Andrew Desperito
Dennis Devlin
Gerard Dewan
George DiPasquale
Kevin Donnelly
Kevin Dowdell
Raymond Downey
Gerard Duffy
Martin Egan, Jr.
Michael Elferis
Francis Esposito
Michael Esposito
Robert Evans
John Fanning
Thomas Farino
Terrence Farrell
Joseph Farrelly
William Feehan
Lee Fehling
Alan Feinberg
Michael Fiore
John Fischer
Andre Fletcher
John Florio
Michael Fodor
Thomas Foley
David Fontana
Robert Foti
Andrew Fredericks
Peter Freund
Thomas Gambino, Jr.

Peter J. Ganci, Jr.
Charles Garbarini
Thomas Gardner
Matthew Garvey
Bruce Gary
Gary Geidel
Edward Geraghty
Dennis Germain
Vincent Giammona
James Giberson
Ronnie Gies
Paul Gill
John Ginley
Jeffrey Giordano
John Giordano
Keith Glasco
James Gray
Joseph Grzelak
Jose Guadalupe
Geoffrey Guja
Joseph Gullickson
David Halderman
Vincent Halloran
Robert Hamilton
Sean Hanley
Thomas Hannafin
Dana Hannon
Daniel Harlin
Harvey Harrell
Stephen Harrell
Timothy Haskell
Thomas Haskell, Jr.
Terence Hatton
Michael Haub
Michael Healey
John Hefferman
Ronnie Henderson
Joseph Henry
William Henry
Thomas Hetzel

Brian Hickey
Timothy Higgins
Jonathan Hohmann
Thomas Holohan
Joseph Hunter
Walter Hynes
Jonathan Ielpi
Frederick Ill, Jr.
William Johnston
Andrew Jordan
Karl Joseph
Anthony Jovic
Angel Juarbe, Jr.
Father Mychal
  Judge
Vincent Kane
Charles Kasper
Paul Keating
Richard Kelly, Jr.
Thomas R. Kelly
Thomas W. Kelly
Thomas Kennedy
Ronald Kerwin
Michael Kiefer
Robert King, Jr.
Scott Kopytko
William Krukowski
Kenneth Kumpel
Thomas Kuveikis
David LaForge
William Lake
Robert Lane
Peter Langone
Scott Larsen
Joseph Leavey
Neil Leavy
Daniel Libretti
Carlos Lillo
Robert Linnane
Michael Lynch, (L.4)

Michael Lynch,
  (E.40)
Michael Lyons
Patrick Lyons
Brian McAleese
John McAvoy
Thomas McCann
William McGinn
William McGovern
Dennis McHugh
Robert McMahon
Robert McPadden
Terence McShane
Timothy
  McSweeney
Martin McWilliams
Joseph Maffeo
William Mahoney
Joseph Maloney
Joseph Marchbanks, Jr.
Charles Margiotta
Kenneth Marino
John Marshall
Peter Martin
Paul Martini
Joseph Mascali
Keithroy Maynard
Raymond
  Meisenheimer
Charles Mendez
Steve Mercado
Douglas Miller
Henry Miller, Jr.
Robert Minara
Thomas Mingione
Paul Mitchell
Louis Modafferi
Dennis Mojica
Manuel Mojica
Carl Molinaro

Michael Montesi
Thomas Moody
John Moran
Vincent Morello
Christopher
  Mozzillo
Richard
  Muldowney, Jr.
Michael Mullan
Dennis Mulligan
Raymond Murphy
Robert Nagel
John Napolitano
Peter Nelson
Gerard Nevins
Denis O'Berg
Daniel O'Callaghan
Douglas Oelschlager
Joseph Ogren
Thomas O'Hagan
Samuel Oitice
Patrick O'Keefe
William O'Keefe
Eric Olsen
Jeffery Olsen
Steven Olson
Kevin O'Rourke
Michael Otten
Jeffrey Palazzo
Orio Palmer
Frank Palombo
Paul Pansini
John Paolillo
James Pappageorge
Robert Parro
Durrell Pearsall
Glenn Perry
Philip Petti
Kevin Pfeifer
Kenneth Phelan

Christopher
  Pickford
Shawn Powell
Vincent Princiotta
Kevin Prior
Richard Prunty
Lincoln Quappe
Michael Quilty
Ricardo Quinn
Leonard Ragaglia
Michael Ragusa
Edward Rall
Adam Rand
Donald Regan
Robert Regan
Christian Regenhard
Kevin Reilly
Vernon Richard
James Riches
Joseph Rivelli, Jr.
Michael Roberts
Michael E. Roberts
Anthony Rodriguez
Matthew Rogan
Nicholas
  Rossomando
Paul Ruback
Stephen Russell
Michael Russo
Matthew Ryan
Thomas Sabella
Christopher Santora
John Santore
Gregory Saucedo
Dennis Scauso
John Schardt
Fred Scheffold
Thomas Schoales
Gerard Schrang
Gregory Sikorsky

Stephen Siller
Stanley Smagala, Jr.
Kevin Smith
Leon Smith, Jr.
Robert Spear, Jr.
Joseph Spor
Lawrence Stack
Timothy Stackpole
Gregory Stajk
Jeffrey Stark
Benjamin Suarez
Daniel Suhr
Christopher Sullivan
Brian Sweeney
Sean Tallon
Allan Tarasiewicz
Paul Tegtmeier
John Tierney
John Tipping II
Hector Tirado, Jr.
Richard VanHine
Peter Vega
Lawrence Veling
John Vigiano II
Sergio Villanueva
Lawrence Virgilio
Robert Wallace
Jeffrey Walz
Michael Warchola
Patrick Waters
Kenneth Watson
Michael Weinberg
David Weiss
Timothy Welty
Eugene Whelan
Edward White
Mark Whitford
Glenn Wilkinson
John Williamson
David Wooley

Raymond York

Acknowledgments

I want to thank the following people for helping to make this book possible: my friend Steve Dannhauser, of Weil, Gotshal & Manges, president of the New York Police & Fire Widows' & Children's Benefit Fund, and his wife, Beth, who conveyed to me what they have conveyed to thousands of New Yorkers – a share of their boundless goodwill and generosity; Tara Smith and Rosemary Rennon, who organized, corrected, and researched so professionally that this book was completed in the allotted time. My lifesaver, Theresa Grogan-Lustberg, and her staff, who made hundreds of pages of transcription appear magically and without complaint. Rob Singer, Eric Friedman, Linda Basile, Roz Goldman, and all the partners of Weil, Gotshal & Manges, who combined thoughtful concern with helpfulness to provide a congenial place to write. Dennis Duggan, William Kennedy, and Frank McCourt, whose friendship knows no indifference, have once again provided succor and sustenance. Peter Quinn, who put me up, and the FDNY library, which never let me down. Clare Ferraro, president of Viking, and Al Zuckerman, my longtime agent and friend, who planted the seed for this book and nurtured it into being. Rick Kot, who is the most inspired editor I have ever met—a man who always knows the difference in writing between what is and what should be. Brett Kelly, who fixed every problem presented. And, finally, my wonderful wife, Katina, who was strong and comforting for four months, and my family, Brendan and Elizabeth, Dennis and Jackie, Sean and Christina, Deirdre, and Aislinn, who provided the support I needed during a time that challenged all of our lives. There are many more whose cooperation and support for my efforts were vital to the completion of this book. They are all found within its pages.

Lower Manhattan

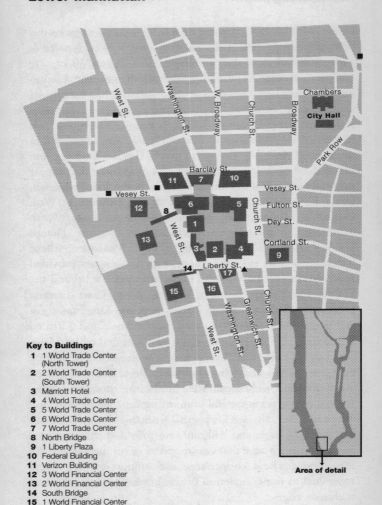

Key to Buildings

1 1 World Trade Center
   (North Tower)
2 2 World Trade Center
   (South Tower)
3 Marriott Hotel
4 4 World Trade Center
5 5 World Trade Center
6 6 World Trade Center
7 7 World Trade Center
8 North Bridge
9 1 Liberty Plaza
10 Federal Building
11 Verizon Building
12 3 World Financial Center
13 2 World Financial Center
14 South Bridge
15 1 World Financial Center
16 90 West Street
17 Bankers Trust
■ Command Center
▲ 10 & 10 Fire Station

Area of detail

Author's Note

The New York Fire Department, sometimes referred to as FDNY, is divided into a command system by divisions, two in each of the city's boroughs, except for Staten Island, where there is one. Each division is divided into battalions. There are twelve battalions in Manhattan, nine battalions in the Bronx, sixteen battalions in Brooklyn, nine battalions in Queens, and three battalions in Staten Island. Each division has four deputy chiefs, and each battalion has four battalion chiefs, who work around the clock.

There are three or four engine companies and three ladder companies in each battalion, but in the larger boroughs of Brooklyn and Queens a battalion might have five or six engine companies and four or five ladder companies.

Every engine company has a captain, three lieutenants, and twenty to twenty-five firefighters. Each ladder company also has a captain, three lieutenants, and an average of twenty-five firefighters.

The engine company is responsible for hose work – extinguishing the fire. Each working tour consists of an officer and four or five firefighters.

The ladder company (the men on the bigger truck, sometimes called the hook and ladder) is responsible for forcible entry, rescue, ventilation, and overhaul (deconstructing a fire area sufficiently to ensure there is no heat or fire in the walls). Each working tour consists of an officer and five firefighters.

The fire department also has specialized companies, like the rescue companies, one in each borough, that require special training for specialized tools and techniques for extraordinary emergency operations. There are seven squad companies designed to provide specially trained manpower at fires and hazardous materials situations, three fire boats, a hazardous materials company, a field communications unit, and others.

Report from Ground Zero

Prologue
September 11, 2001, 8:48 A.M.

For decades to come people will ask of each other, where were you . . . ?

I am sitting with Arnold Burns, the chairman of the Boys & Girls Clubs of America, talking about the needs of youth. I have known this former deputy attorney general of the United States for twenty-five years, and because I am on the board of a Boys & Girls Club in the South Bronx, I am also a member of his board of advisers. We happened to run into each other the way New Yorkers often meet, in unlikely places – in our case, sitting in a laboratory anteroom, waiting to have our blood drawn for annual exams. Suddenly, a nurse enters and exclaims that a plane has crashed into the north tower of the World Trade Center.

I look at my watch.

It is 8:48 A.M.

In Manhattan most people arrive at their offices ten or fifteen minutes early. Many have carried little white bags from local coffee shops, and are just now taking seats at their desks, anticipating the day's work. How many people are already in the World Trade Center? It must be thousands and thousands. What a terrible accident this threatens to be.

I picture the firefighters now responding to the alarm. I know what they are thinking. They are thinking of conditions. For every fire, however, the conditions can never be truly imagined until you arrive on the scene. I remember that from all the alarms I have responded to in my own life. Thousands. Each time there is the anticipation. I never knew what would meet us at the alarm location, never knew what had to be done, just as these firefighters now do not know. A report of a plane's going into a building is just the first part of the story.

I am a retired New York City firefighter, and an honorary assistant chief of department. Engine Co. 82 and Ladder 31 in the South Bronx, where I worked from 1967 through 1973, were then among the busiest fire companies in the city. I am on the board of directors of several fire-related not-for-profits, and I am chairman of a foundation that seeks to improve the health and safety of firefighters. I have never lost contact with my friends in the fire service since my retirement in 1981, and I feel still very much a part of 'the brotherhood.' Indeed, once a member of its ranks, I don't believe one ever leaves them. Friends call me 'brother' all the time.

At 9:03 I arrive home to find my wife, Katina, staring in shock at the television. Another plane has just gone into the south tower.

The second crash leaves no doubt that we have been

attacked, so many thoughts begin to flow through my mind. This is no accident. As every American must be doing at this moment, I wonder: *Who would do this? Who could pull off something this horrendous?* This is not as simple an act as parking a truck loaded with 7000 pounds of fertilizing chemicals in front of a federal building in Oklahoma City. No, this could only have been the result of the carefully plotted efforts of a group, and it would have to have been supported by a government or some organization with very big money. And people would have to have been willing to effect their own deaths. Like the kamikazes. But where were the kamikazes today? Just one place that I know of. Only the terrorists of the Middle East would do this. They had tried before, in February 1993, and now they have come back to try again.

'I will give it a little time,' I say, 'and then I have to go down there.'

I know the drill, for I have been to similar emergencies. I can picture what the scene will be like at the crash site. It is a major disaster, and will be crowded with professionals – firefighters, cops, emergency medical technicians, nurses, and doctors. During the first hour of operations fire department personnel will be rushing from place to place. To an outsider it might seem like pandemonium, but it will be controlled and orchestrated by people who know fully what they are doing.

Pumpers will be connecting to hydrants; ladder trucks will be positioning their aerial platforms; roof men, can men, irons men, and engine men will all be helping, lifting, carrying people out of the buildings. Chiefs of every rank will respond. There will be battalion chiefs who wear golden oak leafs, division chiefs who wear golden eagles, division commanders who wear one gold star, deputy assistant chiefs who wear two gold stars, and assistant chiefs who wear three gold stars. Each will have a specific job, each will direct personnel to pre-assigned duties in evacuation, rescue, or firefighting. The chief in charge, probably Pete Ganci, the chief of department, a circle of five gold stars on his collar, will be setting up his command post in accordance with pre-fire plans at the fire control panels in the lobby of the first building hit. He will begin to divide the emergency into grid sectors, and assign areas of responsibility. His aide chiefs will be generating blueprints, computer images of specific floor plans and elevator banks, prefire evacuation plans, and personnel assignments on each alarm transmitted. There will be lists of each of the three engine companies and two ladder companies that will respond to each alarm as it is transmitted. *How many alarms will there be?* I ask myself. Five alarms is normally as big as it gets. But this is not 'normally.' A situation requiring more than five alarms used to be transmitted during my years on active duty as a 'borough call.' But now the department's Starfire computer system not only makes the first five alarm assignments, but also tracks and moves the nearest ninety engine companies and forty ladder companies to the fire – a force equivalent to twenty-two alarms. Whatever it is called, it will be like a borough call, I think, and companies will come from the Bronx, Brooklyn, Queens, and Staten Island.

I wonder about the conditions. What is the heat like below the fire, going up to it? Heat rises, but there is also radiated heat to consider. How hot is it above? Will the men be able to do a search above the fire? What is the integrity of the stairwells? Can people get down from the floors above? What is the strength of the ceilings and floors?

There are thousands of people working in those buildings. Will the evacuation be orderly? Is there enough room to carry equipment like stretchers, hose, Stokes baskets, generators, ropes, and ladders up the stairwells while hordes are coming down? These are all questions that must be going through the mind of the chief of department, if he is there, or the chief in charge, in addition to assessments of firefighting tactics. Is the standpipe system in working order? Is there electricity? Are there fail-safe generators? Are any of the elevators operable to at least get the hose and heavy equipment up? How will the wind affect the burning? Do we go in from the north side of the building or the south? What is the possibility of an inverted burn, in which fire travels downward in high-rise buildings? At which floors do I place my command stations? How many chiefs do I have? Where will I put the staging areas?

The irony of 'twin' towers does not escape me, for they present two major incidents to the fire department, one in each of the buildings. I know that rescue will be the paramount concern. Will this prove to be the most serious rescue operation in our history?

Helicopters are certain to be dispatched to pick people off the roofs, but I wonder about the thermal columns. Can a helicopter get close enough to drop a rope and a man? If the fire is too great, they can't place the people on the street at risk if the chopper falls.

I can see now on the television just how much fire is coming out of the buildings. Entire floors are involved. I've been in the World Trade Center many times, and I know precisely what is involved. There are many large areas with no interior columns, lots of open space for a fire to burn freely. This would be a five alarm call even if only one building were involved. Besides the helicopters, all the special equipment companies will respond – the hazmat team, the squad companies, the rescue companies, the mask service unit, the high-rise unit, the field communications units, the foam unit, and the fire boats. There will be no need of volunteers like me. This is a fire and rescue operation, and New York has twelve thousand firefighters. Within an hour, I am estimating, the operations will have been planned, the defenses put in place, and the rescues and evacuations will be under way. Thereafter, there might be a need for assistance both in caring for the injured and cleaning up.

I am attached to the television as if every friend I ever had is about to cross the screen. The helicopter shots are mesmerizing. In my mind's eye I try to visualize what is happening in the towers as I see the flames shooting from their sides. How many have been killed by the impact of the planes, and how many have been burned by the fires? How many will be living, and how many will be dead? How many will be trapped on the fire floor, above the fire, and below the fire in collapsed areas? And the firefighters? How many flights of stairs will they have to run up, and how many minutes will it take them to reach the fire floors? How many firefighters are at this very minute racing into the buildings to get to those who need help?

How many will be caught above the fires? It looks like there are ten stories above the fire in the north tower that haven't been hit directly by the plane, and maybe twenty-five stories above the fire in the south tower. *Oh, my God,* I cry to myself. How many people are on each floor? An acre each floor. Maybe two hundred or more? Thousands. And what will they do? Did the planes take out the fire stairs? The fire will go right up the wells, like a chimney. No one will be able to get to the roof. Maybe the only way they can be saved is by extinguishing the fire and bringing ladders in. That will take hours, maybe days. It is so hot above a fire. I have been there many times in the tenements of the Bronx, hot and dangerous. And if the stairs are out, the situation will be untenable. How many are trapped at this very moment, trying to come to grips with their own end?

How many? It is a natural reaction to contemplate the numbers. But I know that I must think only of the individuals affected. Even one person is too many to be in the presence of such mortal danger, yet I know that fundamental to this terrible incident will be the numbers.

The heat must be extraordinary, generated by

airplanes with fuel-filled wings. I remember a question on the lieutenant's test of many years ago: *What is the expansion factor of a one-hundred-foot steel beam as it reaches the inherent heat level of 1200 degrees Fahrenheit?* The answer is nine and a half inches, and I try to gauge how hot this fire before me is burning. Is it intense enough to bring the steel to 1200 degrees? The smoke is first very black, indicating the burning fuel, and then white as it rises, indicating great heat. It is not a good sign. If the steel stretches, the floor will collapse, and that will only make the rescue effort more difficult.

We are at the beginning of a war, I think as I begin to change my clothes. No one could send two planes into our largest buildings without a grander plan, and I fear there will be more to follow this disaster.

I find an old Engine Co. 82 T-shirt, an FDNY sweat-shirt, jeans, and heavy black hiking shoes. This is just about the same kind of uniform I wore when I used to work on Engine 82 in the days before bunker gear, neck protectors, sixty-minute air tanks, and personal alert devices. I make certain to bring my badge with the department's picture identification card, certain there will be tight security everywhere I go today.

At 9:45 a third plane goes into the Pentagon, and I begin to make my way to a firehouse.

The notion of an enemy's attacking us in the innocence of our early morning is repellent. The Pentagon – the very heart of our military strength. How could they have stolen these planes? The question vibrates within me. How could they have gotten onto

24

our planes with guns or weapons? Why didn't someone notice what they were doing, or suspect them? But it is not like us, we Americans, to be suspicious. It is our optimism that prevents us from attributing evil intentions to others; it is our need to protect the rights of everyone that leads us to think the best of people. And that is our strength, actually; this basic freedom to walk around freely without suspicion makes America what it is. But it is that very attitude that also leaves us so vulnerable. One might argue that the price we pay to protect our freedoms is to be tolerant of strangers. All any American needs to do is to look around his community to recognize that we are indeed a nation of strangers. Our cultural diversity is so great that even our good friends and professional colleagues have cultural traits and assumptions that we do not share, or even understand. That is the way it should be in an open society. But how do I explain to myself, and to my children, that our freedoms have led to this horrible event?

It angers me; I want to know whom to blame. I am certain, deep in my heart, that this attack is connected to the first attempt on the World Trade Center back in 1993. Ramzi Yousef, who planned that bombing, and his cohort, the blind sheikh Omar Abdul Rahman, are today in federal prison. The sheikh has been in Springfield, Missouri, since 1995, serving several life sentences. Yousef was sentenced to 240 years in a maximum security prison in Florence, Colorado. I remember how defiant and remorseless they were; and that memory only serves to anger me even more.

I can shake that feeling only by picturing the fire-fighters climbing up those stairs in the highest buildings in New York. It is a tough climb in any circumstance, but they will have mask tanks on their backs and hose and tools in their hands, equipment that will make them about sixty pounds heavier. It takes a person with exceptional commitment to do this, and I know it will not be easy for them. I am thinking about this as I go down the elevator in my building. How easy it is for some of us, and how difficult for others. On Lexington Avenue the streets are packed with people. The trains have stopped, the buses are not stopping at the bus stops. I must find some way to get down to the World Trade Center.

A killing storm of terrorism has transformed our lives. We have been swept from a peaceful Tuesday into a calendar of war. New York, Washington, D.C., and a quiet green meadow in Pennsylvania have been attacked, and our fields are now strewn with the remains of heroes. All of us in the Western world are shocked, awestruck, puzzled, and furious.

There is no center to this day, no middle or end. All its remaining minutes and hours will be collapsed into that single instant at 8:48 A.M. when September 11, 2001, became the saddest day of our history.

Testimony

Retired Firefighter Jimmy Boyle

Former President, Uniformed Firefighters Association

Jimmy Boyle is, without a doubt, the most beloved man within the New York Fire Department family. He is the kind of man who looks you in the eye, and when he does you know he is interested in you. *How are you? How is your family? How are the kids, and do they need help finding a job, or the right doctor, or with admission into a college, or with a transfer from one firehouse to another or one precinct to another?* Jimmy found early on in his career that he had a gift for getting things done, and for only one reason: People trusted him and would do anything for him. He never had to tell a fabricated story or stretch the truth. His honesty and reliability got him elected to the presidency of the firefighters' union, the Uniformed Firefighters Association, in the early nineties, and he served his men tirelessly, going to every one of their company dances, company picnics, weddings, baptisms, twenty-year parties, retirement rackets, hospital beds, rest homes, and funerals in the five boroughs and surrounding counties.

Jimmy retired as president in 1994 and then worked for the fire science program of the John Jay College of Criminal Justice until, just a few months ago, he was offered a job with the Brooklyn district attorney. This morning he is sitting in his corner office in the D.A.'s operations center, and he notices smoke rising from one of the twin towers. He can see them across the East River perfectly from his chair, and they are as much a part of his office environment as the photographs of his grandchildren.

There had been an 8:35 A.M. message on his phone from his son Michael, a firefighter with Engine 33, saying that he was going out to Queens to work on his cousin Matty Farrell's campaign for City Council. It was election day, and Farrell was running in a tight race, with Michael as his campaign manager.

'As soon as I saw the fire, though,' Jimmy says, 'I knew Michael would get on the rig. I knew he would be there even though it was election day. A plane crash in the World Trade Center? What firefighter would miss that job if he wasn't off in Europe somewhere? They will all be there. It will be a tough job, all that fire so high up in the building, but the firefighters would measure their energy and get it under control, no matter how long it took.'

Then Jimmy sees the second plane coming in, and he hears a thud, followed by a huge burst of fire from the south tower. The telephone rings, and a friend explains that there is a big job in the World Trade Center. Jimmy knows he has to get down there. He meets a few fire-fighters in an engine and a truck coming up Jay Street, and ponders for a moment flagging them down. But, no, he will walk across the Brooklyn Bridge and into down-town Manhattan, not wanting to stop them for even a moment.

There is no doubt in his mind who is responsible for the crashes. He remembers that, after the bombing at the World Trade Center in 1993, the terrorists who master-minded it went to the second-floor window of the nearby J&R Music store to watch the rescue workers try to deal with damage they had inflicted. Jimmy decides to make his way to the store to see if there are again any suspicious-looking people lurking about, but the streets are packed with thousands of men and women hurrying north, rushing to get out harm's way. He instead heads to Broadway and begins walking south until he hears the inconceivable rumble of the collapsing south tower. He runs for cover into the lobby of the closest building, 250 Broadway, where a crowd that has gathered is frightened and apprehensive. Jimmy reassures them, 'It will be all right, you are safe here.'

When the cloud lifts a little, Jimmy walks to the corner of Barclay and West Streets when, suddenly, a great, growling roar is upon him. It is the second collapse, and the cloud of black air in its wake rushes at him and knocks him off his feet. He struggles to get up, and he thinks to run, but he cannot see an inch in front of him. He reaches the edge of the sidewalk and the wall of a building and begins to feel his way east. He knows that District Council 37 has an office just up the street here, and he keeps moving carefully along the wall until he comes to a side entrance on Murray Street, where he seeks shelter with other firefighters who have run there for cover.

After a few minutes he heads out into the now dissipating cloud of dust, and returns to West Street. 'I see all the burning cars, and the buried fire trucks. Everything is quiet. I get a gut feeling that Michael is dead. No one is around. I go to Vesey Street, and I see Chief Hayden. He is standing on a truck directing a bunch of firefighters. I search for the rig of Engine 33, and I find it. I look on Lieutenant Pfeifer's riding list, but I don't see Michael's name. Maybe he wasn't on the pumper when it left Great Jones Street? Maybe he got out to Queens to help on the election? But then I meet the company chauffeur, and he tells me, yes, Michael made the rig and was inside the building. But I keep searching because there is always that possibility, and I go down West and up Vesey.'

I meet Jimmy, here at the corner, where he is searching for his son. There we also run into Kevin Gallagher, the president of the Uniformed Firefighters Association, who is also searching for his son, Kevin, who is with Ladder 49.

Both young men responded to the alarm, and I think of how very hard it must be for these fathers to look for their sons, to remember everything about them, to fear that they might be dead, while at the same time

32

desperate for any sign that might indicate they are still alive.

Captain Dennis Tardio
Engine 7

Engine 7 and Ladder 1 are housed in one of the largest firehouses in the city. The Duane Street quarters in downtown Manhattan are famous within the fire department because it is here that its museum collection was housed for many years until the Spring Street firehouse, just a few blocks away, was converted into a modern exhibition building called the New York City Fire Museum. For many decades, firefighters brought their children and their schoolmates to the old museum to view its many antique steam pumpers, ancient hand-pulled hose carts, and its musty exhibits. Every child and adult who walked through this museum, I'm sure, will remember how crowded with artifacts were its rooms, and how creaky and splintered were its nineteenth-century wooden floors.

Dennis Tardio, the captain of Engine 7, has been fighting fires for twenty-two years. He is an intense, serious man, though good-natured, and runs his company with a benevolent efficiency. He is the kind of captain the line firefighters love to work with, for he is entirely 'into the job.' Today he is on the apparatus floor of the firehouse, looking over the pumper, much as a pilot examines an airplane before a flight.

It is a wide apparatus floor, maybe the widest in all of the firehouses, and he begins to walk over to the housewatch area at the center of the space when the telephone rings. The firefighter on housewatch duty answers and listens as the dispatcher reports a possible gas leak in the area of Church and Lispenard Streets. Engine 7 is being sent to investigate, and Battalion 1 is to respond as well. The housewatchman hangs up the phone and hits the loudspeaker. 'Engine goes, truck goes,' he says, then adds, 'battalion goes.'

This is a common run for engine companies in New York, and Captain Tardio thinks little of it as he climbs into his pumper. Someone has reported smelling gas, and it is probably a faulty Con Edison connection or an accumulation of sewer gas. This sort of call is typically not much of a problem, though occasionally it can be a very big one.

The intersection of Church and Lispenard is just seven blocks north of the firehouse. It is also just fourteen short blocks directly north of tower 1, the north tower of the World Trade Center.

Captain Tardio checks that the men on his riding list are on board the fire engine, and as they are pulling out of quarters he notes that Battalion Chief Joe Pfeifer of Battalion 1 will be responding directly behind him. It is all SOP. Cup of tea, as they say.

It is 8:43.

■ We get out of the pumper at Lispenard, and walk

across the street to Church to investigate the report of a gas leak. I notice that Chief Pfeifer has a television crew with him, who are making a documentary about a year in the life of a firefighter. Then there is a big noise, and the shadow of a low-flying jet crosses over us. I can see that, since it's so low, that it must be in trouble or something. It is going right into the north tower of the World Trade Center.

We don't make a determination of a gas leak. My three firefighters and I, with the chauffeur, head downtown immediately, siren and air horn. I see everything all the way. There is a lot of fire on the floors up near the top of the building.

In two minutes we are in the front of the building, and the chief orders us to begin a firefighting operation. Each of my men grabs a rolled-up hose, and I decide to take a fourth one, just in case. We have masks on, and air cylinders on our backs.

People are jumping. Crowds are coming out of the building as we enter.

How proud I am of the men. At no time does anyone turn around and say, 'I ain't going up there.' And if anyone says it, I would have no problem with it. They just keep going on and on. Nobody ever says we shouldn't go up after what we see in the lobby. Three people are burning, their clothes are burning. I know people are caught up there in that incredible heat. We start up the stairs. There are people coming down the stairs burned. Not on fire, burned. We have no idea what floor we were going to encounter the fire on, so we were asking, 'Does anybody have any floors that have fire on them?' And people were coming down from the sixtieth floor and they said no. So we know we have to go past the sixtieth floor. It is tough making all those stairs. We get to the thirtieth floor. I see Chief Picciotto from the 11th Battalion. He tells me that he just got an order to evacuate forthwith, and so we drop the hose and go down. I'm thinking, *What if I hadn't seen Chief Picciotto; what if we had been faster up the stairs and missed him altogether?* Carrying the hose slowed me up.

We make our way down, we hit the lobby. I actually – I actually joke about it. You know our famous command board that we have – our command post – that opens up, with those magnets that they put up for the building? It is just standing there abandoned.

I say, 'This is not a good sign. They abandoned the command post.'

I am totally unaware of what has happened. I just walk out, make a right, instead of a left or going straight. Just make a right turn up to Vesey. I hit that corner and I hear an explosion and I look up. It is as if the building is being imploded, from the top floor down, one after another, *boom, boom, boom.* I stand in amazement. I can't believe what I am seeing. This building is coming down. I start to run west. I get half a block, and turn around. I don't see anybody behind me. *Keep running,* I tell myself. I make another half a block, turn around, and I realize, I can't outrun this. You know, the big clouds –

I can't outrun this, so I just dive in the street and cover up. I think about going behind a car, but I am afraid something will hit the gas tank and it will explode. So I just lay in the middle of the street and just cover up. I think for sure my chauffeur is dead, because he was under the building looking up. And then I think again, if I didn't run into Chief Picciotto, who was on another radio band, I wouldn't have known that they had ordered the evacuation of all firefighters. And I would say that, one more floor higher, another thirty seconds in that building, I don't think I would be here.

It's amazing I'm here. The first thing I do now, naturally, after I made it out of the cloud of dust, is to try to find my men. I am concerned for my men, and it takes me a while to find them. But finally I find them in the crowd. We need some medical attention now; our eyes need to be washed out.

After an hour, we go back to the site. Do what we can. Dig where we can. We keep going. My men are great. We stay most of the night. Then we go back to the firehouse, but I don't sleep for three days. ■

Dan Nigro
Chief of Operations, FDNY

'You name it, he's been in charge of it.' This is what many firefighters will say about Chief Nigro, for in his more than three decades with the department he has headed the personnel division, medical division, the department of purchasing, the emergency medical services division, and the operations division (which had formerly been known as division of fire). He is as well rounded a fire chief as exists in the entire country, and he carries his expertise in a quiet, even manner. He is known for taking the time to understand both sides of any problem that is placed before him.

■ I am in my office on the seventh floor of department headquarters in downtown Brooklyn. It is around a quarter to nine, I hear a thud, and I think it might be somebody upstairs dropping a file cabinet. Just a loud thud and a little bit of a shake to the building.

I hear Pete Ganci, the chief of department, yell, 'Dan, look out your window; a plane just hit the World Trade Center!' I look out, and the north tower is clearly visible. I see gray smoke, quite a bit of it. So we instantly know this is bad. There is a group of us here in headquarters, as is always the case in the firehouse with a change of tours. Donald Burns is still here, who had worked the previous twenty-four hours. We have Joe Callan, who is just starting his tour as tour commander; Sal Cassano; and Jerry Volger – which means there are four system chiefs in the building, plus Pete Ganci and myself. We respond. Because I want to talk to Pete about the operation, I send my aide to take my own car over, and I ride along with Pete. Steve, his exec, is driving.

On the way there, from the Brooklyn Bridge, we can see the building, see the amount of smoke and fire. We realize we will have to wait until we get there to determine a course of action, but I said to Pete, 'This is going to be the worst day of our lives!'

Just judging from the number of floors that appeared to have been involved in this building, I can't think of a more serious fire that I have ever been to in thirty-two years. We arrive in just a few minutes and set up a command post, close to Vesey Street, on the west side of West Street, on a ramp – a ramp that leads to a parking garage under the World Financial Center. While we are setting up at that site, we hear the roar of a plane, which causes me to look overhead. I don't see the plane, only a blur, and then the impact comes. I know now this isn't an accident, that we are under attack. I turn around and see the mayor and a group of people, [Police Commissioner] Bernard Kerik and some other familiar faces from City Hall, who join us at this position.

We have the second fifth alarm transmitted, a fifth now for each building; Joe Callan takes charge of the north tower and Chief Barbara and Donald Burns take the south tower.

Ray Downey [chief of rescue operations] arrives and takes most of the companies that have joined us there at that post into the south tower to assist in evacuating the stairs. The order of the day is, 'Let's get the people out as fast as we can!' Ray leads a lot of units across the street, carefully, because there is the constant sound of people jumping from the north tower, and hitting the plaza below. I think if the day ended right there, it would be the worst day I have ever experienced in all my years of firefighting. I don't think anyone is prepared to have to witness this, such a large number of people choosing to jump from that north tower. It's not an easy thing to experience.

Once we feel we have the radio frequencies and the commands in place in each tower, I tell Pete Ganci I want to quickly circle the buildings in order to see what kind of damage we have. We are looking up at the west side of the north tower, and this is probably the side of the building that shows us the least.

The first plane hit straight from the north and did most of its damage on the north side; the second plane, which hit from the south, also did a lot of damage to the east side of the building, which was not clearly visible from our position.

I walk around the north tower, and from its north side I see the damage is more than we sense. I get around to Church Street, and I'm proceeding south, and again I see a lot of damage. I intend to go to Liberty, make a right, and get back to the command post, but someone stops me. Ordinarily, I would have said, 'I really don't have time to talk to you.' But he was a man I knew who used to work in our head-quarters building – a young guy, Gabe. He says to me,

'Chief, could you tell me what's going on? My wife works on,' I think he said, 'the ninety-second floor of the south tower.' I felt pretty bad for Gabe, and I said, 'Well, you know, maybe, there was some amount of time between the first and second strikes, maybe she walked down, I don't know . . .'

I'm trying to get the people out of the building, and I spend a minute or two just talking to him because it is such a personal sense of risk that he has in this. I remember when he left working for us, his wife had just had a baby. So, he is a new father and is so concerned. It is this conversation that keeps me from getting closer to the south tower. This is just the way life is . . .

I am a little past Dey Street, on Church, when suddenly the south tower comes down. I am just able to grab my aide, who is also my nephew, by the arm, and we find a doorway. It is a Starbuck's. We press ourselves to the doorway, and though things hit all around us, we come out unscathed. After the building falls, there is the dust cloud. I remember things I have seen on TV, like a volcano in which people perish from the dust, and I think this might be our fate. For some amount of time, fifteen or twenty seconds, I keep filling up with this huge quantity of dust, and I am unable to breathe.

I tell my nephew, who is getting a little agitated, 'Try to breathe as slow, and as easily as you can.' And, then, it stopped. We survived it as well.

Once everything settled a little, I still couldn't see

anything at all, like I couldn't tell if the whole building was down or not. I looked up and I saw it coming down on us, so I knew what was happening, but I didn't know if the other buildings around it were damaged, or what else was going to fall down.

So I say to my nephew, 'We are going to walk a little east and a little south, and we'll make our way back around in a little bit of a loop. And we'll find our way back to the West Street command post.' As we proceed to do this, there is that sound again as the other building falls, and another dust cloud, pushing us, pushing us forcefully down the street.

We found a building with a lot of people in it, and just stood there for a few minutes. Once the cloud lifted a little, I said, 'Okay, let's get back out there.' My nephew didn't even have his coat and helmet. In the first collapse I just got him under me, to protect him. He just had a short-sleeve shirt and pants.

Witnessing this in lower Manhattan is horrendous. So many people, dust-covered people, asking us what to do, so we just tell them, 'Keep walking, walk north and east towards the Brooklyn Bridge.' And then we proceed to find our way down to West Street.

Radio traffic is very spotty, and I don't know who has survived in the command post. At West and the Battery we meet Doctors Kelly and Prezant, who were both chased down West Street and were injured, and two firefighters, and the four of them look like the last survivors on earth. These were the first people we saw from the department after the collapses. And two members of the communications unit. So I ask them, 'Where's the command post? Who's manning the radios?' They answer, 'A command post was established just south of City Hall Park, on Broadway.' This gives me a place to head to. Chief Harring is now in charge, and he hadn't even been there when I left the first command post. This information gave me this terrible feeling that everyone I knew had been killed. I just thought, *Where's Bill Feehan, where's Pete Ganci; where are all the other staff chiefs?*

Once we get to the new position, we learn the command post was moved north, to Chambers and West. That's where they are running the now multiple fires, and the rescue attempts. I get up there and assume command. I am told that Pete is missing. I learn that Commissioner Feehan, Chief Burns, and Chief Barbara are missing. Chief Cruthers had gotten there from home; Chief Fellini and he are running various parts of the job. We move on from there. Harry Meyers comes in, Butler, other chiefs.

I am not sure how many survivors we might find. I am taken immediately by the height of the rubble that we have, for it isn't as high as I think it should be. This does not give me a comfortable feeling about our ability to find people. I have seen enough buildings collapse to know what should be left of a six story building, and what is left of a seven story or four story. It certainly appears to me that if buildings reaching into the sky collapse then it should leave a larger pile. We know about all of the levels beneath. Perhaps people were at the concourse levels, and perhaps we will find a few there, unless we are unable to get down there.

So we have some amount of hope going into the night to find people, but this is certainly nothing like any collapse any of us had ever seen and, privately at least, it was hard to be optimistic that people could have survived such horror. Publicly, everybody wanted to believe that we would find some of our friends, but looking at what was left . . .

At some point in the night I feel I have to . . . well, I have so much dust in me, and on me . . . I think I'd like to just go home, try to get as much of this off of me as I can, and get changed, because I know we are going to have a very busy week. I want to actually show my family that I'm . . . let them touch me, know that I am alive. And I do that. My car got crushed, and so I take Harry Meyers's car. I say hello to my family. I sit with them for an hour. Then I take a shower, and take another shower, and get as much dust off me as I can. I get changed into some sturdy clothes.

I then pick up Al Turi at his house, which is right near mine. He had gone with Steve Mosiello, Pete's exec, to tell Kathy [Ganci] that Pete wasn't coming home.

I pick Al up, maybe it's eleven or so, and we go back to work. We go to headquarters and organize a few things, and then sometime in the middle of the night, I go back over to the site.

Everybody did his duty that day. I'm sure I wasn't the only one who felt like joining the general populace, and walking over the Brooklyn Bridge, thinking, *I've had enough of this.* But, you come back just like the other firefighters did, and you do your job.

That night or the next day the commissioner mentioned that he would want me to be chief of department. I said of course I would. At any other time this would be the culmination of my career, being the number one man in the number one department in the world – what an achievement. But now, it felt like just a responsibility.

I feel that this department is so strong, and had the building also fallen on me, it would be just as strong today as it was. There is just a wealth of people here that are great and will move up. We lost a tremendous amount of experience and talent, but we are able to recruit such good young people. I don't want to say it's like wartime, but if you look at the record of people in the Second World War, and at their ages, you will say, wow, look at what that person accomplished, how young he was. Because they had to.

This disaster has, of course, dominated me. I wake up thinking about September 11. I go to sleep thinking about it. And, if I happen to wake up in the middle of the night, that's what I am thinking about. It is that way, but not a terribly morbid way. I have moments of terrible depression over losing these people. I consider myself very fortunate – very lucky, for some reason, to have survived. And I wouldn't begin to imagine why. I don't know if there is a way, but I can't even think of speculating on why I happened to survive and someone else didn't. ■

Dispatcher John Lightsey

John Lightsey has been a volunteer firefighter for twenty-five years in Hampton Bays, a small town on Long Island. He has also been a dispatcher for the last five years for the New York Fire Department, and the morning of September 11 is his turn to be up on the radio. He is working in the Manhattan dispatcher's office, which is located in a landmark building on the 79th Street transverse of Central Park. Dispatchers are assigned various types of duty at the start of each working tour. Today, as every day, some are working the phones with the 911 system, others the voice alarm system with the firehouses, and still others the DD desk, which is where the dispatching decisions are made – which fire company will be sent to investigate a complaint, for example, or which will respond to a report of a water leak, or a gas leak, or some other emergency that does not require a full first alarm assignment.

A call comes in reporting a smell of gas in the downtown district, and the 1st Battalion is called, as is Engine

7, and they are sent to investigate.

Not much happens in the early morning, at least not before nine. John knows that soon after that, many companies will be leaving quarters to go to company medical exams, training appointments, or regularly scheduled building inspections. They might also be attending a recently started program at the Brooklyn Headquarters Building where companies get outfitted with the newly mandated fire protective outerwear called bunker gear. For whatever reason they leave the firehouse, though, they must go on the air and inform the dispatcher when and why they are leaving quarters. The dispatcher will give them the necessary approval and note the time of departure for the records.

■ Suddenly, [at about twelve minutes before 9:00] Ladder 10, or it could be Engine 10, calls. Ladder 10 and Engine 10 are in the same firehouse, just across the street from the south tower of the World Trade Center. The officer does not wait for an acknowledgment: 'A plane hit tower 1 of the World Trade Center. Transmit the box.'

Just then a department voice alarm comes on. It is [also] someone from 10&10, saying, 'A plane just went into tower 1.'

I transmitted Box 8087, the building box for the north tower.

All of a sudden every phone in the dispatching center began to ring.

Not long after I transmitted the box, Battalion 1,

on his way over there, came on and said, 'Transmit a third alarm for this box.'

Just after that, a battalion, I think Battalion 1, said 'Transmit a fifth alarm.'

And then, when the second plane came in, we got the order to transmit a second fifth-alarm assignment. We had at least twenty-five engines and sixteen trucks, maybe five or six battalions. Here in Manhattan we dispatched everybody, every company, from 125th Street on down. We sent them right on down to the World Trade Center – everybody on the west side, we sent them to #1 World Trade Center, and everybody on the east side we sent to Box 9999, at 2 World Trade Center. There was a lot of chaos. There was so much going on I couldn't tell you exactly what the names were, or who was transmitting what. At the time, we sent everybody out there instructions that they were to bring every piece of equipment they had in their quarters.

People caught in the towers were calling in, from above the fire floors. From what they were telling me, these people were very excited. You could see what they were dealing with on the TV. Normally we would call the division chief on the air, and give him the floor numbers. We realized that no one was going to get to those people above the fire. So we had to decide, do we transmit this on the air, or just put the information on the side and transmit everything going on below the fire floors?

I knew I had the biggest event in my life when

those towers came down. No one would ever, ever even dream about something like that. Marine 6 was transmitting over the air, and everyone started yelling, *Urgent!* over the radio – tower 2 had just come down.

The worst thing was when everything got to dead silence after the collapse. For at least fifteen or twenty minutes there was dead air. We tried to raise anyone at the location. We tried to raise Division 1. We tried raising field com. Some of the companies. But nothing. We kept trying and trying. We had turned the TV off because we all had a lot of good friends in there, and it got to be too much. A few of us still can't come to terms with this. Just the thought of it all, trying to accept it.

All during the day we gave each other little hugs, and supported each other. My tour had started at 7:00 A.M., and I stayed until eight or nine that night at the radio. I stayed there for three days straight. We just spun out on the floor and got some rest.

I get teary eyed because of the people I knew who we sent down there. Just thinking about them being down there. Or feeling guilt about assigning all those companies at the same time. We didn't follow the rules; we went above the rules. We went ahead and assigned more than was necessary because of the instinct. A chief, Joe DeBernardo, told me, 'You have to look at it this way. You are not going to stop a firefighter from going in,' and he said, 'what you guys did down there in sending as much manpower as you did, you ended up saving a lot more lives than what they lost.'

Hardly anyone here talks about it. We go to the wakes and the memorial services, but we don't talk about it much. A few of us leave the room when we do the four-fives [the traditional fire alarm signal for a death in the line of duty]. They get upset. Everyone is still pretty stunned. ■

Police Officer Steve Stefanakos
Emergency Service Unit, Truck 10

■ As usual, at 7:00, Tommy Langone is down on the truck, checking the load – Hurst tools, power saws, ropes, all that. Tommy is a rappel master, and he just put together a high-rise rescue kit, a rope package with pulleys, carabiners, special ascender and descender connections. He is very proud of this and has it with him. The morning goes on, and my partner Richie Winwood and I go out on the air, and at 8:48 it begins. We have the emergency radio on, and a voice comes on and says, 'Clear the channels.'

We know it is bad, and as we go down Richie and I talk about it. We have been partners for eight years. He was the best man at my wedding, I am godfather to one of his children, so we say we will just watch out for one another. As we pull up to Church and Vesey, my first thought as I got out of the truck was to call on God. I go to the bin that holds our ropes, and I get the equipment. We put our Scott bottles on.

There is a little bit of confusion at the start as to who will do what. Tommy Langone and Paul Talty had pulled up just before us, and Tommy said, 'We are going into the building with this team, and they need a helicopter to be prepared for roof operations.'

'Maybe we should go with you, Tommy?' I say.

'No. We got it, bro. Be safe. I'll see you in a little while.'

So Tommy goes into the building with Talty, and Richie and I get ordered to go over to meet aviation, and to prepare to rappel down to the roof. We get into the back of Truck 4 and drive over to a playground where aviation is waiting for us. There are a few men from ESU there, and they had prepared the helicopter at Floyd Bennett Field, rigged it with enough rope to get down onto the roof of the building. I am worried [by] this time. I am not scared about doing a rappel, but the fire looks like it is rising up now just a few floors from the roof. It just looked dismal. At this same time there is a report that comes over the radio that there will be no roof rappels, no roof operations – it is just too risky. So we stand by for another few minutes, and then a crack, like an echo, and then it is like thunder magnified a million times, [and] the south tower comes down. I look up, and it is like watching it come down in slow motion. On the radio, there is a lot of noise and frantic calling: *Everyone evacuate the building!* ∎

Deputy Chief Pete Hayden
Commander, Division 1

Pete Hayden is one of the most respected fire chiefs I know. He is a deputy chief, but since he is also the division commander of the 1st Division, he wears one gold star instead of the usual gold eagle that his title would warrant. He has the bearing of a cardinal, very secure in his elevated position, yet humble, and every moment conscious that he is also a member of the flock. This morning he is sitting at his desk in the headquarters firehouse of the division on Lafayette Street, a sheaf of papers in his hands. It is the oldest division of the department, and Chief Hayden is responsible for some of the city's most valuable real estate, all of Manhattan south of 34th Street, an area that includes the Empire State Building, the World Trade Center, and the World Financial Center. He peruses carefully the papers before him, evaluating each manpower log for the seventeen engine companies, fourteen ladder companies, and five battalions of his division.

He is good at paperwork, although it is certainly not his first love. He went to law school for a year when he was a young firefighter, but he decided he would be happier studying fire department material than various legal codes.

Suddenly, he hears something out of the ordinary, an approaching sound, and he looks up. As the sound gets closer and closer, it begins to resemble, inconceivably, the roar of a jet plane taking off on Lafayette.

Chief Hayden jumps from his chair and goes to the window, opens it, and looks up. It *is* a plane passing by. *My goodness,* he thinks, *that plane is awfully low.* It is lower by ten times where it is supposed to be.

Because of the surrounding buildings he cannot follow its southward course.

Better get ready, he thinks, hustling down the stairs to the division van. And, then, almost as if he expects it, an explosion . . .

He sits in the car as his aide turns the key in the ignition. 'I think,' Chief Hayden says, 'a plane just hit the World Trade Center.'

■ Joe Pfeifer was the first arriving battalion chief. I came in two or three minutes after. The all-hands chief came in, Battalion 2. Then [Assistant] Chief Callan came in very quickly. The whole staff, it seemed, civilian and uniform, showed up shortly thereafter: Commissioner Von Essen, Commissioner Feehan, Commissioner Fitzpatrick, Chief Ganci, Chief Downey, Chief Burns.

Chief Callan was the tour commander working that day. He came in and took over tower 1, the north tower. And when the second plane hit, Chief Burns was actually the tour commander. He took over the second building.

We responded to the lobby command post. But all the building services were out. The elevators weren't working; there were no communications lines established. Basically, we had no communication at all. No elevator controls, no cell lines, no landlines. We were really at a distinct disadvantage. Relatively early – I haven't really seen the whole crime sequence, but at one point in time – we discussed moving the lobby command post out of the building. It was not a good place for it to be. At that time there was a discussion with all the staff officers there, and they agreed they would move to West and Vesey. What we were concerned about, in the lobby area in particular, was some of the elevators crashing down. We thought they were going to come crashing down a hundred floors and then just blow out into the lobby. If you take out the core in the elevator shafts, you have no emergency brakes. We didn't know what kind of damage was done up there, but if some of the elevators failed and came down, we would have a significant problem.

I heard that a number of people were burned in an elevator because of falling jet fuel. It could be true that that happened because there were at least six or eight people right outside the building when we came in that were badly burned. So there must have been some type of a blast somewhere. I was wondering how they got burned, and I remember saying to myself, *How did they get down the stairs so fast?* Remember, [the collision was] eighty floors up, and so they probably were in an elevator. We were getting reports of the strong odor of jet fuel on the fifty-something floor. I was thinking that maybe that jet fuel was pouring down into the elevator shafts. We were in trouble. We were in really a lot of trouble.

We had begun evacuating the north tower soon after the second plane hit the south tower. It was approximately – in fact the first call, because we have it on video, was about twenty-six minutes prior to the collapse of the north tower. We started calling people down. Come on down, come on down, get out.

Commissioner Feehan, Commissioner Von Essen, and Commissioner Fitzpatrick and their staff all left, and Chief [of Department] Ganci – that was the last time I saw them. I never saw them again. As a matter of fact, I thought at the time, when I heard that Commissioner Feehan and Chief Ganci were killed, I thought Commissioner Von Essen was dead also, because he was with them. But as fate would have it, he made a detour – he met up with the mayor and took off in another direction.

Chief Ganci and Commissioner Feehan started out [toward the new command post] and walked southward along West Street. That would put them walking towards the first collapse, south along West Street – I guess to better view the building, and then the second one came down. I was told they were covering the area on the other [west] side of West Street by the Winter Garden.

We had stayed in the lobby [of the north tower]. We were still getting a number of distress calls coming in and we were assigning units to respond to them. People [were] trapped in the elevators, [and] we were in contact with them via the communication in the elevators. You have to remember that there are ninety-nine elevators in [each WTC] building. One of the elevators was trapped right on the first floor. We didn't get those people out until much later, with all the commotion. There were a number of reports of people trapped and burned on the upper floors, which was certainly expected.

In addition to people trapped in the elevators, there were people who were in wheelchairs who couldn't get down the stairs. There was one call for a person who was lying, disoriented, on one of the floors.

What we tried to do to the best extent possible was to address all of those calls for distress, and then also to assign every five floors (say twenty to twenty-five) to a chief and a couple of companies, [who would] search those floors and get them evacuated. I think the highest floor we got to in the north tower was about the fiftieth.

One elevator opened later on in the operation, and a number of uninjured and relatively safe civilians exited. They were in the lobby, but the door never opened up, and I guess with all the noise and commotion somebody finally realized that the elevator was down. We were working with the engineer to get the elevators going, and we were trying to contact each elevator to see where it was. 'What floor are you on, are you trapped?' We were somewhat successful in that. Then we were dispatching companies up to search those floors, and to get those elevators opened if they could.

The magnitude of it was really overwhelming. The problems we were encountering, aside from the number of people that were in need and distress, with thousands of people coming down the stairs. The number of distress calls, and the fact that we had no building systems to assist us, just complicated the matter altogether.

And all of those civilians above the fire. How terrible. They had no way down, I'm sure. When the plane hit, it took out part of the core there, and the areas were blocked off. They had no exit, and when they tried to go down the stairs they were hit by heavy heat and smoke. They were doomed, I think, after the plane hit anyway. All that – I mean, they were doomed from the first.

Dozens of them jumped rather than face the fire. On my side. I mean, it could have been many, many more that I did not see or hear. But I know just on the West Street side, by the lobby, there was dozens of them came down. And they could have been jumping out on the other three sides. It sounded like bombs. Like a bomb going off, when they hit. I mean, it was huge. Nobody knew what it was. It was unnerving.

Once the south tower collapsed, the lobby of the north tower was knocked down. The sound was almost like being next to a jet plane. A tremendously loud roar, [and then] there was a great blast of hot air. That's when you were really kind of like, *Oops, this is it*. But then it subsided. We were very lucky to be alive. The cloud and force of the dust and the debris came into the lobby of the first tower. What protected us mostly was the fact that we were on the northwest corner of tower 1, so we were on catty-corner with the core of the building actually protecting us. So [when] the debris came around that core . . . we didn't get a full blast of it. But we had enough of it. We certainly had enough of it. It actually took me and threw me.

After the collapse of the south tower we lost some lives in the lobby of the north tower because part of the debris fell into the lobby where we were standing. We found Father Judge by the escalators, and though he didn't appear to have suffered any apparent physical injury, he was [clearly] dead. Someone found a board and we laid him on it, and I and a firefighter, someone from the fire patrol, and another person began to carry him out. It became totally dark. You could barely see the brightness of your own searchlight. We went across the lobby and up the escalators to the concourse level. We put Father Judge down outside and on the other side of the building, and then several others put him in a chair and carried him away, down a set of stairs to the street.

When I found Father Judge, I almost started to cry. So many good people. He was such a good man. I helped carry him out.

[By now] there were a number of off-duty firefighters coming in, and there were the change-of-tours firefighters who had gone in without radios.

That was an area where we lost a lot of control. Individual firefighters and groups [were] not necessarily reporting into the lobby command post. They were going straight upstairs. This was a problem, particularly since we had set a staging area up early on at West and Vesey, and that was not adhered to. We got out onto the street, south of Vesey, and we started to look for the [outdoors] command post and then suddenly the second [tower] came down, and it was complete chaos and confusion after that.

There really was no command structure at all – the command had been wiped out, and what you had was groups of firefighters and individuals acting on their own, trying to dig out a pile of rubble. It was only after a certain period of time that we were able to get some type of organization and group together, and say to an officer, 'Hey, you take these firefighters and you search here.' Besides, we still had so many other problems. Even after the collapse [of the towers] we had a full fourth-alarm fire going in 90 West Street, and that wasn't even covered. [Deputy Chief] Bobby Mosier was over there [and Chief Vallebuona]. 'Hey,' I said, 'I'm giving you some partial companies; I'm sending groups of guys over there, and just do the best you can.' It really was more of a defensive position to keep it in certain areas, confined. I know that Chief Mosier didn't have the manpower available to extinguish it, but he held it, and he did a good job up there with whatever he had.

I went down towards Liberty and West, and that's

where I stayed most of the time. That was on the other side from where I had been. I operated there with Chief Blaich and Chief Mosier most of the time, and that was like a sector for us. We had a number of guys, a number of firemen over there, a few hundred operating in and around the area trying to get the rescuers in those areas where we could do some surface exploration. We were concerned about additional collapses, stability of the hotel and the rest of the façade there; plus, we still had 7 World Trade Center, which was burning also. We were worried about that collapsing, and it did collapse, about six hours later. There was a conscious decision to let that building burn and just keep everybody clear.

There was some optimism early on because we did have radio communication with Ladder 6 and Captain Jonas. We were actually talking to him on the radio when – sometime in the early afternoon, probably about 2:00 P.M. – he actually came out with his men and then walked up to me and reported for duty. That was probably the happiest moment of the day for me. In fact, it definitely was.

When I looked at the carnage that was there, the devastation, and what had just transpired with the two collapses . . . I looked around, and my personal thoughts were that *nobody survived this.* This was not a six-story building coming down – this was the equivalent of an earthquake.

When the second [tower] came down, we were heading for the [crossover] bridge [on West Street]. If we had been fifty to seventy-five feet farther south, we would have been under that crosswalk so – there but for the grace of God, you know? There is no reason why I'm alive and anyone else is dead.

Much later on in the night, I realized I was exhausted. I mean, the adrenaline was thick. I think everybody there felt the adrenaline. First, there was a period of shock. It wasn't real. You looked and said, 'My God, what happened here?'

But there was adrenaline going, you know, that kept everybody alert, [doing] the searches. But later on that night, I think – I went home about midnight because I was coming back in the morning – it was starting to settle in. The enormity of the event, and the evilness of the whole thing, hit me. It was evil. There is something wrong with people who do this. Something evil.

I think everybody has some sense of second guessing. If you want to say maybe we should have done this differently? Done that differently? Looking back, you say, 'You know, we should have established a larger collapse zone earlier on – gotten the apparatus out of there – really enforced the staging area. But there was a period of time where [there was] complete unaccountability of who was going into the building because these people were coming into the building – firefighters, police officers, off-duty. And I know they weren't reporting in because I can remember saying to one of the battalion chiefs, 'I want everybody accounted for who is going upstairs –

leaving the lobby.' Because I could see guys coming in, so once we lost that control, I think, that was critical. And it was a chaotic event and difficult maintaining the kind of control we would want.

Joe Pfeifer, the first battalion chief on the scene, lost his brother. And he's working out at the site with us today. John Vigiano, Lee Ielpi, Eddie Schoales – all these guys I worked with. Al Santora's son. There are a number of firefighters that I knew the sons of, or I knew the fathers of, or I knew the brothers or know in-laws and cousins. It is really a terrible loss both professionally to the department and personally to so many people. Terrible loss.

It is very easy to say in retrospect that it was time to leave early on. But, you know, we discussed that. That was all taken into consideration. It wasn't as if we were oblivious to the possibility of a collapse – we were very aware that the building might collapse. I don't think anybody could have forecast that it was going to be so early on and so catastrophic. I think we envisioned a gradual burning of the fire for a couple of hours and then a very limited type of collapse – the top fifteen or twenty floors all folding in, and we would have [pulled] everybody back. We would move the apparatus back. We would move a few blocks north and south, east and west. Certainly, that was discussed.

In fact, [Assistant Chief] Joe Callan came in, and one of the first things he said to me was, 'Are we thinking collapse here?'

And, I said, 'Yeah, we got to. We just saw a plane hit the building. But we have a lot of people coming down and we have a lot of people in the building.' And even after we started calling them out . . . in the time between the first collapse, [and] when . . . the second building collapsed, some of the companies didn't come down. We interviewed them and talked to them, and some companies did hear us and didn't come down. [In other cases] with poor communication and handy-talkies we would call a particular battalion and ask for a response, and we weren't getting anything back.

We tried the PA system in the building. That was out. All the building services were out, which was a distinct disadvantage. Above a certain floor – that was it, they were out of communication with us. So whether or not they ever heard the call to come down, we'll never know. They may have started down, they may not have. There was a strong sense that we were going to lose some people. Well, you know. We were there. We committed. We did what we had to do.

There were so many considerations. We were concerned about a secondary device. I think we recognized early on that this was a terrorist attack, even with the first plane. That was discussed. In fact, Chief Downey brought that up.

We said to Chief Downey, 'We're aware of that, but we . . . have twenty-five thousand people in the building, and we got to get them out.'

There's a video of the lobby command post operation during the entire time. If you saw it, you would say to yourself that the operation was handled in a very professional manner. In fact, it's eerie, how calm it is. There are calm discussions and people talking about issues and information with bodies coming down all around us. A French camera crew had been making a documentary about the life of a firehouse, which I never knew was even there. The entire event is on video, and that video is over at the chief of the department's office now. It was nice for us to watch it because there is some vindication there for us. It shows we are concerned about the fact that there was a possibility of collapse, and it shows us on the video specifically calling guys down on the radio, to evacuate. We're saying: C'mon down. We timed it from the first call – it was twenty-six, maybe even twenty-eight minutes prior to the collapse of the north tower. It kind of vindicated – those of us in the lobby were saying, 'Jeez, we didn't get these guys down in time. We didn't get them out in time.' We were trying for a good twenty-five minutes to come down.

The video shows myself and Joe Callan talking with Ray Downey, and I remember saying to them, 'They're not coming down.' In other words, who's coming down to the lobby. So we weren't oblivious to the idea of a collapse, but once again I don't think anybody could have anticipated anything that catastrophic. We never expected that. We thought we had time.

Everybody out there deserves some type of medal, even the people who were there during the collapse who survived it and stayed and worked. Certainly, having survived that and [being] willing to stay for hours is a testament of [there being] a lot of good people. These guys usually say, 'That's it. I've had it for the day.' Or, 'Are you kidding me; I'm not going back there,' – or whatever. But they were right back in there again. If anything, you had to kind of pull back. We were worried about stability in a lot of the buildings. Guys were searching in front of the Marriott, and you were looking at this thing, thinking the rest of this building could start to fall down again. A lot of guys wanted to get in there. You have to tell them, 'Hey look. We lost a lot of guys here today. We lost a lot of guys. Let's not lose any more.'

You could say that the dust hasn't quite settled yet. We are still busy trying to recover our people. I don't think that is going to be discussed, naturally, for a long time. They want this to go away.

Chief Joe Pfeifer
Battalion 1

Joe Pfeifer has been a battalion chief for five years, and has been working the downtown area with the 1st Battalion. He is a thin, athletic man, with a studious air. If you saw him in a suit, you might take him for a lawyer, or a financial expert. If you mention his name in a group of firefighters, they invariably say, 'Chief Pfeifer? The best.'

On this pleasant morning he is standing in his shirt-sleeves at the intersection of Church Street and Lispenard Street. It is 8:48 A.M. and the day is already summer clear. He has been sent to this location with the men of Engine 7 and Ladder 1 to investigate a report of a gas leak. Ladder 8 pulls up as Chief Pfeifer takes a gas meter from his van, a small device with a long thin neck, at the end of which is a sensor. The chief circles a grating in the street with the meter until it buzzes, indicating a slight presence of gas. Sewer gas, maybe?

Suddenly a shadow falls over the street corner, and a firefighter, Steven Olsen of Ladder 1, looks up. It is accompanied by a heavy, roaring sound that is abnormal and surprising. *You never hear planes roaring over Manhattan,* Chief Pfeifer thinks. It is also much lower than it should be. All the firefighters are now staring upward, as is a film crew that is shooting a documentary about firefighters. It is a plane. Chief Pfeifer's gaze follows the path of the plane, and he has a clear view as it crashes into the north tower of the World Trade Center, near the top, somewhere around the ninetieth floor. There is no discussion among the fire companies as they rush into their trucks and speed fourteen blocks south.

In the chief's van, Chief Pfeifer hears the Manhattan dispatcher announce that a plane has gone into tower 1 of the World Trade Center. The chief reaches for the telephone.

'Battalion 1 to Manhattan.'

'Okay,' the dispatcher answers. It is John Lightsey who is working the microphone this shift.

'We have another report of a fire,' Chief Pfeifer says, calmly and resolutely. He knows these are public airwaves. 'It looks like a plane has steamed into the building. Transmit a third alarm. We'll have a staging area at Vesey and West streets. Have the third-alarm assignment go into that area, the second-alarm assignment go to the building.'

'Ten-four,' Lightsey answers.

He next hears on the radio. 'Division 1 is on the air.'

'Ten-four, Division 1. You have a full third-alarm assignment.' Chief Pfeifer now knows that Chief Hayden is on his way. He and Chief Hayden have been colleagues for a long time, and he is glad this is the deputy on assignment today.

When he reaches the staging area and steps into the street, he sees a large amount of fire and white smoke lifting from the top of the building to the sky. Behind him, Jules Naudet follows with his video camera. Chief Pfeifer pulls his heavy bunker pants over his trousers, and steps into his boots. He puts his helmet on, a leather helmet painted over with white enamel, which has been the traditional style of the FDNY for more than a hundred years. His own has a large front piece that reads, in large antiqued lettering: BATTALION CHIEF. There is a shower of debris falling, and he runs through it into the building, straight through to the elevator control bank, where he sets up a command post. He notices that all the twenty-five-foot-high windows that surround the lobby, at least in this northwest section of the building, have been blown out, and people are walking through the frames. They undoubtedly were broken upon the impact of the plane, which means the building must have shook violently.

A maintenance worker runs to him, saying, 'We have a report of people trapped in a stairwell on 78.'

'Okay, 78,' he answers. People have begun staggering through the lobby, badly burned, while others are running. The firefighters of Ladder 10 appear, as does an officer who reports to the chief.

'I want you to go to 78,' Chief Pfeifer says.

'What floor, Chief?' the officer asks. It is almost as if he does not want to register the floor number, anticipating the length of the climb, the weight of the equipment he has to carry, and the smoke and fire that will confront him there.

'Seventy-eight.'

Just then Chief Hayden arrives. The unspoken transfer of command is passed from Chief Pfeifer, even as they realize they are all in this together. The elevator control bank stands behind a five-foot marble wall, and Chief Hayden and Chief Pfeifer station themselves behind it. As firefighters report in, they speak to the chiefs as customers do over a high counter in a meat market. Civilians are running past the firefighters, leaving the lobby as the firefighters are entering it. The firefighters move more deliberately, carrying equipment and hose, for they know they have to conserve their energy to ascend the highest building in the city.

Chief Hayden and Chief Pfeifer are very focused and exacting. Not a voice is raised in any discussion between firefighters and fire chiefs. Chief Bill McGovern of Battalion 2 joins them. About thirty firefighters have gathered, patiently awaiting assignment, hose and tools at their feet. As their orders are issued they disappear into the building, and other firefighters appear, among them the men of Engine 10. In front of the elevator bank counter, the field communications unit begins to set up a portable command station consisting of a large suitcaselike piece of equipment on four legs, with a magnetic board where the list of entering fire companies can be logged. Just behind them, firefighters are dressing the windows, breaking off the dangerously hanging shards of glass.

Lieutenant Kevin Pfeifer and the men of Engine 33 arrive and report to Chief Pfeifer. The chief seems surprised to see his brother now in front of him, for he knew that Kevin had put in for a couple of weeks off, combining vacation days and mutual trading of working tours with another officer, just to study for the up-coming captain's test.

'The fire is reported,' Chief Pfeifer says, 'on 78 or 80.' There is none of the normal joking between them. Kevin nods, needing no further information, and then the two brothers lock their eyes together for a few moments of shared and worried concern before Kevin leads his men to the stairs. Their names can be read on their bunker coats as they depart: PFEIFER, BOYLE, ARCE, MAYNARD, KING, EVANS.

Rescue 1 arrives with Captain Hatton carrying an extra bottle of air, followed by Tom Schoales and the men of Engine 4. The affiliation of firefighters is designated by the front piece of their helmets, on which is their company and badge number. A group of police officers passes through the lobby and disappears into the stairwell, all with masks, air tanks, and blue hard hats. They are carrying large black nylon satchels filled with their tools and ropes. These are members of the elite emergency service unit of the NYPD.

Chief of Safety Turi has also arrived and speaks with Chief Hayden. Assistant Chief of Department Callan steps into the lobby and takes command, almost automatically. He speaks into his handy-talkie, 'Battalion 7 . . . Battalion 7 . . .'

Captain Jonas and the men of Ladder 6 enter and speak momentarily with Chief Hayden and then to Chief Pfeifer. Just as they take their orders there is a loud crashing sound outdoors. Debris, some large and burning, showers down from the crash floors, hitting the ground with such force that pieces shoot into the lobby like shrapnel. Someone says, 'I just saw a plane go into the other building.'

Tom Von Essen, the fire commissioner, appears, but no one goes to greet him. He is on the civilian side of the department, and his counsel will not be sought by the fire chiefs.

It is 9:03 A.M. All of the chiefs have radios to their ears, and they do not seem surprised by the news of *a second plane.* Maybe it is responsible for all the burning debris outside.

The fire commissioner walks over to the huddled chiefs and listens for a few moments, then moves away. A firefighter from Rescue 1 is standing before the command post waiting for someone to tell him the location of his company so that he can catch up with them. Another firefighter passes with a length of folded hose on his shoulder, ninety feet long and forty pounds.

The field com firefighter says, holding a phone, 'I have Battalion 2 here.'

Chief Pfeifer takes the line to speak to Chief McGovern, who has now gone up into the building. Chief Hayden says to an officer, 'I have a report of trapped people in this tower up on 71.'

The deputy fire commissioners are now on the scene and are standing off to the side with the fire commissioner. The firefighters of Ladder 1 move to the command post, receive their orders, and then quickly advance into the building. Chief Hayden is talking in the middle of the group of fire chiefs when he is suddenly interrupted by a loud report, as if a very large shotgun has been fired. Startled, he turns toward the sound and realizes it has come from the impact of a falling body, the first person to jump on this side of the building.

An African American firefighter nearby looks profoundly pensive as he moves to a wall and scans the lobby.

Deputy Commissioner Bill Feehan stops to assist a group of people in the lobby. One says, 'Please take me out of here.' Bill Feehan was once the chief of department, the highest uniformed rank, but today he knows that he is also on the civilian side. He directs them to proceed in the direction they are going, to keep moving, and reassures them. It is Commissioner Feehan's way, to keep everyone reassured.

The fire companies keep stepping up to the command post, as orderly as in a parade. They stop briefly as their company officer speaks to one of the chiefs to receive instruction.

Few firefighters are in the lobby at one time now, for they are all quickly and methodically dispatched to their work, going up, up, as high as the sixtieth floor. All the while there is an incessant whistling from the firefighter's Air Paks, the crashing of debris outside, and the frightful *bang, bang, bang* as the bodies begin to fall one after another.

Chief Hayden is in a discussion with the chief of safety.

'But,' he is saying, 'we have to get those people out!' He doesn't gesture, he doesn't flail his arms, but the tension in his voice is noticeable. All the chiefs are by now beginning to appreciate the extent of the terrible scene above them, getting worse by the second.

Engine 16 comes into the lobby as more emergency police officers and EMTs move in and out of the building. A probationary firefighter, his orange front piece denoting the short period of time he has served in the department, stands alone, waiting for his officer, looking ambivalent, as if he is trying to gauge the dangers that surround him. None of the firefighters remaining on the floor are talking among themselves. As they wait for their orders, every one of them exudes an undeniable apprehension, as if they suspect, each of them, that an almost certain doom faces this group. But for now, all they really know is that this is the toughest job they have ever faced.

A voice is heard, a chief's command, 'All units down to the lobby.'

There is another sudden crash and breaking glass, as bodies continue to fall from the very top floors, ninety floors up, traveling at 120 miles per hour. The sound has become a part of the environment, and hardly anyone reacts to it.

A field com firefighter stands before the opened suitcase of the command unit. There is a telephone connected to it and lined boxes with titles written across the open lid: STAGING AREA, ADDRESS, ALARMS, R&R, CONTROL. In the boxes are white tags for the fire companies in both buildings.

Father Mychal Judge appears, in his turnout jacket and white helmet. He stands off to the side, his hands on his hips, careful to stay out of the way of the huddles. The loud whistling in the background is now coming in series of fours, *scheeeeee, scheeeeee, scheeeeee, scheeeeee.* It is from the alarm packs of firefighters that are going off, what they call the personal alert alarm, designed to locate them when they are down.

Yelling is the preferred form of communication between men who are not sharing radio waves. Police officers in shirtsleeves and men from the mayor's Office of Emergency Management are all about, mostly on cell phones and handy-talkies. On a balcony above them a crowd of men and women moves forward in an orderly way. Where have they come from, a stairwell or an elevator? These are survivors. The *bang*s of the falling bodies in the plaza are now regular, and almost syncopated.

Father Judge's lips move in silent prayer as four firefighters pass him carrying a Stokes basket. The Franciscan is so focused in his meditation, so completely one with his inner voice as it is connecting with his God, that it appears he is straining himself and that the power of his prayer is outpacing the ability of his heart to keep up. Chief Pfeifer notices that the priest seems to be carrying a great burden in the midst of this disaster, and he is struck by his aloneness, praying so feverishly, like Christ in the Garden of Gethsemene. Finally, a man goes over to Father Judge and shakes his hand. 'I'm Michael Angelini,' he says. Father Judge is pleased to see him, for Michael's father, Joe, is a member of Rescue 1 and also the president of the department's Catholic fraternal association. 'Ahh,' Father Judge says. 'Your mother and father were recently at my jubilee celebration. I will pray for your family, Michael.' Father Judge taps him on the shoulder as well. It is the human thing. A handshake is not quite sufficient. Michael leaves the priest's side, and Father Judge is again left alone with his prayers. He has no one to counsel, no one to console, no one to shepherd. It is only him and God now, together, trying to work the greatest emergency New York has ever seen. It is obvious that Father Judge is trying to make some agreement about the safety of his firefighters.

A Port Authority police officer in white shirtsleeves has the attention, momentarily, of the chiefs. Chief Pfeifer tells a firefighter to write TOWER 1 across the panel top where they are standing, the elevator control bank. Then Chief Pfeifer begins to call in his radio for 44, when a loud noise is heard, a new and odd noise, a rumble. No, it is a roar.

It is 9:59 A.M.

Chief Pfeifer looks up to the ceiling as if to concentrate his hearing. Suddenly, he turns and runs toward the escalators behind him. Four fire patrolmen in their red helmets, from the insurance-industry-supported fire patrol, are just ahead, and Chief Callan and Chief Hayden are following closely behind.

And in the midst of the roar, a great cloud chases them, and then envelopes them, until, almost immediately, all is black. It is black as midnight. The roar has lasted only sixteen seconds, a great rumble wave, flinging a tidal wave of ash against the building. And now, just as suddenly, all is quiet. It is a profound quiet, in a profound darkness. Within this stillness the men are one with their thoughts, and asking themselves, *Am I alive?*

After a few minutes pass there is a faint rustling. People begin to call out, and flashlight beams shine vaguely through the whirling dust. It is reassuring to realize they are a sign that men in the lobby have survived. It seems as if it is snowing, thick clouds of pulverized concrete mixed with pulverized marble, and computers, and office chairs, and teapots, and . . .

A voice asks, 'Is everybody all right?'

'Yeah, I'm okay,' someone answers.

'Hey guys, we need a hand here.'

'Right. We got four guys.'

'Top of the escalator.'

'Yeah. Go.'

'Joe?'

'Pete, where are you, Pete?'

'Where are the stairs?'

'Everybody join hands. Keep together.'

'We gotta get out of here.'

'Now.'

'Here are the stairs.'

Chief Pfeifer takes the lead when the group sees Father Judge on the ground, and they all rush to him. Chief Pfeifer loosens the priest's collar, but he isn't breathing. The prayers have consumed all his energy, and he wasn't able to survive the shock of this catastrophe, this enveloping cloud. They work on him now, mouth to mouth, but the environment is dangerous. They can't just stay here, and they look for something to place Father Judge in, a stretcher, a Stokes basket. Michael Angelini finds a wooden board, and they carefully lay the priest's body on it, and they lift him up and out.

The powder begins to lift as the sound of crashing continues. Chief Pfeifer reaches a mezzanine and heads down a corridor, completely gray with ash.

A firefighter who is following close behind asks, 'Where are we headed?'

Another *bang*.

It is a long wide corridor, the pedestrian walkway over West Street, two stories above ground.

Chief Pfeifer is thinking, *The whole building came down and we have to get everyone safely out.*

Mayday, Mayday, Mayday.

He notices the blue tape strung between the columns of the corridor, as if someone had prognosticated an emergency scene.

'Battalion 1 to Division 1.'

'You okay?'

He has raised the division and learns that they have evacuated safely through a window on the east side of West Street. He turns and walks the long corridor for the third time until he comes to a set of stairs that take him down to the street.

On West Street the rigs are parked up and down the road, the apparatus of Rescue 1, High-rise Unit 3, pumpers and ladder trucks from Manhattan and Brooklyn.

The chief hears someone say, 'I can't believe this can happen.'

There is an eerie, otherworldly nature to the surface of the street. It is like the soft earth beneath the boardwalk at Coney Island. But here the ground is strewn with millions of pieces of paper. It is quiet, a heavy quiet like the quiet beneath the boardwalk at night.

Just a sliver of the Vista Hotel is standing. Could it be? Tower 2, the south tower, has come down, and the entire tract across West Street that had once been the tallest building in the world is now a field of rubble and raging fire, and . . .

And people. How many people were in that building? How many cops? How many firefighters?

And Kevin, Chief Pfeifer thinks. At least Kevin is in the north tower. He lifts his radio transmitter to his mouth.

'Battalion 1 to Engine 33,' he asks, searchingly. 'Battalion 1 to Engine 33. Lieutenant Pfeifer?'

But there is no answer.

The air above the pile that was tower 2 is a haze and now an impenetrable cloud and gold with fire.

Chief Pfeifer is standing beneath the overpass on the Vesey Street side, the overpass that crosses West Street and connects the Financial Center to the World Trade Center. He surveys the site before him, and he is planning strategy.

Two trees are lying in the street, still green, but a gray green.

He walks to Vesey Street and meets with Chief Cassano and Chief Hayden, still uncertain about exactly what has happened, for smoke obscures everything. The south tower has collapsed. But all of it, half of it?

It is 10:28 A.M. when that sound, that terrible roar erupts again, as people scream, 'Run, run!'

Everyone is rushing in a different direction; Chief Pfeifer heads toward the river, still accompanied by the cameraman who is following him around. He doesn't see that just behind him a policewoman has been hit by a piece of flying glass and has been thrown to the ground.

Someone stops to help her as Chief Pfeifer darts between a car and a truck. The cameraman hits the ground and the chief, in his bunker gear, falls on top of him, covering him from the debris crashing all around them.

As it had before, the cloud changes from a brown haze to one of complete blackness. There is just one thought in Joe Pfeifer's mind now: *Oh, God, I want to see my family again.* He closes his eyes and waits for something to crush him.

But the din stops, and the smoke begins to lift a little, lightening in color back to dirty brown. Then, suddenly, loud gunshots ring out, just down the street at Vesey and West. For one fleeting moment the thought of an invading army crosses his mind. *Now they are shooting at me.* His eyes are crusted, and he can barely see, but he makes his way back to the corner of West and Vesey.

It is almost as if snow had fallen, for the gray dust mutes all ambient sound. But, eerily, there is no sound, no screaming, not even crackling from his two radios.

He has no idea how many people are lost, or where they are, or where his brother might be. *I am out here,* he thinks, *I am more visible. My brother will come looking for me. He'll find me easily enough.* He sees the rig of Engine 33 on Vesey Street and goes to check the riding list. He knows Kevin is on the list, but, still, just to make certain, he will check everything.

Each chief has stepped out from the place he has sought for cover and heads for a different area of the devastation. They seem to develop a sectoring plan without articulating or drawing it, without the formal convention. They consult with one another by radio, remaining in their separate locations, directing work, any work that advances the idea of rescue. Get the men out. Get the people out. There is no actual command structure.

Chief Pfeifer receives a communication that several firefighters are trapped in the north tower, Chief Picciotto with Captain Jonas and Ladder 6, between the second and the fifth floors. But where could what used to be these floors even be? That operation lasts for several hours, and it is reassuring when the men are rescued along with a civilian man and woman.

The chief tries again and again to place a phone call to his wife, but every line he gets goes dead. He repeatedly circles the entire site looking for his brother. He doesn't see Kevin, or anyone from Engine 33. He tries to ignore the pain that is pressing at the center of his stomach, the feeling of everything sinking away.

It is 11:00 P.M.

He can't walk anymore and looks for a car, a bus, anything to take him back to the firehouse. But there is nothing, so he sets off on foot. On Duane Street, the firehouse is dark. He doesn't stop to change clothes but gets in his car and drives home to Middle Village. Because all the bridges, highways, and tunnels are closed, he has to keep presenting identification.

It is near midnight when he arrives home, covered, head to toe, with garbage and soot. He goes upstairs where his wife, Ginny, rushes out, hugging him and crying, followed by his two children. The four of them stand in the hall, locked in their embrace.

He is exhausted and can barely keep his injured, beaten eyes open. He thinks of his brother, his last thought of the day. *We will find him tomorrow,* he thinks. *We will find Kevin tomorrow.*

Firefighter Dan Potter
Ladder 31

Dan Potter likes to make his wife, Jean, a royal breakfast. 'She is a princess,' he says, 'and I like to take care of her. She's always rushing around, and I want to make sure she gets off to a good start, so that day I made her an asparagus omelet.'

'My sweetie is so good,' Jean says, 'and I had to get to work early that morning. My boss's bosses were coming up from Charlotte for a breakfast meeting, and I was responsible for making sure everything was ready in the dining rooms.'

Jean is a thin and attractive woman, a redhead with a kind and professional manner. She is the executive assistant for Lou Terlizzi, the head of operations of Bank of America, and she leaves you with the conviction that she can get things done. Her office is on the eighty-first floor of tower 1, the north tower of the World Trade Center.

The Potters have been married for less than two

years, and the air of the honeymoon is still about them. They hold hands when they are together. They kiss in public. He walks her to work and picks her up whenever he is not on duty.

Dan is a nineteen-year veteran of the New York Fire Department. He has recently transferred to Ladder 31, which is housed on Intervale Avenue in the South Bronx. The firehouse is much less active than it used to be during the heavy fire days of the 'Burn, baby, burn' period of racial unrest in the late sixties and early seventies, when it responded to forty or so alarms every day, a record still unmatched in the department's statistics.

Dan asked to be transferred there for a change in fire duty after working in Ladder 5, in the heart of Greenwich Village, for seven years. He had made a personal commitment to study rigorously for the upcoming lieutenant's test, and the change would facilitate a schedule determined by studying periods.

But recently Dan had heard the department was looking for volunteers to take a ninety-day temporary duty assignment to 10&10, as the quarters of Engine Co. 10 and Ladder Co. 10 are known by firefighters. It is the firehouse directly across the street from the south tower, tower 2, of the World Trade Center.

Compared to Ladder 5 in the Village and Ladder 31 in the Bronx, Ladder 10 is a slow company. The down-town area of the city essentially closes down after five each afternoon, when the hordes of financial district workers return to their homes uptown, or across the bridges and out to the boroughs. In the World Trade

Center alone some 50,000 people are employed, and the two buildings host more than 140,000 visitors daily. Aside from the patrons of Windows on the World, one of the most successful restaurants in America, only maintenance and cleaning workers, a corps of not much more than 1000, can be found in the towers late at night. Ladder 10 would be an ideal place to work for the next ninety days, Dan thought, particularly since he lived up the street from the firehouse, in Battery Park City, and the assignment would cut his traveling time to almost zero. By taking the detail he could pick up a few more hours of intensive studying each day.

This morning he has driven out to Staten Island, where he is registered for a study class to prepare for the lieutenant's test. He is seated in his chair, next to Harvey Harrell of Rescue 5, when another firefighter, in civilian clothes, bursts into the room. 'Holy God,' he says, 'did you hear what happened?'

■ So this firefighter tells us a plane went into the towers. And immediately I jump up and go to the front of the school, from which you could see the World Trade Center. I knew Jean was in there. I got onto a telephone; I had to know where she was. I just needed something because it was going to be a hell of a ride back to the city, and my heart was already in my throat. I can't really tell if it's two towers that are hit, or one; I could just see a lot of smoke. When I reached her office, I got a recording, so I said to myself, *Okay, she must be evacuating, because she always picks the phone up on the first ring.*

With that I ran. I got into my truck and was able to go through a couple of lights and right onto the Staten Island Expressway. I just flew right over the Verrazano Bridge, weaving in and out of traffic, trying to get by, because people were rubbernecking. As I came into Brooklyn I got into a police lane, where I almost got stopped, but I held my badge up to the windshield, and the cops just flagged me through. There were no buses. It was when I got into Brooklyn that I could really assess and see that both towers were burning. I took the Brooklyn Battery Tunnel into Manhattan, where they had a backup and were trying to move buses. I was able finally to get around to West Street.

I parked my truck and got out and started running up West Street, and as I am running I am noticing there are body parts in the street, different body parts. I'm figuring they are from the plane.

By the time I got to the firehouse, to 10&10, I realized how severe this was. I can now see where the fires are going as I look up, and I am thinking, *Jean has got to be on the roof. If she is on the roof, she'll be okay.* That was my whole preconception. I didn't think she could get out of something like this. I was just trying to approximate where she might have been.

Things were still raining down on the street, and it was an unbelievable, just an unbelievable sight, like a Bruce Willis movie. But this wasn't stopping.

Lieutenant O'Malley was going to start putting a

crew together. 'I'm with you,' I said. 'Let me just get my stuff, get whatever I need. Let me change.' I wanted the fire department shirt on. I ran upstairs, got that, and when I came downstairs, I didn't see him. But I knew there was a fireman who was hurt about half a block down between the firehouse and the south tower, and that four guys had gotten the Stokes and taken off [to help him].

There was an ambulance in the firehouse that was just sitting there. I looked around and I can see an Asian man with a broken leg, sitting on the floor. There seemed to be nurses there; I remember nurses. But no one was helping him.

Right in the front there was a fire on the ground. That's when I just started putting on my own bulky gear, and I saw Pete Bielfeld, from Ladder 42. He was sitting there in the firehouse, trying on different gear to get a good fit – the boots were either too big or too small.

I say, 'How are you doing, Pete?'

He looked at me, and said, 'Oh, hey 31, how are you?'

I said, 'Good. I am going to be right behind you, soon as you are ready. We'll just go.'

Pete said, 'Okay'. He had an unlighted cigar sticking out of the side of his mouth.

He started out, and I told him I wanted to grab some tools first. I knew where the tools were on an upper floor, but they were already wiped out. Captain Mallary enters the firehouse and tells me to go to his locker, which was on the second floor. He set me up with an ax and a Halligan.

When I came down, just as I get to the front of the firehouse, a guy – I believe it was another fireman – says, 'Holy crap, here it comes.'

It's coming down. With that, the nurses were just panicked. They scream, and a couple of women run towards the back of the firehouse. When I look down I see the Asian man I had seen before, and he is yelling, 'Ow!' It is just like a jet engine's roar up in front of us. I hear things crumpling and coming down. His hand is out, and he is reaching for anyone, and he is going, 'Ow, ow, ow.' I just grab him and pull him back, as far back into the firehouse as I can. Right behind us there was a little brick wall, and I laid down behind it, and just covered myself, and covered him, the best I could. And as I'm trying to cover him, suddenly all the heat, the rocks – whatever – was falling, was coming in at us like the back of a jet engine. I peeked over my shoulder and saw it flying, hitting the doors, breaking things, rolling into the firehouse. The air pressure is unbelievable, and black, just black, black smoke, or whatever it is. I laid right up against that wall, just to sort it out. And the stuff was all just passing, passing right past me. It blew all the windows out.

You could not outrun this collapse, and I thought of Pete Bielfeld, who had just left me a few minutes before to go into the building. If I was a permanent firefighter in 10&10, I might have known the engine office didn't have any tools. I knew that there were tools somewhere in the firehouse, and I don't like to be anywhere without tools, with empty hands. But that Captain Mallary brought me upstairs for tools, that was the difference that saved my life. Otherwise, I would have been right with Pete Bielfeld.

The rumble stopped, and the pressure slowed down a little bit, and I told the Asian man, 'We're going, we're going,' and I just grabbed him and pulled him to the back door, by the kitchen. I wanted to break these windows out, to ventilate the blackness, but the windows were already gone. I take him outside, and it is a black curtain there. I said to myself, *Why would there be a black sheet here?* I thought I wasn't behind the firehouse at all, and that maybe the guys had done something, put up a big black sheet of wallpaper or a black blanket or curtain. I wasn't thinking right and went to break it, and it was right then that I realize it is all black smoke, and that I am in the street.

In the back, it is like a cotton gray surrounding. Papers are blowing and there are a couple of cars that are pushed up against each other, and small fires by the cars, and just dust everywhere. I remember inhaling felt like I was inhaling cotton balls, big cotton balls. The EMS people came to take care of the Asian gentleman, and I said, 'All right, I have got to go.'

My first thought, again, is: *I want to know where my wife is.* All the while I kept telling myself, *She is up on the roof, she is up on the roof.* And now I start to think,

I don't even know what building came down. There was no time to see anything. I have to get over there and find her. But it's long blocks to the north tower. I remember a high-rise fire I had worked where the glass came down and destroyed the chief's car – completely destroyed it. I was aware that anything coming down, even a small piece of glass, can really do damage or kill you.

So what I wanted to do is go to the nearest covering, and that was Deutsche Bank in the Bankers Trust Building. There I ran into a woman who was toward the back. I told her, 'Just stay here, you're okay,' and then I came upon five or six employees of Deutsche Bank. I saw blood on a couple of them and said, 'Look, just go to the back of the building.' One guy says, 'We have a fallout shelter.' 'Then go to your fallout shelter,' I told him. 'As long as it is in the back of the building, go to your fallout shelter and just stay there.'

Then I saw there was a nursery room, and I had to check this quickly. I opened the door, and I said 'Ah.' Then I went through to the front, which is when I saw for the first time that the whole [façade of the building] was wide open. It was also the first glimpse I got of which tower had gone down, and I realized it was the south tower.

I was able to catch my breath and told myself, *Okay, she is still on the roof of the other one. Everything I know is still okay.* So I went quickly by the escalators in the lobby of the Deutsche Bank building to see if there was anybody who was needing help. As I came out I saw Mel Hazel on the street.

He says 'Hey, 31.'

I say, 'Dan Potter, how you doing, Mel?'

He goes 'Okay, good, good, good,' and then he realized it was me, that we've known each other a long time. Mel is one of those guys from Intervale Avenue, with Engine 82 and Ladder 31 in the blaze days. We just hugged and we held each other for a few minutes. I said I have to get over there; my wife is on the eighty-first floor. I have to see if she is okay. He tells me he wants to go to the command post, and we head that way together. As I am talking to Mel, a fighter jet flies overhead, something more to make me realize how huge this thing is.

Just as we decide to walk up Albany Street, two police officers suddenly come running down the street, yelling that the second building is going to go. There is a lot of rumbling. We both looked at each other. We can't see it, but I can hear it, and I think it is going to hit back into the south tower. I hear that and I say to Mel, 'My wife is on the roof of that building.'

You could hear the rumble in the same thunderous roar, and then that crash, and it all starts to come down. I grab Mel and I say I know where to go. We run back to the Bankers Trust Building and just as we get inside the front wall of that building, standing against it, everything comes tumbling down again. I couldn't believe that I was in another one of these things, but I kept thinking about Jean. Is she on the roof? Is she out? Did she leave before my call?

Everything came in through the opening where the nursery is. We were blasted right there, and we were just huddled as if we were in an upright fetal position. Everything was rushing over our heads, and hitting the back of my head, making my hair stand up. And with that rush once again, the rumbling noise stops. ■

Mel Hazel
Retired Fire Marshal

■ I have been teaching forensics since I retired a few months ago, at Mercy College, and I got here as soon as I could, just after the first collapse. I saw the south tower, and wondered how could it be, how many fire-fighters were in that building? I saw all this devastation on Liberty Street, and I kept thinking of what I learned so many years ago when I was in Ladder 31 and working with Willy Doyle in La Casa Grande, the night Larry Fitzpatrick of Rescue 3 died. I couldn't figure out why Fitzpatrick had to die, just being on a rope, doing a rescue, and then the rope breaks? But Willy Doyle told me, 'Remember this, Mel, it's part of the job.'

And I am looking at this, and saying, no, this is not part of the job, it can't be. Just the day before, on

September 10, at the firehouse of Ladder 42, I was with Father Judge, and he said to me, 'Keep the faith, Mel.' That's the part, now, that's important, keeping the faith.

There were so many fires when I got there. I was starting to report to the command when I met up with Dan Potter. I didn't recognize him at first because he is all gray with ash. He tells me his wife is in the north tower, way up on one of the high floors. 'She's okay,' I keep telling him. To me it doesn't sound good up [that] high, because it takes that much longer to get out.

We are going to go to the command post together when a cop tells us that the north tower is about to collapse. The cop runs away. So Dan and I go into the Bankers Trust Building, and just as we get there we hear a loud rumbling. To me it sounds like thousands of cobblestones smashing together. We start to run. The ceiling is already partially collapsed when we get in the building. We are looking for shelter, and some firefighters jump through one of the broken windows. The front wall of the building was open, and so everything came in – the wind, the heat, the smoke. It gets worse and worse, and I couldn't get a breather. I lost all reference; I didn't know where I was or what side was up. We got flat on the floor, searching for a wisp of air. It felt like I had a rag stuffed down my throat.

'We have to get out of here,' Dan says to me, and he nudges me. 'Let's go.' We begin to crawl through the darkness, over everything — concrete, steel, anything that came down. We are crawling, and I am wondering, *When are we going to get out of this building?* Then I realize, when the dark lifts a little to gray, that we are in the middle of the street. I'm just saying to myself, I'm glad I'm with Dan Potter, to have someone to be with in this.

I'm thinking throughout all of this. I don't know what hit me, and the guys in the firehouse will never know what has hit them, this thing is so big. ∎

Firefighter Dan Potter
Ladder 31

∎ Somebody tells us that the command post got wiped out. We have no radios, so we don't know for sure if [that is true]. With all the dust, it is still like we are in a snowstorm. It is strangely quiet. And then somebody would come by and say something to us. One guy says we must have lost one hundred guys at the command post. We didn't know where the command post was.

I can't stop thinking that Jean was on the roof. In this kind of emergency, would she go to the roof? It's faster than going down. I just feel empty and desolate, I don't know what to think, it just feels like my head is spinning.

I always told her, because she was always

concerned with [being in a] high-rise, just go to the stairs. Get to the stairs, and come down, don't go to the elevators. I said you'll be all right. But with this situation, I mean all bets are off.

I thought then to go home. I could call around. There was no point in being in the streets. I said good-bye to Mel, and headed home. But I was feeling so down, I saw a bench and just sat for a minute, to try to think of what I am going to do here. What's my next step? You know, I can't do anything. I have got to get some sort of composure. I just felt so empty, I guess my thoughts were with her, such a sweet woman like Jean. Why, why, how can this happen to such a sweet woman, not a mean bone in her body? And a photographer came over and began to take pictures. I told him to stop, and waved him away.

I went inside my building and got one of the custodians to help me. I then broke my door down and I was able to get to the phone, where there was a message, just one. It was my dad telling me that Jean had called.

I called him. Dad is crying, and tells me Jean is at Engine 9, Ladder 6 on Canal Street. He said, 'You know where that is?'

I said, 'Yeah, Dad, I know where that is.' My father forgot for a moment that I had worked for so many years down here at Ladder 5.

I take the back streets up and around, through Pearl, and the Bowery, then up to Canal Street to the firehouse. And there she was. In Chinatown. She was all gray and caked up, like a china doll, and wet. She was still all wet from something. I was still all gray, completely gray. We were a match, another match made in heaven. I think she was more surprised when she saw me, because she never thought I was going to be there.

I said, 'Where do you want to go?' Her parents live in Pennsylvania, and Jean suggested we go there. I said, 'Let's go; I'll take you there. We can clean up. There is no way we can get back into our apartment.' I think she just wanted to be with her parents where she knew it would be safe. So we got in the pickup, still covered with the gray dust.

I went back to work the next day. I wanted to get back to the firehouse. I had a pretty bad pain in my back, but I wanted to be on the piles. I worked in the pit for the next twenty-four hours. Then I brought Jean back to the apartment, for I wanted to get her situated. I had to get her together first. We had to evacuate our home now, and that wasn't easy. We went to a hotel that was taking in people like us. We had some things with us, but we had to leave everything there when there was a bomb scare in the Empire State Building. And so we [next] went to the Helmsley Hotel, and they took us in there. We walked in looking like vagabonds. We left everything behind at the first hotel, and didn't even have a tooth-brush or a change of socks.

The following day I couldn't move. My back was hurting too much, and so I called in sick.

The next day there was a list that came out with all the guys that were missing, and I looked at it real quick. There were so many names I knew — every name I read I felt like I was getting punched in the face. There were seventy names that I knew well on that list. When I went to the 10&10 the next day they were talking about a guy who left his keys, his wallet, and a note to his wife in his locker. That was Pete Bielfeld of Ladder 42.

His wife subsequently called me about a week later because I was the last one to have seen him. I told her I had seen Pete, just before he went into the building, that he had a big, beautiful cigar in his mouth, a brand-new cigar, and that's how he left the firehouse.

Through all that day, 9/11, there was so much going on, I didn't know who was on first, and who was on second. I didn't know this would be Pete's last run out the door, and the guy I was sitting with at fire tech, Harvey Harrell, too. I went to my truck, and he went to Rescue Company 5 to get his gear, and he got on that truck. I passed them on the way to the city that morning. And he's missing. And my good friend Brian Hickey of Rescue 4. And I heard about Captain John Vigiano losing his two sons. I worked with him in Brooklyn. He's a marvelous man, very professional, very smart, and my heart breaks for him. I know he's at the site. He's there until he brings his boys home.

I know the Higgins brothers. I know Joe Higgins,

Mike Higgins, Bobby Higgins. I did a detail down at Ladder 103 for a while, and worked with them. Their brother Timmy is missing. And their father is a retired captain as well. It is so sad. ■

Jean Potter

■ I loved going to work. It was such a wonderful group of people. I absolutely loved my job. The space [in the north tower] was just beautiful. We were in the clouds, and on a rainy day you could practically see the rain forming before your eyes.

But the building always made me a little nervous. It would take four or five minutes to go up on the elevator, to the seventy-eighth floor, and then transfer in the sky lobby to an elevator that would go to 81. I never felt the building sway, but I always had a picture of the towers exploding. Every so often this picture would come into my head. And I would think, that's bizarre. When Dan was detailed down to Ladder 10, it was the first thought in my mind. I hoped he was not coming back downtown because something terrible was going to happen.

I was at my desk, reading e-mails. The breakfast meeting was going well on the other side of the floor in a conference room, and all of a sudden we felt this impact and the building literally shook. Things started falling from the ceilings, and there was a smell of flames, smoke, something – very strong smell. Everything rocked; the building rocked. The light fixtures were swinging back and forth. I think, *I'm on the eighty-first floor, and something terrible has happened to us. Something terrible.* My voice said, *This is not your time; we are with you, and your brother is with you.* I lost my brother two years ago. I get chills when I say this. That's the first thing that came to me as soon as that building was hit.

Everybody just jumped up. I was going back to get my purse and I had a stair buddy, Ben. He grabbed my hand and said, 'You're going.' He pulled me along, and I mean, minutes really mattered.

We immediately proceeded to the stairs, and I'm thinking, *Dan.* We had no instructions. You know, you go through fire drills, and I was the floor fire warden. I'm supposed to call downstairs for instructions. In a situation like this you don't look to call downstairs to see what's going on. We just knew we had to get out. Everyone was very calm. We heard we were hit by a commercial plane. We did not know it was a terrorist attack. That was also something that saved us because no one was panicked.

Our staircase was pretty good. We were in the opposite section of the building where it hit. At one point they made us cross over to another sky lobby on 44. We got to that sky lobby, which has windows, and we just stood there. They [told us to] stay here.

That's when all of a sudden – I don't know if it is when the second plane hit, or just that we heard another explosion. We can see out the window, [where there is] flame and debris flying. We ran back to our original staircase and proceeded down.

We moved pretty fast in the beginning. We got slowed up as we got farther down by the many people coming in the stairs. The emergency lights were on.

People were even making jokes. We were completely calm. I prayed – I prayed the whole way down because I didn't know what was going on. I prayed for everyone. It was just an unbelievable situation. I guess we started seeing the boys come up in the twenties. It was the fire department that [gave us] a sense of relief. *Okay, well, they'll figure out what is going on.*

I also met Vinnie Giammona, a man Dan worked with in Ladder 5. I grabbed his arm, and I said, 'Vinnie, be safe.' And I prayed for him.

As we were getting to the second or third flights, people were getting slow. But after walking eighty-one flights your legs are [tired, and] your mind is telling your body to move, and I just started yelling, 'Go. Move. We're almost out of the thing.' I don't know why, but I had this sense of urgency. 'Let's go. Move it. Move it. We're almost there. Go.' I was screaming. And that's not me. But there were literally minutes to spare.

And then, when we got out of the World Trade Tower, we were taken down into the concourse where water was pouring from the ceilings. There was glass and water all over the floor, and I'm glad I hadn't taken my shoes off. People had been saying, 'Women, take your heels off.'

We were completely soaked, and they took us up and out by Waters, [where] there is an esplanade. I began walking toward Broadway. In the middle of the block I saw one of our doormen. I saw for the first time that both towers were on fire, and I said, 'What? Richard, what's this? What is going on?' There are certain pictures that I will never forget, certain feelings. The initial hit I will never forget. When I came out and turned around and I saw both towers flaming, [that] I will never forget. Walking to Broadway I was talking to a police officer. 'I have to get word to my husband,' I say to him. 'He's going to think I'm in there. He's a fireman. I have to get word to him that I'm okay.'

Now, I am out of the building literally five minutes, and I have walked one block, and then the building collapsed. All of a sudden, I hear rumbling and people screaming.

Oh, my God! I thought, *I'm not out of this yet.*

I remember just looking at the ground, thinking, *Maybe it is my time.* And this cop suddenly grabbed me by the arm and pulled me into the Dey Street subway station, [and we] went downstairs. All that cloud came in, rushing in, thick stuff, and the people kept walking. And then I met someone I worked with, who took my hand, and we were walking deeper and deeper into the subway. I was uncomfortable with that, I said, because we could get trapped in here. And as soon as the noise stopped I left him, and I walked out. I went to the right, and then I just kept walking. People were yelling at me to cover my mouth. I was walking toward Chinatown, towards the seaport. And I can't see. I am in this cloud – it wasn't dark, black was closer to it, and then it was a gray cloud. And I was soaked [because] the water had been just pouring. I had slacks on; my pants were wet and still had pieces of broken glass on them.

I'm thinking that Dan is okay. A cop told me to go across the street somewhere, and again I was very disoriented. I just kept walking, and then the other tower collapsed, and people started running again, now up past the Brooklyn Bridge. I met an FBI person who was great. She helped me call Dan's partner. But, again, we couldn't get through. All I had to say was my husband is a fireman, I have to get through, and the FBI and the cops with phones were wonderful people who were helping me. And we couldn't get a line out.

Then I met [a man named] Jamie, and he gave me water and put a chair outside of his house. He brought the chair out because I was afraid to go into his house. He had just stopped me and asked, 'Do you need something?'

I said, 'Phone? Do you have a phone? And water.' He gave me water. And I got through to 31. And he said, Jean, we want to bring the firefighters a sweet basket, and we brought the guys in Ladder 6 and Engine 9 pastry to say thank you. Jamie took me there, and I was there for a few hours. And Jamie said, 'Listen, I thought you were so distraught I had to do something.' He has me on video walking down the block, so that would be scary to see.

If I had known for one split second that Dan was headed to the firehouse, for 10&10, that is where I would have gone, even if they told me not to. Really, Jesus was with both of us. If I would have gone in another direction, and if Dan hadn't had a class and was at home, he would have been the first one in the building. We live just one block away from the south tower.

For us, it is a miracle, and we are so grateful. We are so upset, and so heartbroken for everyone, but for us it's such a miracle that we were both there, and we are now both here.

So many good people are missing. We had just had dinner with Brian and Donna Hickey a couple of weeks before. What a beautiful couple. I would never have thought he would be there, because his firehouse is in Queens. They were so wonderful together, a beautiful love story. They had just moved into their dream house, a house that they have always known in the neighborhood that was never for sale, but all of a sudden it came up for sale, and they got it.

I started to learn all these things just two days later. We were at the hotel, the Park Lane on Central Park South — we arrived there at 11 o'clock at night — with a T-shirt. All our bags were in another hotel because we had to get evacuated. They took care of

101

us, which was another amazing thing, all of these little blessings. I had no credit card or anything, no bags, and they gave us robes and checked us in. And when they found out our story, one of the head guys, John Moore, came over to us. We were in the lobby, and he says, 'Excuse me? Are you a fireman?' And Dan said yes. And he said, 'Well, you are staying here free of charge. You are not to worry about a thing.' We started to cry.

We went to the apartment when we came back, picked up some things, because that was such a sight. Oh my God, it was the day after. Battery Park City was inches of this soot, people, buildings. Our apartment was covered. There were shoes thrown all over. Papers. Baby carriages. Our beautiful building was black. It was so frightening. The windows were open just a little bit, but it was unbelievable. Covered. We had to get a professional company in to get our apartment cleaned. And then it took me two days to finish up.

It has been a lot. I think it was probably good that we were both in it, and experienced the same thing. We talk about it to each other. So that's very helpful. We just get a lot out. And for some reason, I think we're doing better than people who weren't in it, and viewed it from afar. It is strange. For the two of us to have been a part of this history-making event is a load to carry, a burden. We talk about it, but first we just talk about the load.

And now we are looking to the future. We'll get through it. We have good people all around us. I feel so specially blessed since I met Dan that I have been brought into this world of firefighters – to get to know these men and their families. It is just such a special blessing to be around people like this. ■

Joe Dunne
First Deputy Police Commissioner

Joe Dunne might well be the tallest man in the police department. Everyone in the Dunne family is tall, and Joe's brother, Tom, who is a vice president of Verizon, likes to say that he is, at six four and a half, almost a full half inch taller than his younger brother. They are a Brooklyn family, and both brothers went to St. Francis College. As in most Irish families in New York in which the siblings are all successful, they are proud of one another.

Joe is among the few men in the police department history who have risen through every rank on his way to becoming the first deputy. He was until a year ago the chief of department, the highest ranking uniformed police officer in the country, when Mayor Giuliani asked him to take the first deputy commissioner's office. Joe accepted, and now his friends say, out of his earshot, that the police department couldn't run without him. Joe is a modest but a very tough man, a spirit and demeanor he learned during summers on the athletic fields of Rockaway, where he grew up just a few streets away from his friend Pete Hayden, the chief of the fire department's first division.

On the morning of September 11, Joe got up earlier than usual to vote in the Democratic primary election. He noted that he was getting the first taste of fall, that it was clear, that the temperature was just a little bit down from hot, and the humidity was low – that it was just a beautiful New York day.

After voting, his driver took the Midtown Tunnel into Manhattan, as they did most mornings, or whenever he did not have a meeting scheduled somewhere in one of the other boroughs. Joe's large office in 1 Police Plaza, overlooking City Hall, is no more than fifteen minutes away. He came out of the tunnel on the Manhattan side at just about a quarter to nine, as he listened to the chatter on the police radio, on the special operations division frequency, which is reserved for citywide emergency services. Any serious jobs are broadcast over that frequency, which makes it a better choice than listening to the individual police division frequencies when he is traveling. That one channel gives you all the action.

On that morning he was riding in an unmarked Winstar, because he had recently broken his Achilles tendon and his left leg was in a cast from the knee down. The Winstar gave him room to get in and out of the car, and to stretch his big frame out. He had spent nearly three months in the cast and the Winstar, and he was looking forward to getting his old car back at the end of the week when the cast finally came off.

He takes a call from a friend in his office, about a

minor and forgettable administrative issue. Meanwhile, he is thinking that, since it is such a bright and beautiful day, he should put his sunglasses on. He loves those sunglasses, and at $125 they are the most expensive pair he has ever owned. He remembers that he left them in the glove compartment.

But he suddenly puts the sunglasses out of his mind, for after thirty-two years on the job, he has developed a cop's ear for the radio. He picks up something, a tension in the voices ordering cars to the World Trade Center; after apologizing to whoever it is on the phone, he hangs up.

He turns to his driver, and asks, 'Dennis, what's going on?'

■ As God is my judge, the first thing I said was that this was a terrorist act, and that we better get down there fast. I just knew it in my bones. We headed down 2nd Avenue like the hammers of hell, trying to get down to the Center, and as we were approaching it you could see that the north tower was ablaze and smoke was coming out of it like a chimney. We then heard it was a commercial plane and that it had hit someplace in the high eighties. My objective was to get to the Office of Emergency Management, which is on Vesey Street, building 7 of the WTC. We worked our way down through the maze of fire trucks and police cars and people staring at the crash. I pulled up near #7 and I was told that the building was being evacuated. So I made a U-turn and went up to the corner of Vesey and Church. I stopped the vehicle because there was fire hose in the street, and some of our police vehicles were there.

I got out of my van and tucked my crutches under my arms. I told my driver to get a jacket out of the car that would identify me. It had my name on it, and people would know who I was. Because I travel back and forth to work in a suit I don't wear a uniform. Just as Dennis was pulling up the hatch in the back of the vehicle, I heard a huge noise and explosion.

I assumed it was a secondary explosion in the north tower, not knowing that it was in fact the south tower being struck by the second plane. Almost immediately debris was flying through the air from the explosion and crash in the south tower. One piece fell through the hatch and shattered the window, and other pieces were falling on the car and all around me and Dennis as well. I was able, on my crutches, to run underneath the overhang of building #5, which is right on the corner of Church and Vesey. That building provided some protection for us, and we flattened ourselves against it until the falling debris stopped – I guess the better part of a minute or so. We saw an airplane tire fall to the ground on fire, and a lot of flaming material all around. I worked my way around then with Dennis and a couple of off-duty cops.

We went around the corner and into the lobby of building 5, though I didn't know at the time it was building 5. People were streaming out, and we began helping them. They were in shock, and we were just directing them out. There was another group of police officers there in uniform, and we formed a line on either side of the door, and we told the people to go east, go north, you're safe now. We were also telling them to get off the damn cell phones. Everybody had a cell phone in his ear. I guess they wanted to call people to tell them that they were safe, or that something had happened. But I knew that becomes a problem for communication, so I told them, please, get off the phone. Pay attention to what you're doing.

We did this for about fifteen minutes or so. Then I noticed that the chief of the department, Joe Esposito, had arrived at the corner of Church and Vesey, which became one of our forward command posts. So I went over – hopped over, actually, with my crutches. The chief of patrol, Bill Moran; Barry Maun, the assistant director of the New York office of the FBI and his aide, Ken Maxwell; and several emergency service vehicles were also there. It's at that point in time that I learned that it was a second plane that hit the building.

I said to Barry, 'How many planes do they have?' – meaning the terrorists.

'Joe,' he said, 'I don't know. We think two others.'

I said, 'Where are they going?'

'We don't know.'

I said, 'Do we have air cover?' And as I'm talking to him I can't believe that I'm standing in downtown Manhattan, New York City, speaking to the New York director of the FBI, asking him if he could get us some air support. Because we had already begun thinking of the other buildings in the city as potential targets, and we had already ordered an evacuation of [places] like the Empire State Building and of all high-rise [structures].

Our emergency service people were [already] organizing themselves into rescue teams. There were several emergency service unit trucks at the corner, and they were very methodical in what they were doing: putting on their protective gear, their air tanks, helmets. In fact, a couple of emergency men came up to me and said, 'Listen, Boss, put a helmet on.' This was a good idea, and I told some other cops to put helmets on, [including] Lieutenant Terri Tobin. I told others, 'If you're not needed, get out of the area.' I was concerned for their safety because [there was still] stuff coming down, and it wasn't getting any better. It was getting worse.

At that point in time, we were standing at the corner, just looking up at the building. At first, I thought I was looking at debris falling, and then it became painfully clear it wasn't debris. It was people falling, or jumping, and they were jumping by the dozens from what I could see. It was horrific and I'm looking at them. It was so painful to watch.

I turned to Esposito and asked if we were going to put a helicopter up on the roof. He said, 'We can't; there's too much smoke up there,' and I told them [not to attempt it]. Our guys had been looking up there, but it [seemed] too dangerous, and they said they were not sure they [even] had a place to land. The smoke was just coming up so intensely that it wasn't a feasible or correct thing to do, and I'm sure there was a lot of heat there as well. I think any attempt to land a helicopter would probably have caused the helicopter to explode, or to crash to the ground.

I don't think anyone was seen on the roof, though I can't vouch for that. But people in our helicopter took some close-up pictures of people standing ready to jump. Dozens of people in the portals where there once were windows – just standing there and holding on for dear life, and then just jumping because of the intense heat and the smoke that was engulfing them.

We learned, and I don't know how, that the mayor was at West Street near the fire command post, and so I had my guy bring up my car. It had been damaged, but it was still running. We found our way down to West Street – I, with my driver and several people from my staff who had responded from 1 Police Plaza to join me at Church and Vesey.

On West Street I saw Pete Ganci and Father Judge, [though] I didn't speak to them. As soon as I got over there I learned that the mayor had just re-located to 75 Barclay Street. It is now around 9:30. We turned the vehicle around and went to 75 Barclay, where I joined the mayor, who, I believe, was on the phone to the vice president, or somebody at the White House. He was with his entire staff, and Chief

Esposito had found his way into the building, too, [as had] the fire commissioner, Tom Von Essen.

The mayor got off the phone, and he began to discuss quickly that he wanted to have a press meeting so that he could reassure the public, and give some direction. He felt that it was important that he get out in front of this tragedy. But just as we get up out of our chairs we hear a roar. A tremendous roar. And then someone yells, 'Get down.'

So we all kind of hunkered down and stood still, and you just heard this roar like a train coming into a station, when the express is coming through. And then nothing but dust.

And then somebody said the tower came down. We were all in disbelief.

In my mind's eye, I thought that maybe the top of the building had fallen over, or something – I had no idea that it was the catastrophic fall that it was.

We began to look for another way out of the building because the way that we had come in was now blocked with debris and with broken glass from the force of the fall of the south tower. I stayed back and made sure that everyone was out of the room that we were in – there were a good twenty people – before I would leave. Someone said, 'You don't need to be last, go ahead, go ahead.'

I said, 'Listen, I will be last because I'm on crutches.' People attribute certain qualities to you that aren't always accurate, you know. In my mind, I wanted to be the last one so that I would not be delaying someone, or getting knocked over by someone who maybe got into a panic. *I'll get myself to where I have to go,* I thought. And it's a good thing that I did, because they went to one location, and that location was blocked. Then they went to another location that was blocked – so there was a lot of up and down stairs going on, and I just had the luxury at this point to stay where I was. Finally they hooked up with somebody from the building, somebody in the janitorial staff or engineering room, and they got us out back onto, I believe, Church Street. I think we were in the basement, but we wound up getting out on street level. Those buildings, as you know, are interconnected, and even though [an address] says 75 Barclay, you might actually have an entrance or an exit on Church Street.

The mayor left, and he went with his staff and the police commissioner north to make his way to City Hall. I said to myself, *The hell with that. I'm going back down to see how my people are doing.* And so I went back down to Church and Vesey with some of my staff.

I knew we had casualties. When I got outside, what I saw was a nuclear morning, as the mayor described it. All of that dust, all those papers flying in the air. The concrete dust was already several inches thick on the ground. It was like we were walking through moonscape dust, very much like we saw when the astronauts were on the moon. It had completely changed from what it had been just fifteen minutes before, when I had left there to find the mayor. The visibility was [so] clouded that I really couldn't see that the building had fallen, but they told me that it was completely down, which shocked me.

I saw people walking around in the street, and they all had the look of shock on their face. They were walking slowly, and many of them were walking into harm's way. They were walking south on Church Street. Many of these people were coming out of 7, 5, and the other surrounding buildings that had suffered collateral damage. In fact, at this point in time, those buildings were now on fire.

We began telling them to turn around, to go north. We told whatever police officer we met to urge them to go east. And then I began to warn everyone, 'Look, run if you can because you're in danger here.' I felt that we owed them that message.

[By now] my staff was trying to encourage me to get out of there. They said, Boss, we should be leaving, it's dangerous. And they were reminding me, you're on crutches, there's debris and stuff all over the place, you're stumbling. But − and I don't want to sound like a hero, because I'm not − I just had this overwhelming need to stay there. I can't even now understand why I did not want to leave that spot.

Another ten or fifteen minutes or so later, one of my guys said to me, 'Listen, the north tower is making noise, we're not safe here, that building is going to come down, too. Let's get out of here.'

But I [still] didn't feel in danger. I don't know why. Maybe because I was out on some planet there at that time. I just felt that there was work to do. And my car was just down the street, a few feet away. I could get there in a minute if I needed to.

Just then a bomb squad Explorer truck happened to be passing, and one of my guys commandeered it. They grabbed my crutches, and they grabbed me by the arm, and one of them said, 'Look, you're out of here.'

The bomb squad truck is fairly large, and my guys literally threw me into it. My guys [got in, and] we started to travel, because of all the action, at a very slow rate of speed north on Church Street.

And then *rooaarrrr,* we heard the second building go down with that same horrible rumbling sound.

But what I saw at this time that I didn't see when the first building went down is the night smoke, black as night. People who were enveloped by it all described it the same way. Even with my eyes open, I could see nothing. It's as dark as anything I've ever seen. I'm grateful that I got into that truck when we did because as we were moving up the block, one of my guys said to the driver, 'Step on it.' We were going fine enough, but he said, 'Take a look.' I'm hearing the roar now, I'm looking back, and I can see this nighttime chasing us up the block.

It caught us, finally, and wrapped over us. But we were safe; we were in the vehicle. We continued north on Church and wound up by City Hall through making a few turns here and there. Now this dust cloud that had been lingering in the air since the first fall was all over downtown Manhattan. At City Hall, I learned that the seat of city government had been abandoned as unsafe. City Hall is only nine blocks away, and on top of everything else, it was a target.

At this time I didn't know where the mayor had chosen to relocate, and I decided that I didn't care, that my place was in 1 Police Plaza. And so I came back to the command post center here, which is located on the eighth floor. I reported immediately to that room, which was up and running, and had some conversations with the commanders there. I got a call from the chief of department, who was in the field, and we had a brief discussion about whether we should leave this building.

We knew that we lost a command post, the Office of Emergency Management at #7, and we were up and functioning. All the agencies reported here, and we were doing pretty well. So I said, 'Look, let's just stay here, we'll hold our breath, and we can do business here.' I sat down in the situation room, and I stayed there for the next fifty-six hours or so. During that time we made a thousand decisions – little decisions, big decisions – there was so much going on that we were just trying to move the ball along.

We ordered everyone in, in a general call, and we suspended our duty chart and went to twelve-hour tours. We began mobilizing right away. We stood down all of our nonpatrol enforcement stuff, and everybody came to various command post locations in the city for assignment. This agency is pretty good like that. It's automatic; everybody knows what his role is, and they don't sit and wait for somebody to make a call. We've practiced events in the past and we know how to handle them in terms of personnel resources and how to mobilize. So that was going like clockwork.

We began quickly to shut down the city – almost immediately when the first plane hit, we closed the Brooklyn Battery Tunnel. The Port Authority already closed the Holland Tunnel. The Manhattan Bridge, the Brooklyn Bridge, and the Williamsburg Bridge were closed to incoming traffic. After the second plane hit, we closed down the city altogether. All the bridges and tunnels into the city were closed both directions, except for emergency vehicles, and emergency control plans were put into effect throughout the city – each borough has a plan to deal with these things. This is stuff you practice for the day that it's never going to happen, but now it had happened. Most of those plans worked very, very well, and we also began to get the calls from the different state and federal agencies which were offering help.

I spent most of my time in the command post center and [among the] things that I remember doing is promising myself that I wouldn't raise my voice. And I didn't; I didn't for a week. I did not want to let anybody think that I was cracking under the pressure or to put undue pressure on anyone else by yelling or making demands. It would accomplish nothing. We stayed very calm and deliberate in what we did and we tried to analyze as much as we could. I was also very conscious of the fact that image was very important, and at one point in time I was sitting downstairs and took a look at myself in a piece of glass. I looked awful — I had lost my jacket, and though I had a shirt and tie on, the tie was askew and the shirt was filthy. I had a pair of blue pants on, and they were covered with dust. And that's not an image that I wanted to project. I just had this need, so I came up here and I cleaned myself up. I had a suit here, a fresh shirt and tie.

Before I left, two firefighters, Dennis Conway and Pete Ganci's driver, Steve Mosiello, came to my office very upset and crying. They came to tell me that Pete was dead. Pete was a friend, a very decent and good man. We cried a little here, and Steve said they were going to go out to see Mrs. Ganci. I said he shouldn't go out looking like that. I made them get in my shower and clean themselves up. I think they walked out of here with police uniforms on, which is probably one for the books. We should have taken a photograph of that.

We began almost immediately thinking about how we were going to remove the debris. And it's funny how your training comes in. We were in the middle of this tragedy, and we must have thousands dead, yet I'm thinking about the next step — what's the next step? The next step is rescue. How can you perform a rescue if you can't get the debris out of the way? We were strategizing about getting the equipment in, how it was going to respond, how we were going to move debris out. We talked about securing the trucks. We called the coast guard to get some tugboats ready for us in case we needed to use barges to remove the stuff. We decided immediately that it was going to be Fresh Kills where we brought the stuff.

The state police, the national guard, and all the state security officers were mobilized almost immediately by the governor. They began sending in resources – it seemed like almost right away, although I know it wasn't, but everything was time condensed. We spent the next two and a half days or so trying to get through the many issues of managing the disaster, not only our plans for shutting down the city, but the plans for moving personnel around. The hardest thing was dealing with the missing police officers. Fourteen of the twenty-three missing were emergency service officers. Two of them were brothers of firefighters who were killed, too. Can you imagine? Two of our twenty-three had brothers who were killed in this as well – Joe Vigiano and his brother John, who was a firefighter, and our Tommy Langone, whose brother, Peter, was a firefighter.

The best assignment I ever had was when I was chief of department – leading the men and women of the New York City Police Department: forty thousand cops in uniform. I'm very proud of that, to have been a part of the uniformed force for over thirty years.

And now we have twenty-three of them missing. I cannot tell you how hard our men worked at the site digging for our guys, and for Moira Smith, the only woman officer lost. She's so beautiful, this Moira. She responded from the 13th Precinct in her car. We heard her – we think she was on the third floor and we have her voice calling for help. She had called saying that she had a victim who was an asthmatic and she was giving him first aid, and then she called a 10-13 because the building started to collapse around her and that was it, the last time she was heard. And John Perry was in this building filling out his retirement papers. When the first plane struck, he heard a woman screaming in the office. From his position there, he could see the towers. He saw the towers aflame and left his paperwork on the desk. He ran to the Trade Center then – ran. He went inside the north tower, and he never came out. A great man – sergeant major in the United States Marine Corps. You know how rare that is, to be a sergeant major in the U.S. Marine Corps?

A number of our officers, including Pat Lynch from the Police Benevolent Association [the largest police union in the country], were on the west side of West Street when the south tower came down. The debris was literally flying across the street. Several people who were there said they were fortunate enough to get behind the round pillars that were holding up the crosswalk that goes from the World Financial Center to the World Trade Center. So the debris was falling, literally, on either side of them, and they saw other people being consumed by it. But the debris kept falling and the dust kept coming, and they were so overcome that they were pressed up against the window of one of the buildings there. One of the police officers had the good sense to pull out his automatic and shoot the window out so they could get into the building, and away from the acrid smoke. You sometimes hear of people shooting their way out of something, but here they actually shot their way *into* something.

One of the first things that we did, painstakingly, in an effort to [learn the] last known whereabouts of our missing, was to listen to tapes of the radio communications, and interview other officers and survivors who might have seen them. It was a Herculean task to listen to all those transmissions and pick out voices. We were able, with pretty good accuracy, to determine where our people were when the buildings went down, and a lot of that work proved to be [helpful] in terms of where we found our people, or the few we found. We're doing DNA [testing because] we didn't find much of them, [though we did find] their guns. Actually, firefighters found their guns. The protocol that has been established between the fire department and the police department is that, if we find bunker gear, we step down and call the fire department, and they do the same thing if they find guns or badges.

It was a good thing that I had met Lieutenant

Tobin and told her to put a helmet on because when the first tower came down, she was thrown thirty feet. It was a heavy bulletproof helmet, and after losing her wisdom tooth and fracturing her ankle, she was lying on the ground when a concrete piece fell on her, cracked the helmet in two, and embedded in her head. She could not have survived that without a helmet, and she continued working down there. It took forty stitches to close that wound. Then thirty minutes later she was running towards the Hudson when the second building came down. This time she got hit between the shoulder blades with a shard of glass. It took another forty stitches to repair that injury. She has a good mind, a good heart, and a good soul. Her dad was a cop, and three brothers are cops. Her cousin, a firefighter, was killed on September 11.

I feel lucky and blessed. There was a rumor throughout the site that first day that I had been killed. I think that they confused me with my friend Bill Feehan, the first dep of the fire department. And then to add to it, they found what was left of my car. My driver had parked it close to where we were at Church and Vesey. They were getting ready to take me to the car. I was resisting them, but I probably would have gone with them in another couple of minutes, and we would have gone right down Vesey Street. If that bomb squad truck hadn't been passing by just then, the one they pushed me into, we would have gone down to my car just at the time the building came down. Of course, the car was completely destroyed by the collapse and the fire, and about five days later I got a call from the Fresh Kills dump saying they found my sunglasses – I'll never spend so much money for sunglasses again – in the glove compartment of the car. I'm just glad I wasn't in them. I just have to look at my sunglasses and I will always think about how fortunate it is that we didn't get there.

No one made the right move. No one made the wrong move. No one made a critical mistake. No one made an ingenious decision. We were just in the hands of God, or fate if you prefer, and those that got out of that place were fortunate and blessed and those that didn't are with God now. There's no rhyme or reason why people made it and why people didn't.

Everyone has an incredible story. You can see why it hurts so much. You may think we are hard and tough, but we bleed and weep just like you do. ■

Lieutenant Mickey Kross
Engine 16

It was a normal night tour for Engine 16 and Ladder 7. There were a few runs, but nothing particularly memorable. George Cain's sister, Erin, came in to chat with her brother at the East 29th Street firehouse, and she stayed for dinner. She is a looker, and has an ability to wisecrack as good as the big leagues. Lieutenant

Mickey Kross enjoyed it all, as he enjoyed every one of his tours in the firehouse.

Every morning Mickey looks forward to his first cigarette of the day. He knows smoking is a health risk, but if there is anything Mickey hates in life, it is people telling him what to do. And so he smokes as much to proclaim his independence as to satisfy the addiction.

A quest for independence motivates much of what Mickey Kross does, because he believes that in the firefighting business you must be able to think independently in an emergency situation, to think outside of the box of expectation, if you are to survive. This self-reliance is a trait that carries over equally to other spheres of his life, and he is well known in the firehouse for carrying his own weight in any situation, even when he simply wants a cup of coffee.

Some years ago Mickey had a cup of coffee in the firehouse that sent him to the gent's room for a week, and he vowed to never again drink the firehouse brew. And so on the morning of September 11 while sitting at the desk of Engine 16 sorting out the cards for the building inspection duty the engine would undertake later in the morning, he opens the thermos of coffee he had made in his apartment the night before, and pours a cup for himself. He lights a cigarette just as the phone rings. It is his good friend of many years, Christine, and as they are talking he hears a loud bang in the background.

'What's that?' he asks.

'I don't know,' she answers as she carries her cell

phone out the door. When she reaches the street, just three blocks from the World Trade Center, she sees the north tower on fire. 'Mickey,' she cries, 'the World Trade tower is burning!'

■ I am on the phone with Christine, and the third alarm comes in [to which] we are assigned. I tried to soothe her a little bit [before leaving] by saying, 'Don't worry. It's a plane, but it'll be okay. We have to go now. I'll talk to you later.'

I hung up the phone and off we went. I was monitoring the department radio during that time also, but I still wasn't 100 percent sure what was going on.

I told the driver, 'Please don't use the tunnel that goes under Battery Park,' because I was afraid of God knows what – that we could be flooded, or that a building had fallen at the other end – I didn't know what. So we got off the East River Drive and went south against traffic, and when we reached the front of the World Trade Center there was absolute chaos. I assumed that it had been a small plane from what I heard on the radio, but I was quite shocked, when I first saw the building, at the immensity of the damage.

It looked like a movie scene, where the monster was coming from the river and everybody was screaming and running and tripping. It was exactly then that the second plane struck. We were directly under the tower, on Church, coming from south to north, and we got showered with debris. It was like [being in] a meteor storm, and I didn't think we would get off the block, that we would get killed on the block. Things were hitting – *bing, bang, boom* – over your head.

The siren was going, and I had the radio to my ear, so I didn't know it was a plane that hit it. I thought it was the original building [exploding]. It was chaos. I told the chauffeur, 'Ronnie, let's just get out of here and find a hydrant.' I figured we'd go to a safer area, so we went about three blocks away and we saw an open fire hydrant. I had a very good chauffeur – very experienced guy.

We got off the rig and I had a little pep talk with the guys right in the street. It was something to the effect of, 'Okay, guys. This is going to be tough. It's going to be tough, but it's another fire – the main thing is, we just stay together and watch out for each other.' I was trying to impress upon them to watch out for each other and try to keep our heads about us. You know, I was afraid of panic. That was a thought in my mind.

We were ordered to the lobby command post of 1 World Trade Center. In the confusion I forgot which [building] was 1 – I didn't know if it was the north tower or the south tower. As we were approaching, things are falling from the sky. Maybe a hundred yards away I see the number 1 on the northernmost tower, and I thought, *Well, thank God. At least I now have a place to go. I know where I'm going. There is number 1.*

My mind was directed on getting in that lobby. I

didn't even look around. The men told me things were coming down — bodies were falling around us. All I [wanted was] to get in that lobby safely with the guys. I did see a body in front of the building. I think it had no feet; I remember that. I tried to ignore that because I realized that he was gone. He was burned also, and we couldn't help him.

I don't remember who was there, except I remember seeing Commissioner Von Essen, and he looked quite distraught. I think I said something to him quickly in passing, I don't remember what. I tried to say something uplifting, but I think I said something stupid, like, 'You're gonna have a tough day.' I remember feeling embarrassed. In all the craziness that is going around, I'm going through embarrassment about what I said.

A chief told us to team up with the next engine company that entered the lobby and go up together and take three lengths of hose between the two companies, one officer in the front and one in the back of the two companies. I thought it was a very good idea.

He said to report to the twenty-third floor and find the command post that was set up. There was supposed to be a battalion chief there. Immediately, Engine 1 came in, which happens to be my old company, so I knew all the guys except for the officer, all friends of mine. So that was kind of cool. My new company and my old company together.

As we headed to stairway B, which was in the center of the building, the core of the building, the men told me that the elevator shafts were open, that the doors had blown out, and the elevator pits were full of bodies. I never saw that. I saw injured people in the hallway coming out of the stairs. On stairway B mobs of people were coming out, [some with] skin hanging off, clothes disheveled. A lot were just very calm and walking with a determined look to get out of here. But it was pretty hectic, and I did have a concern about how we were going to get up those stairs.

When I was told where to go, I just kept my mind on that one point. I didn't want to distract myself with anything. It was hard enough keeping a sense of grace under fire, you might say, and that was what I was trying to maintain. As we went up I found it wasn't so bad. The people moved out of our way, and it was one line going up, one line coming down. It worked out pretty well. People were very good about it, and were encouraging us as we were going up, saying nice things. Offering us water. We were encouraging them, too, telling them they were doing a great job, just keep the pace going.

Well, we got up to 22, but we made a few stops along the way. We'd go maybe six flights and take a breather, then go another six. When we got to 22 the men stayed there with their equipment, and I went up to 23 to find the command post chief. I didn't want to bring everybody upstairs to clog up what would be the command post. The guys were pretty tired at that point, after carrying up the helmets, gear and boots and all that equipment, and the hose, and extra air bottles and all. They were pretty exhausted, and we hadn't even begun to fight yet; we weren't even near the fire yet. Which we truly believed we were going to attempt to put out.

When I got up to 23, it was kind of quiet on the floor – very dark. I saw some firefighters moving around, but not any command posts or chiefs. The hallway was full of debris where I was walking, debris about three feet high covering a whole section of the hallway. I was puzzled as to what that was, because I knew that the plane had hit way above us. So as I got there I realized it was the elevator shafts. The doors or the walls had probably blown out, and I had to climb over this debris to get to the other side of the hallway. Here, I was a little concerned about falling into the pit, the elevator shaft. I thought, *Oh my God, what if I get sucked in for some reason? What if there is some kind of low pressure in the shaft that creates a wind coming down, and it would suck me in because of the venturi principle.* I had been told that the command post was on the other side of the building, so I went there and looked around but didn't see anything. And then I went back, and because there was a hallway heading in the other direction I said, okay, I'll go back and I'll try that. Meanwhile the radio is going; there was a lot of communications. I didn't hear anything in particular, but as I was going by the second time, climbing the rubble, the section of floor I was on started shaking. And I said, *Oh God – it's the elevator shaft.* I thought the elevators had cut loose, or whatever, and something was coming down the shaft. So I thought this section of the floor was going to collapse, and that really kind of shook me up.

I didn't know that it was the south tower going down; I thought it was a localized thing, and that was that. But it was [soon] over, so once I got beyond it, I felt okay again. I saw that everything was all right, and I pursued looking for the chief. I finally found him, and he said, 'Evacuate the building.'

This was serious. There was a fire up there on top; there were people. And [yet] we are told to evacuate.

So I [begin heading down] staircase C. There is a large African American woman coming down the stairs with a gentleman. I wish I knew who he was; I wish I could find out what happened to him. He was a civilian and he had a suit on. He stayed with her. He was quite a heroic figure, this guy, he deserves a pat on the back. He could have just taken off, but he didn't. He stayed with Josephine. They were all alone, and there is nobody else in the staircase. She was barely able to move because she has something wrong with her feet, and I was looking for the chief now for direction. I had her and him, and I realized that they were never going to get down these stairs like this. She may have a heart attack or something on the way, or go unconscious. He will not be able to handle her. So I say to her, 'Go to stairway B with him.' I knew there were people there that could help her. She wasn't going to get help if she stayed in C; she probably wouldn't have made it. They started arguing with me, I remember that. And I got pretty angry because there was enough going on, I had something to do, and here they are arguing with me. And so I yelled after them, something like, 'Shut the hell up and listen to me, and get to stairway B. Right now.' I remember they looked at me and then proceeded to stairway B, where eventually I guess they got picked up.

Then I saw the chief – he was not a big man, and I don't know which chief he was, but he told me, 'Okay. Evacuate the building.' He said, 'Everybody, out of the building.' And then I headed back to the staircase B – which I had originally come up – wanting to tell the firefighters there that we had to evacuate.

I got back to stairway B and told the guys that were there that we got the order. Some of them may have already gotten it; I think it came over the radio. I remember seeing one of my guys in the hallway [and telling him], 'Get out of the building.' He had that look, you know, like 'What do I do? What do I do?' It's just that moment in time where it freezes. We all started down, and somehow I got separated from the guys. I ended up with the officer of Engine 1, whom I didn't know. There were a lot of people in need of assistance; a lot of people were hurt. I think there was a guy in a wheelchair, and there was supposed to be a blind guy, with a dog – all these things I don't clearly remember. I was on one side and [the Engine 1 lieutenant] was on the other side and somebody [we were assisting] was in between us. The person was going very slow, very slow; it was like time was standing still. I was just praying this person would move faster. You wanted almost to shove her down the stairs. It was a woman, and she was going one step at a time – she was hurt or something. We were trying to figure out some way to do this faster, maybe getting her a chair, but then we changed our minds, and other guys came to help.

At some point, [because] I didn't know who he was, and we were kind of stuck in this predicament, I reached over and [said to the lieutenant], 'What's your name?'

I remember him looking at me with a nice smile and saying, 'Andy.' And I say, 'I'm Mickey.' And we shook hands across this woman. That second where you smile at each other – it was kind of neat. It was the last thing I remember until the rumble started.

Then suddenly, there was this loud, loud noise overhead. There is a sense of tremendous energy, like being on a locomotive track with a train coming at you, that kind of feeling. Just tremendous energy. I think I surmised that the building was coming down – though I wasn't 100 percent sure.

I don't know what happened at that point. My next memory is being on a staircase next to a railing. The railing was to my left, so I must have been on the inside of the staircase, and I remember crouching down as low as I could get. I was thinking of my situation – what should I do, what can I do? What do I have that is positive? What tools do I have? Anything that had been in my hand was gone. My big light was gone, the survival rope was gone, and the only thing I had was what was in my pocket or clipped to me. The main thing I had was my helmet. I remember thinking how important it was to have had that helmet. That was the biggie: the helmet, and holding on to my helmet because I had forgotten to snap my helmet.

And the wind is powerful, trying to lift it off my head. There was a fierce wind. It was actually almost lifting my body, and pulling the helmet off my head. It is almost moving me. So I'm holding on to the helmet with my fingers, and I'm thinking, *Damn*.

I taught probies at the probationary school for two years, and I always made a point of saying to them, 'Snap the helmet. You don't want to lose your helmet. You really don't want to get in any situation without the helmet. With no head protection, you are in big trouble.' So here I am with an unsnapped helmet. Here I am, the teacher, and then I started getting hit with things. Something heavy came down on me because it cracked my flashlight, so something was mighty heavy. I figured, *Well, that's over. I'm a dead man.* And I remember feeling that I don't believe this is happening. I'm going to die in the World Trade Center. Another feeling, you know, I was angry – I was a little angry.

I don't know what happened to Andy at that time. We were together, but then we were separated. There was a lot of changeover [because] a lot of guys were helping. It's not like you took one person all the way; there was a lot of switching on and off. That's why I guess Josephine ended up with other fire-fighters. A lot of other people were helping because she was very big. Ladder 6 ended up with her.

I'm down in this little crouch position, and there is this noise and debris and fierce wind and the next thing was silence. No more debris. No more noise. No more wind. Total silence. Incredible silence. And then darkness within a few seconds, eight to ten seconds. I didn't know that at the time. I wasn't counting, but they told me it took ten seconds to come down. [After all] that tremendous energy and noise, silence in ten seconds, and there I was, okay.

I was waiting for the next shoe to drop.

I said, 'This can't . . . there's got to be . . . something's got to happen.' But nothing happened. I was alive and I didn't feel injured. What happened? Was there a partial collapse? I wasn't really sure. So I stayed there for a few seconds, trying to assess my situation. I wasn't hurt. I was breathing, but my eyes wouldn't open.

I think I thought the darkness was because my eyes were covered. So I took off my gloves, and I put my finger in my mouth to get some moisture, just a tiny bit because my mouth had dried up. I just kind of peeled away at the corners of my eyes, worked my eyes open. And then I tried to get my light on, but the switch was covered with dirt. So I cleaned that off, and it went on. I saw the light – and it felt good. My eyes worked. I mean, they were itchy and dirty, but they worked.

But there wasn't much to see. I saw just a lot of powder and steel and a bunch of crap lying around. But there was a lot of smoke. Then I thought, *Oh, man, now we're going to die here,* you know, that kind of picture. That was troubling. I was just crouched up in a ball, and I thought, *How many hours or days will I be here?* That was a scary feeling. I was thinking of Viet Nam, where POWs crouched in the fetal position in tiger cages, and I knew it was a very painful position, and so I was thinking it was like that. So, just for the hell of it, I said, *Let me just push around and see what happens.* I didn't think anything would move. But the rubble moved.

I then began to feel things hitting, and I figured that was the beginning of the end. I remember that once I started feeling things tapping on me, I said, *Well, this is it. It's over.* I was just waiting to die. I remember hoping that it would just be fast. I didn't want to be lying there hurt, with a broken back or something. I didn't want to be crushed in that way.

I tried to climb out of where I was, but I couldn't get out because the mask was holding me back. I just took the mask off – I wasn't using it anyway – and it just kind of fell off, rolled over my shoulder and disappeared. I didn't worry too much about it. One of my thoughts was, *Well, I'm probably better off without a mask. Why prolong this?* There was a lot of carbon monoxide around – that would be the best of all the situations. Maybe I'll just go unconscious and die out – that would be a blessing. Rather than be here and die of whatever you die of when you're trapped, dehydrated, whatever.

I climbed over where I had been, and it led me out and over on top of this pile. I knew as I looked around that the staircase was separated from where the wall had been. There was no wall to my right, just piles of ashes. So I was cantilevered. I was on a cantilevered staircase and disconnected from the left wall. I thought, *This is not good.*

I was afraid to move because any movement might bring down whatever was holding this up. Whatever was holding this up was probably not much, and I didn't want to go down where the mask had gone because it sounded like there was a little distance there. I remembered seeing the number 4. Maybe there was nothing below the fourth floor?

There was a great silence. It might have been five minutes. It is hard to tell. That was a kind of existential isolation. That would be a good term. You are the only one on earth.

Then I started hearing some movement. I hear things moving around, and I hear a little *rrrrrrr,* you know, grumbling, and 'How are you?' and 'Are you okay?' I heard guys talking, and I realized I wasn't alone.

That is very inspiring, when you find out there are other people alive near you – that felt very good. So we started calling out to each other, and found we were actually pretty close to one another. I didn't realize that. I could see the guys after a little while, moving around. I could hear guys moaning, guys saying they were hurt. I think, though I'm not sure, that Andy was one of them moaning. I have a memory of that, but it's not clear. I had an image in my mind of this '4,' just like when I saw the '1' of the World Trade Center. Certain memories are very clear – images of a rope, images of a 4, images of 1. Things like that are clear, but the sequence of events is not. That's where I'm having a problem remembering.

So then below me somebody said he found a doorway. And they were yelling, 'Does anybody have any tools?' Now, remember, the staircase above me is intact, and the staircase below me is intact, and I'm kind of in the middle on this heap. So somebody passed me a tool – I think it was a Halligan. It was hard to hold on to, for someone in the position I was in, and I remember I had an awkward time with it. I passed it down to somebody below me, and then the guy passed the rabbit tool, the thing that they force doors with, and then I was saying, 'Oh, man, I hope they find a way out.' Because there was a slight movement of air. I was feeling a very slight movement of air on my cheek.

There was smoke in the air, but there was a breeze. I knew that air was coming into this place from somewhere, so there must be an opening – a big opening. There has to be an opening or we wouldn't be breathing. Right? So I said maybe it is that doorway that they found. When they open that doorway, there will be a passage way out of here.

Then they forced the door open, and I heard them say, 'Nah, there is too much debris.'

I felt bummed out about that. Then I said, 'Let me get off this pile of things.' I'm assuming I'm secure, so [I'm going to] try to get on the other staircase to my left. I'm not doing any good over here. Let me get down to where the guys are. So I kind of leaped over the staircase. It was all right, nothing moved but me. And then I climbed up. And the chief was up there – there was a chief. Guys were on a landing. There was a lieutenant there, too. They were just right next to each other, huddled up. And that's when we found a bottle of water. It was just like in the movies. You open up the bottle, and everybody takes a sip. It was like a movie. It was wild.

The chief there, Picciotto, he did his job. He handled himself real well. He wasn't heavy-handed. He said, 'Guys, we ought to turn our radios off except one and shut our lights off except one. Then we'll look around, see where we are, and assess our situation.'

He kept up the focus. When somebody tells you something positive, it gives you a little sense of direction, you have something to do and to think about.

Then Chief Picciotto started handling the communications. At this point, Maydays are coming out – we are hearing Maydays going out on the radio while Chief Picciotto was giving Maydays, but there was no response to anything. We heard guys down below us sending out Maydays and guys above us sending out Maydays, but no response. Some guys started peering around – there wasn't much room to crawl, but they were crawling around, feeling [their way]. [I started] talking to the guy next to me – we were kind of cheering each other up a little bit:

'We'll get out of here.'

'There are guys looking for us – we'll get out of here.'

You don't really know in your head if it's true or not, but we kept telling each other that. It did help. Then we got a response on the radio. Apparently Chief Picciotto had made communication with somebody on the outside. They were talking on a first-name basis, like they knew each other. I thought this was a real good thing; now they know we're here. We were trying to give our last known location – stairway B, fourth floor. Little did we know that there was nothing out there but a pile of debris, so that didn't really mean too much.

So they sent teams out looking for us. It went on for quite a while, [because] they were having a lot of trouble finding where we were. It was just chaos on the top. When I got out and saw it, I realized why they couldn't find us. You had no idea where you were; you could have been on the moon. You saw nothing but piles of debris and razor-sharp steel. It was like a forest, and there was just no sense of direction whatsoever.

After a while we were talking on the radio when

I saw this great light up and to my left. A very dull, bluish-gray light. And so I'm looking at it and wondering what that could be. Maybe it's an emergency light with a battery left on; maybe it's some guy up there with a flashlight. And then the light went out, and back on again. Then it went away, and eventually it started getting a little brighter. So now I'm kind of just crouched down looking at it, and I'm thinking, *Maybe I'm hallucinating.* But then, within maybe fifteen minutes of this going on, all of a sudden sunlight broke through. It was just the most amazing thing to realize – you are under a collapsed building that was a thousand feet tall in the air, and sunlight came right through. A beam of sunlight maybe about six to eight to ten inches – very small and narrow, but it came right into where we were – right into this. It looked like a dirty laser beam. It was all full of soot and garbage – debris. It was sunlight, but it was all peppered. But it was clearly sunlight. It was one of those moments like you see in the big dramatic movies where light comes in, and Hallelujah, and all this stuff.

I said, 'Holy Jesus – there is a hole up there. There's light. No wonder we're getting air. And now there is light.'

And I thought, 'We're getting out of here. Maybe.'

Well, maybe we'll get out if nothing else bad happens – if there is no secondary collapse, if the fire and the fuel don't get to us or something. So we started all cheering each other up. And then we got the word from Jay Jonas, upstairs, to come, so we started working our way up, which was not easy because there was no staircase. There were a lot of beams we had to climb, and we went one at a time on them. It was like being a kid on the playground in the monkey bars. I remember when it was my turn, and I said, 'Oh, man, I hope this beam holds.' So you get up there, and you got on the landing, and there was the opening, and then I think I saw Josephine there – she had a purple dress on.

And there were two guys there, outside, on the pile. Everybody was trying to get out. There were the two of them, on their own. And they just happened to be where we were, saw this opening and realized what was going on. They came down with ropes, and they set up a rope thing. And they told me when I came out I was completely dazed and I couldn't get my hand on the rope. I couldn't find a rope that was right in front of my face. I don't remember this at all.

I went through a lot of different things in my head. I remember feeling a tremendous peace and acceptance of the situation I was in, and that whatever happens, happens. I had to accept it. I remember sometimes feeling really scared and terrorized, and other times just focusing on what I had to do to survive. I went through different things. There was never any panic by anybody. Maybe it says a lot for our training or something – that with so many people trapped, nobody panicked.

There was a pile out of the opening, the initial

pile when I climbed out of the void. The outside was horrible. It was far from over. I thought that once I got out of here it was gonna be over. I was shocked by what I saw. I didn't know the construction of the building, but these columns were standing there, fifty to sixty feet, [almost] a half wall of them, and they were slightly canted over. There was this crackling sound around, fire. There were small hills of steel that looked like they were ready to come down. And we had to work underneath them, because we went towards them initially, crawling towards them to figure out the best way out. But the chief said, 'Turn around and go back; we cannot get out this way.' He said there was a fire behind those columns. And I said, 'Oh Jesus, we got to go back.' We did all this work for nothing. We were in the precarious situation, and now we have to go back, and that's when I said I wish I knew which way was the right way. I wish I had a compass on me because you had no idea what direction you were going. It was impossible to tell. You couldn't see any building. Just these columns, and piles of steel that were still hot and smoking.

We could only see maybe thirty feet. Nothing but pieces of concrete wall, bent columns, steel twisted up, metal torn up. A lot of sharp pieces of metal around, and a forest of crazy wire besides that, so it was an ordeal getting down and out. Finally, we went over a little mound of debris, and I saw these fire-fighters coming up. I felt like I was imagining that I was soldiering, and my troops were coming up the hill towards me. That's what it felt like, when I finally felt safe. Here they are, they are coming to get us and they are coming from safety.

I got over to West Street, where there was a temporary command center set up. I saw guys sitting there, officers at a desk and table. I went to check in. I remember walking up, and the first thing the officer said was, 'Give me your riding list.' I knew that my riding list was accurate, but I wasn't certain. It was the change of tour when we left the firehouse, so I puzzled about having the complete day tour on the riding list. I remember my nose was bleeding at this point.

This officer looks at my list, and said, 'These guys are found, he's found, he's found, he's found.' And the only one we couldn't account for was Paul Lee. I tried to remember where the rig was parked. Maybe Paul went to the rig, so let me try to find it. Little did I know that everything was buried at that point, and the streets were all covered with debris, and God knows where the rig was – covered up at some point. So I went to where it was clear, but I was way north of the rig. It was south of me, on one of those streets that was piled up.

And that's when I ended up in front of my girl-friend's apartment. I was looking for the rig, and I looked up and saw Christine's apartment at Independence Plaza, and I wanted to tell her I'm all right. I better go up there. But officers were standing around, a sergeant and a bunch of cops. They saw me going towards the building and very gently stopped me. I could see their concern, and I realize, *Oh boy, they think I'm bonkers. How am I gonna get out of this one?* I just wanted to get in that building. But, I remember, I have a bloody nose and I'm pretty cut up.

So I said, 'My name is Lieutenant Mickey Kross. I am with Engine Company 16 and searching for a firefighter; his name is Firefighter Paul Lee, who's missing. I happen to be in front of my girlfriend's apartment in this building. Her name is Christine Gonder. She lives upstairs. I want to leave her a message. She probably thinks I'm dead.'

I think they realized that at least I had half my marbles. So I leave a note for Christine, and then I report to the triage center. All along I'm trying to grub a cigarette, asking everyone I see. Son of a gun, nobody smoked. Finally, I find a guy who gives me a cigarette, and sure enough, it's a lousy lite or something, a crappy brand and I'm thinking, *Where's my cigarettes?* They're on the rig, and I can't find the rig.

They wanted to take me to the hospital. I said, 'No, I don't want to go to the hospital.' I'm embarrassed. I felt like, there are people dead, trapped, body parts falling all over, and here I am going with a bloody nose. So I said, 'No, I don't want to go to the hospital.' The doctor then took my pulse, they washed me up, they took my temp, they gave me a quick once over, and the doctor says to me, 'It looks like you're okay; I can't find anything wrong.'

So I decided to walk around a little bit, and I

thought then about Josephine. I know she got out. Well, I hope that guy who helped her got out okay. That guy was courageous, and he deserves a big thank-you.

There was no transportation set up, and I spent quite a while going around trying to get a ride back to the firehouse. I was waiting around two hours for the bus and no bus came. So I finally hitched a ride with some engine company that was covered with debris, like you see in the army. We were going up 3rd Avenue; we were heading uptown. They dropped me off on 29th Street, and I walked down the one block. I felt like going for a walk because I still didn't know what was happening yet. I didn't know what happened to the guys in the truck, Ladder 7.

I found out then they were all still missing, all six of the men.

I have thought a lot about surviving this. The flip side of 'Why me?' is 'Why *not* me?' That is probably the healthier way to look at it. Oddly enough, I just read some writing by Seneca, and he deals with that very thing. It's the kind of book that applies to your everyday [life]. I bought it on our baseball trip to Baltimore, the firehouse trip. That was just a couple of months ago, and then this happened. The guys went out drinking. I drank myself years ago, but I stopped. I'm not ashamed of it; I just don't drink. So the guys were all going out drinking, and I went out to have a taco. I was wandering around and I saw a Barnes &

Noble, an absolutely beautiful building. And I went in and I saw that book. Seneca. Seneca writes, 'Why not you? What do you think, you're special? Bad things happen in the world. Bad things are going to happen to you. It doesn't mean you're a bad person. In the same way, good things happening to good people doesn't mean they're good.' I found it interesting. This bad thing happened, but I lived through it. A good thing happened. I don't know. I have no answers. I try not to go there and question it. ■

Firefighters Billy Butler and Tommy Falcon
Ladder 6

■ BILLY: We see some people outside the north tower; they are on a curb, and badly burned, hair burned off and all. Captain Jonas talks to Chief Hayden, who tells us to go up and see what we can do.

■ TOMMY: I'm seventeen years in the company, and I was here for the explosion in '93, so I know these buildings. I see two cops come over with blankets for the burned people as we leave to go up. People were running south, toward tower 2, and then when the second plane hit tower 2, they stopped and began running the other way. It was terrifying for them.

■ BILLY: Somebody said, 'They're trying to kill us.' So

we start up. On the stairs, too, people are coming down, skin just hanging off. One woman had no clothes on, a piece of a skirt and panty hose, everything else burned off her. Someone had thrown a jacket around her. So we are going up to the fire floor to do searches, [as if it were] just another high-rise fire. We paced ourselves, and I'm carrying tools. Someplace high up, 25 or 26, there is a chief with a bullhorn, telling everyone to evacuate. There was a guy there from Engine 9 who was one floor ahead of his company, and I was one floor behind Captain Jonas and my company. So I tell him to go catch up with his company. He says, 'No, you come, too.' I say, 'No, I have to wait for 6 Truck to come down; you go down – get out of here. We'll be all right.' So my company, Ladder 6, comes down, and we start down together. Then we come across this woman, Josephine, a large woman, a couple of hundred pounds. She has something wrong with her feet, and she's slow, but we do everything we can to help her along. But she slows us up a whole lot, and we don't want to separate the company.

■ TOMMY: So then the collapse. The rumble, and the noise. . . . One of the guys, Matt, is the last guy coming down the stairs and behind us, but somehow during the collapse he goes flying over everybody, over me and Captain Jonas and Billy, and he's now two floors below us. Captain Jonas takes an accounting of everybody; there are about twelve of us there,

145

including a civilian and a police officer, Officer Lim, from the Port Authority. He gives a Mayday over the radio. Chief Picciotto was there, and then Matt was talking to a couple of firemen from Engine 39 who were below us. It was black; we couldn't see anything.

■ BILLY: Several of us had cell phones, and we kept trying to call the dispatcher, the firehouse, the police, anyone, but nothing worked, and we couldn't get a live line. Then I tried my wife, who is working up in Orange County. And I get through. 'Are you okay?' she asks.

'We're at the World Trade Center,' I say, 'and we're trapped.'

'What?' she says.

'But we're okay,' I say, wanting to throw that in real quick. But she starts whimpering, and I say, 'Listen, you can't cry now. You have to remember some information for me here. We're in stairway B, fourth or fifth floor, tower 1. Get through to the fire department.'

She said later that she was crying so bad, she couldn't do anything for a few minutes, but then she got through to the firehouse, and Chris Fisher of Engine 9 just happened to come in from home. He was the only one in the firehouse, and he called Steve Godette of Ladder 6 who was at the scene, and Steve finds Chief John Sulka and tells him where we are.

In the meantime, the smoke is lifting, and we can see a vague light off in a distance.

■ TOMMY: I keep thinking, there are six stairways in these two buildings, and just these four floors of one set of stairs are left? It seems like everything else fell perfectly straight down. So why am I here?

■ BILLY: So Captain Jonas says, 'Let's set up a rope operation.' And so we set up a rope operation even the department doesn't recognize because it is so complicated – a tensionless hitch on the newel post, with a butterfly knot halfway up. So Tommy wrapped it three times, snapped it on itself, and I did a butterfly at the top at the next landing and then clipped it to myself there. Then we did a Munter hitch – this is all climbing stuff, but, you know, Captain Jonas teaches rescue operations, and I teach ropes to the New York State rescue school at Camp Smith, and I was also in the rescue company in Fairfax, Virginia, before I got this job. This allows us to climb over to the light, holding on to the rope. There are voids just next to us that go down forty, fifty feet, and so we have to step carefully every inch of the way.

■ TOMMY: So then I see a lieutenant from Ladder 43 come out of the rubble from the Church Street side; he comes in through the small [area of] light. He is real intense and says, 'The way we came in is how we'll go out.'

■ BILLY: One of the hardest things on the firemen right now is that the families are calling the firehouse

147

and talking to friends, people the wives know, and asking them, 'Please bring my loved one home.' And the firemen are stressed beyond belief. The biggest thing we do is save people, and we bring them out. We bring them out of buildings, and now a lot of guys want to go down because the families want them to bring the guys home, and because they want to go down, but the department is telling them, you need to go back and cover your own company. We need your company to be in service. And so it's like a shoving match, the job says this and the families say that. . .

Lieutenant Jim McGlynn
Engine 39

■ As we were going down 2nd Avenue we could see the thick black smoke, and I knew it wouldn't be a regular high-rise fire, that there was an accelerant involved. When we went across Houston, the hole in the building looked to me like there might have been a huge explosion. I saw Chief Ganci in front of the building [the north tower], and he had this look of fear in his eyes that said we have a major problem here. His aide told us to watch out for jumpers, and we got ourselves into the building. We were directed to go up the B stairwell, and we went up, stopping every ten floors for a rest. We made it up to the thirty-first floor and were regrouping there when we all felt this tremendous rumble, where we all hit the floor and the building began to shake. Not knowing that the south tower was hit, we thought that it was the north tower causing this vibration. We found out later that this was caused by the south tower collapsing. There was a chief there who was working on a command channel, and he said, 'We're evacuating the building. We're out of here.'

I told the guys to take the masks, to leave everything there – tools, roll-ups – and we started down. On the way down there were people we were helping, older people, people with injuries. John Drumm was helping a woman, and when I got to the lobby of the building, I noticed that he wasn't with us. I asked Bob Bacon when the last time he saw John Drumm was, and he said on the second floor. So we went back into the building; I went into the stairwell, and we started walking up to look for him. We went to the second and then the third floor, shouting his name, but we didn't see him, and we went up another floor. We thought he must have made it out, and we decided to go back down. As I was between the first and second floor, and Bob Bacon was between the second and third, that's when the whole building began to shake and come down. I heard and felt those vibrations, and I knew it was the building coming down. Then I started to hear the floors coming down individually, the pounding of one floor collapsing into the next, in pancake fashion, and this sound started to get louder and louder and worse and worse, and I knew it was getting closer and closer. I just hit the floor and rolled to the nearest wall, to get under some type of structure that might offer some support. The noise gets louder and louder, and I am waiting there for the one noise that will come and end it, that something will hit me and end it. Then the noise suddenly stops, and I say, 'I guess I'm going to survive that part.' Then I heard the rest of the cement come tumbling down, the rubble. So I survived the collapse of the steel beams, and now I'm going to get covered up by rubble, and I said, 'This is not the way I want to go.' Then I felt this rush of air, and it was going up. That stopped, and I said, 'Can I pick myself up?' I took a quick inventory of myself and realized I was alive and in one piece. I was shocked to find this out. I immediately began giving Mayday messages, 'Mayday, Mayday, Mayday.'

Bob Bacon was hanging on one of the I beams. He fell a couple of floors and ended up on this beam. The stairwell is now something out of a horror movie, the beams twisted like pretzels and the stairs pushed against the wall, not even stairs anymore. The half landings were half there and half gone. The landing beneath me was totally blocked. So Bob and I crawled up to the next floor, and there was a doorway there, to the second floor. I get no response to my Mayday messages, but I understand that there is a reason for this – how can anyone hear anything the way the radio is going? I didn't yet know the condition of this building, that it had fully collapsed, or even that the south tower had collapsed.

I now get contacted by Chief Prunty, who told me he was in the lobby. And I said, 'Chief, hang on, we'll see if we can get to you.'

He told me that his legs were pinned by an I beam, and that he was feeling numb and starting to lose consciousness.

I said, 'Chief, hang on, try to stay awake. We'll do what we can. Try to stay alive.'

At that point, he just said, 'Tell my wife and kids that I love them.'

We went up to the next floor where someone from Ladder 6 was, Matt, who had a rabbit tool [a hydraulic tool that forces open doors at the frame]. We forced the door to the second floor hallway. It was just a maze of debris and fallen I beams, and led nowhere. I realize now that there is no way to get to Chief Prunty. There is no way at him from below, no way out from my floor. Ladder 6 up above tells me that I should stay where I was because there was no way out from above, at least not that they could see.

Now, Firefighters Coniglio and Efthimiades were on the first floor, and they tried to talk to him and keep him conscious. I asked them if there was any way they could reach him. They said, 'No, we're trapped in here.'

They were on the first-floor landing, and what saved them is that they kept the door opened, and the debris and the ceiling from the landing above were now being held up by the door. On their right there was just this big hole. They flashed their light into it and couldn't see a bottom. They dropped something into it and never heard it land – that must have been the subfloors of the World Trade Center, where it goes down another seven stories. They were on this three-by-five landing, with nowhere else to go. They asked if they should try to get up to me, and I asked them if they felt safe where they were. They said, 'We're okay,' and I told them there was nothing going on up here with me.

Everything is just debris and dust and smoke, and we still did not know where we actually were, just where we thought we were. We did not know the condition of the outside, and we did not know that this building, all 110 stories, had come down, or that the south tower was down. So I am thinking there are other units operating around us, and they will just come in and get us from the ground floor.

We stayed there for quite a long time until we heard from the men of Ladder 6 above us that they had found a way out. This was a great relief, but I knew the place was still very unstable. There was a large crack in the cement on the half landing just below me, and I was hoping it wouldn't fall because who knows what would have happened if something else failed and crumbled? In the meantime, Bob Bacon was bleeding from the mouth, and I told him to go up and out with Ladder 6. I thought I had better stay to guide whoever showed up where Efthimiades and Coniglio are below, and where Chief Prunty was located.

Finally, Glenn Rohan of Ladder 43 came down, and I was glad to see him.

'Jimmy, how you doing?' he asks.

I say, 'Glenn, I'm okay, but how many men do you have with you?'

'It's just me, Jim.'

'Glenn, we need about five guys because I have two guys beneath us, and then Chief Prunty is below them. I'm scared to death to touch anything, for I'm afraid that the floor is going to fall in on top of these guys. I need some extra units in here to reshore the floor. I'm afraid I'm going to lose them after all of this, after four hours in this thing.'

'Well,' Lieutenant Rohan says, 'it's just me. And, we've got to get them out.'

So together we went down one more flight, and I shined the flashlight to the guys below. They could see it. I knew the only way we were going to get them out was up through a hole we would make in the floor.

They were looking up, and said, 'It's about sixteen feet up, but we can climb up some utility pipes and get there.'

So I asked Efthimiades, and he said he could fit through the hole. Coniglio said, 'Lieutenant, can the fattest man in the fire department fit through that hole?'

[Though I was] ready to lose it, at the edge of keeping things together, Coniglio made me laugh, which I appreciate.

So I backed off a little, and [left] Rohan on the

landing [alone] because I didn't want our combined weight on the concrete. He was able to enlarge the hole just enough to drop down a rope and a Halligan tool to get the two firefighters out.

We hadn't had contact with Chief Prunty for two hours and we knew it didn't look good for him, but Rohan and a firefighter from rescue and another from Engine 54 got down there. Now, there was no reason for me to stay, and I climbed up to the fifth floor, or wherever it was, and I saw the guide ropes leading me down and out of the collapse. It was only now that I saw the great destruction all around me, like in movies of the Second World War. ∎

Officer Dave Lim
Port Authority Police Canine Unit

∎ My office is in the first basement level of the south tower, where I have my desk and a cage for my dog, Sirius. He is a yellow Lab retriever, and he might have been the brightest dog I ever have seen, which is why I named him after the brightest star in the Canis Major constellation. He is a great dog, trained in explosives detection, and he is also the family pet. Our job is to check every truck that comes in and around the building for explosives.

I jump when I hear the plane go into the north tower. It sounds like a bomb has gone off, and I say to the dog, 'Maybe they got one by us, Sirius.'

I leave the dog there, and run over to the north tower. I see a dead body next to the bandstand where they have set up for a noon concert, and I call it in on my radio. 'WTC, I have a DOA on the plaza.'

And WTC radios back, 'Is that DOA confirmed?'

I am about to answer when another body lands just fifty feet from me with a very loud noise, [and again] I jump. I look over and see that the skin has been forced away from the flesh. 'There is another,' I say, as I run into the lobby of the north tower and up the B stairs. I see people coming down, and I keep saying, 'Go down, down is good.' On the twenty-seventh floor I come across a large man in a wheelchair. He will be difficult to assist, but I call it in on the radio, and proceed up to the forty-fourth floor, where there is a sky lobby, a large, open space where people change elevator banks. I look out of the window and see this huge fireball rushing out of the north side of the south tower. This is the second plane.

Suddenly, all the windows on the east side of my sky lobby are blown out by the concussion, and the wave hits me and several people around me, and we are thrown to the ground. We get up and start down the stairs immediately. I go from floor to floor and try to make a quick search, but I don't stop at any floor where I see firefighters because I know they are searching.

Somewhere between the fortieth floor and the

twentieth I feel the building shake, and then the radio says that the south tower has collapsed. Obviously, if that building collapsed, we are now in a lot more danger. There is a call for immediate evacuation. I am thinking about Sirius in the south tower, but I think he must be safe because he is in the basement, and the collapse would be above. So, I get down to the fifth floor and there I meet Billy Butler and Tommy Falco from Ladder 6. They have a woman named Josephine with them. I begin to give them a hand carrying her, and then the building starts to collapse.

It is an incredible sound, like the combination of an oncoming locomotive and an avalanche, with a huge windstorm right behind. Everything is shaking, like in an earthquake, and it feels like an eternity, that it is never going to stop. I know what I am in, and I just want to see my wife and two children again.

But now there is a dead silence, and it stays dead until I hear someone begin to take a roll call. 'Who is that?' I ask.

'Captain Jonas,' I hear. We cannot see anything.

There is a man, Mike Meldrum of Ladder 6, sitting on the landing. He has been hit hard, and has a concussion. He is pretty woozy, and he ties his personal rope around his waist, and I tie it onto a handrail to keep him stabilized in case he falls. Mike is asking for something to drink, anything, because he has all the debris and dust in his lungs. I guess we all do. Amazingly, I find a can of soda on the steps, and I pass it to him. He takes a big swig.

I hear a radio message from Chief Prunty, who is a few floors below us. It is a very sad thing to hear this. 'Listen,' he says. 'you have to get here quick.' The firefighters try to get down to him, but the whole second floor is completely entombed, and there is no way to get down to him.

We smell jet fuel, and I wonder now, having survived this, is everything just going to explode into fire? I have been on the job for twenty-one years, and I just think in terms of emergencies.

We climb up to the sixth floor, and there the smoke and dust begins to clear, and suddenly there is a little light, and then a big light, and I realize it is the sun. We seem to be on top of the collapse, if that is possible. I say to anyone, 'What are the chances that we have survived this?'

Chief Picciotto is there, and he answers, 'One in a billion, Officer,' he says. 'One in a billion.'

I try my wife, Diane, on my cell phone, and I get through to her. We talk for a while, but she doesn't want the conversation to end. She wants to stay on the phone until she dies or I die, till the end. And I have to convince her to hang up so that I can give the firefighters a chance to call home, too.

Captain Jonas has communication with people in the street, and finally we see Lieutenant Rohan come in the distance. Chief Picciotto and Captain Jonas make a rope guideline, and we send Meldrum out first because he is very dizzy and hurt. I go with him, and we have to stop a few times on the climb out, so that

Mike can rest and keep it all together. Rohan sends two firefighters to help Josephine and prepare her for the Stokes basket that will carry her out, and he tells another firefighter to show us the way out of the collapse area, [via] the way he came in.

All of a sudden I started to hear gunfire. These rounds popping off, and I think of our enemies. I know they attacked us, and maybe they landed on the beach. Maybe we'll have to fight. I have forty-six rounds on me. And then I realize that the Secret Service has a firing range in the next building. I look at my watch. It is now 3 P.M., and it is the first time I get to see the scene I am in. It reminds me of the war scene in the first *Terminator* movie, everything destroyed. We get led out – it takes a long time and a lot of climbing. The ambulances come to meet us, and sitting in one, I say to Mike, 'I have to go get my dog. He's downstairs.'

'Good luck,' Mike says.

I go over to tower one, and there are several cops and firefighters there who will not let me pass. It is too dangerous. I tell them about the dog, and they said I could not try to go down in any case. I had Sirius for eighteen months, and I truly believed he would survive. He's a good dog. He'll just wait there in his cage until somebody comes to get him out. If we could survive, he will survive.

They take me to St. Vincent's Hospital and admit me. I'm beat up, and my breathing isn't right. I can't see. The Port Authority Canine Unit brings my wife and my children, Debra, 14, and Michael, 12, in to visit. This is a very emotional time for me, and I am very happy when I am with them.

After a week, I have to realize that Sirius isn't coming home, and I have to have a long talk with my son, Michael, about that. He has bonded with the dog, and it is a sad time for us to know he won't be coming home, and he won't be there on the small rug by the bed where he sleeps every night. He was a good working dog, and he was a great family pet at home. The rest is history. ■

Captain Jay Jonas
Ladder 6

■ I'd come downstairs after getting cleaned up for the day tour. I worked the night before. I shaved and had gotten dressed, and I was in the kitchen having a bowl of Wheaties and a cup of coffee. The guys keep telling me I should call the Wheaties company because I must hold the record. . . .

All of a sudden we hear this loud boom, and I say, 'What the heck was that?' It sounded like a big truck driving off the Manhattan Bridge, which is only about a block away from the firehouse.

And then, my housewatchman starts banging on the intercom and yelling into the intercom. 'A plane just crashed, a plane just crashed into the World Trade

Center.' So we go running out the front of the firehouse, and I'm standing on Canal Street, and I can see the black smoke wafting over, and I said, 'Oh, my God.'

I keep my bulky gear in the office, which is right off the housewatch, and as I'm pulling on my bunker pants I hear on the department radio, 'Engine 10 transmitting a third alarm and a 1060 signal [a catastrophic emergency, like a plane or train crash].'

By the time we get on the fire trucks and are heading out the door, our alarm kicked in. The teleprinter starts going off. Both of our companies are second-alarm units there, so I knew we weren't jumping the gun. As we head down Canal Street, right by the Manhattan Bridge, I have a panoramic view of Lower Manhattan, and I can't believe what I am seeing. The top twenty floors of the World Trade Center are on fire. From that time of me sitting in the kitchen to getting my gear on, and us pulling out of the firehouse, there are twenty floors of fire? What did that take? All of thirty-five to forty-five seconds? I can't believe what I am seeing; it is something out of Hollywood. I've worked in the Bronx. I've worked in Harlem. I've worked on the Lower East Side. I've seen some unbelievable fires in Chinatown, and this is the most amazing thing I'd ever seen. I just yelled to the back of the truck, 'Buckle up. We're going to work.'

We get to West Street, park the truck, and begin gathering all of our equipment. We're right by the north pedestrian bridge that goes across West Street by Vesey Street.

We're keeping an eye out because there is stuff that is crashing near our truck, big pieces of the building coming off. So we're all seeking shelter under this footbridge, and once we get everything gathered, I turn to my guys and say, 'All right, is everybody ready?'

'Yeah,' they answer in unison.

I look for a time when I don't see anything falling and I say, 'Okay, run!'

And we all sprint to the front of the building. We make it to the lobby. The first thing I see there is two badly burned people, right at the lobby entrance door. I also see that people were starting to come to take care of them. There might be a thousand people upstairs that we have to take care of. I report in to Battalion Chief Joe Pfeifer.

Deputy Chief Pete Hayden was there already, and a couple of officers were in front of me. So, I am waiting my turn in line, waiting to get my orders. I see Paddy Brown from 3 Truck who says 'Jay, just come on upstairs; they're just going to send you upstairs.'

'No,' I say. 'Let me check in. Let me get on paper that I'm here.'

That was the last time I saw Paddy Brown.

I report in, and just as I'm about to get my orders we hear another loud explosion. I'm not sure what this is, whether it was a few tanks blowing up on our floor, or what. But a guy comes running into our building and said a second plane just hit the second tower. I looked at Chief Hayden, and he says 'Jay, just go up, and do the best you can.'

I say, 'All right, chief.'

So we knew that things had radically changed from 'Oh, what a horrible accident,' to 'Oh, my God, they're trying to kill us.' I saw it. I am standing next to Jerry Nevins, from Rescue 1, when the plane hit. Terry Hatton is there, Dave Weiss, all those guys and we just look at each other, and Nevins says, 'Oh, man, we're going to be lucky if we survive this.'

I am not thinking collapse. I am thinking eighty stories up, for I have in the back of my mind that we have to make the eightieth floor before we will hit anything major. I figure there will be fuel dropping down, and vapors. I figure that will be when we will start hearing things. 'This is going to be a long climb,' I say to my men, 'and I know it's going to be hard, but that's what we're doing.'

Maybe a lower bank of elevators is working that might take us to the fifteenth floor. But I'm not comfortable with that, especially after seeing the people in the lobby that are burned. So that's not a good option.

There is a line of firemen waiting to get into the big stairway. We jump on line and start heading up. I tell the guys, 'We're going to take frequent breaks, so we have something left when we get to the upper floors. We're going to rest every ten floors, catch your breath for a little bit, and then we'll push on.'

We immediately start encountering civilians coming down. They are to our left and we were on the right. It is enough for two people to be comfortable side by side. Every once in a while we stop a civilian trying to come down in the middle. I say, 'Stop, stay to your right, and this way everything will work.'

For the most part the civilians are very calm. They are on every floor where there is a vending machine, breaking into them, and taking out bottles of Poland Spring water to give to us. That was terrific.

We are staying nice and hydrated on the way up. Every once in a while we see somebody who is burned coming down, but I can't believe how remarkably calm they are. And they are shouting out words of encouragement to us. 'God bless you,' and 'You guys are remarkable.'

We just keep going up, and up. We meet a member from Engine Company 9 who is having chest pains, so we stop, and we take care of him until his company gets with him.

Going up a little farther we find a fireman from Squad 18 who is also having chest pains, and we stop to take care of him quickly.

We move on again. We get to the twenty-seventh floor. I am ecstatic that everybody has stayed together. Each floor in this building is like an acre.

'This is the mother of all high-rise buildings,' I say, 'and I can't afford to be looking for you guys. So stay together.'

When we get to the twenty-seventh floor, I'm

missing two guys. I tell my other guys to stand fast, and I go down to look for the other two. They got separated due to the civilians coming down, and I find them. So we are all on the twenty-seventh floor, and taking a quick break. I notice that I am no longer seeing civilians. I'm on the twenty-seventh floor with Andy Fredericks from Squad 18 and Billy Burke, the captain of Engine 21. And all of a sudden we hear the rumble, an earthquakelike rumble, and the loud rush of air like a loud jet engine kind of a thing. I look at Billy Burke and say, 'Go check those windows, and I'll check these windows.' We ran our separate ways, and then came back.

'Is that what I thought it was?' I ask.

'Yeah,' Billy answers. 'The other building just collapsed.'

I just look at my guys now, and say, 'It's time for us to go home.'

If that one could go, this one could go. As we are leaving, one of my men wants to ditch the roof rope, but I say no, we might need that. We don't know what we might encounter on the way down.

We start heading down. There is no immediate call for evacuation of our building, at least I don't hear one, and I am a little nervous heading down. Are we doing the right thing? This is a hard climb up, and I would hate to cover the same territory twice if there is no evacuation.

But I say to myself, *This is not good,* and we just have to get out of here. I realize right away that our situation is pretty grave and that we have to get a move on.

We are in stairway B, and somewhere around the twentieth floor we run into Josephine Harris.

Tommy Falco and Billy Butler find her on the stairs. She can barely walk because she has a serious case of fallen arches or something. She has already walked one step at a time down from the seventy-second floor. Falco says, 'Captain, what do you want to do with her?'

We will have to bring her with us, I think. Tommy Falco and Billy are big football linebacker types, and I told them to keep with her. But she was very slow.

Somewhere close to the twentieth floor, I run into Chief Picciotto. I am surprised. He and I have been friends for a long time. We are in the same study group. We socialize together. But I work in Chinatown, and he works on the Upper West Side, so chances of us running into each other are pretty remote.

'Hey, Rich, how are you doing?'

'Jay, all right,' he says, and starts heading down with us.

I see Chief McGovern's aide, from the 2nd Battalion, Faustino Apostol. I know him pretty well, for he drives me whenever I am working as acting battalion chief. I look at him and say, 'Faust, let's go. It's time to go.'

He answers, 'That's okay. I'm waiting for the chief.'

So he is just standing in his post. He doesn't want to abandon his boss.

As we are going down, I see a few guys from Ladder Company 5 taking care of a man who is having chest pains. They are on a landing, and I know the officer, Mike Warchola.

I say, 'Mike, let's go. It's time to go.'

He said, 'That's all right, Jay. You got your civilians, we got ours. We'll be right behind you.'

We are moving down very slowly. A few times we stop to let other firefighters go past us. One company that passes us is Engine 28. I used to be in that firehouse as a lieutenant.

When we get to the tenth floor, I think that I am glad we brought the rope because it will reach the sidewalk from here. I say to the guys, 'Well, we can single slide with the rope now if we have to.'

Finally we get to the fourth floor, and Josephine is saying she can't go any farther. So the guys try to keep her spirits up.

I wonder now if I am getting a false sense of security. Maybe this building is not going to collapse because I look at my watch, and it's been almost twenty-three minutes now since the other building collapsed. I say to the guys, 'Well, maybe this isn't going to come down.'

But now Josephine falls to the floor, and she is crying. She says, 'I can't walk anymore.' Meantime, other firefighters are passing us by. We stop to deal with her for a little while, but finally I say, 'We gotta get moving.'

The fourth floor door is locked, and I break into it, looking for a sturdy chair. We can throw her in a chair, I think, and we can run with her. But there are no chairs. It's a mechanical equipment room, so there isn't much furniture – one desk with a stenographer's chair that won't do. There are a couple of overstuffed chairs and couches. They won't do either. So I start running back to the stairway, and I'm about six to seven feet away from the stairway door when the collapse starts.

I tried the stairway door, and it wouldn't open. But after a second pull, assisted by a gust of strong wind coming down the stairs, the door flies open. And I dive for the stair. I just crawl up into a ball on the stairway landing, and I wait for something to hit us.

There is unbelievable shaking. Almost like I'm being bounced like a basketball. I am literally bouncing off the floor, like if a train derails and the wheels are hitting the railroad ties. It was that kind of *boom, boom, boom, boom* in a loud succession. It's unbelievable, the sound of these massive steel beams and gutters twisting around you like they were twist ties on the loaf of bread. It is a painfully loud screech of steel all around us. Debris is falling on us like a deluge shower.

As it is collapsing, I think, *This is it. It's over. This is how it ends for me. I'm done.*

I thought of Judy and the kids and that's about it. *This is it. It's over.*

And now the sound stops, dead. Just dead silence.

167

My first thought is, *Oh, man! I can't believe I just survived that.* But it's not like we were in this pristine stairway. All kinds of debris is still whacking us all over the place, and knocks my helmet off. I see a couple of my guys, and it knocks their helmets off. It is like we got mugged, we got beat up pretty good in the stairway, but we didn't get hit with anything that was gonna hurt us seriously. Once the collapse stopped, I start to take a deep breath. We are all coughing and gagging from the dust.

And we can't see anything because of the cloud of dust and smoke.

I give a quick roll call to see who we had. I go through all my guys. I say, 'Is Josephine still here?'

'Yup.'

'How about Chief Picciotto?' I knew he was in front of us.

'Yeah, he's here.'

My first thought is that we will continue down. There was probably a collapse on the upper floors, and it can't be that the lower floors are too bad because we're still here.

Josephine can't walk, and we will have to move her. I teach rescue classes for the state, and I have gotten my men to carry one-inch tubular webbing – it's a twenty-four-foot piece [with which] you can make a harness and everything else. I said, 'Put her into a full body harness, and we'll carry her down the stairs.' The guys get down one flight of stairs, and Matt Komoroski, our lowest guy, said, 'It's no good. There is no getting out this way.'

There are two handholds on this webbing, and they pick up her upper body, and her feet are dragging on the way down. She's going down head first, but there is no getting out.

I am thinking that if our stairway was good, we could just walk right out of the lobby. But this isn't the case. So they bring her back up the stairs, back to where we are. Now I hear the radio begin to squawk. 'This is the officer of Ladder 5, Lieutenant Warchola, Mayday, Mayday, Mayday. I am on the twelfth floor, B stairway. I'm trapped, and I'm hurt bad.'

And so I picked myself up out of the rubble, and I started climbing up to the fifth floor.

I had to move a lot of debris [on the way], but when I got there I couldn't move anymore. It was too big and too heavy; it would take a crane. Which broke my heart. When I first heard him I just thought I could climb up there; I could get to him. I've known Mike Warchola for a long time. I used to carpool with him when I was a fireman. And so I press the transmitting key, and say 'Mike. I'm sorry. I can't get to you.' This is such a hard thing to say.

At this time we start making radio contact with Chief Richard Prunty. He is just below us, on the first floor. Matt Komoroski was on the second floor along with Lieutenant McGlynn of Engine 39. Mickey Kross of Engine 16 is between us.

Chief Prunty isn't able to get much radio contact on the outside, and I think that because we are a little higher, I am able to make pretty good communications.

Chief Prunty, in the meantime, radios that he's in the stairway, he's hurt bad, and he's pinned. And he is starting to feel dizzy, which is very hard for us because we can't get to him.

It is so frustrating to know that geographically you're not that far away from someone, that he's slipping away, and there's not a thing you can do about it. But [at the time, we had no] concept of what the damage from the outside might look like.

We see a light that, after an hour or more, grows bigger and bigger as the smoke lifts. There's a small hole in the stairway where I am on the fourth floor. And I can see out. But all I see is a mountain of steel and the big dust cloud.

From where we came in the stairway, I know looking out that wall is east, and I know exactly geographically where I am in the building. I'm not sure if there are 106 floors above me, but there's got to be something dangling over my head here. So I just kind of stay put. Again, I have no concept of what it looks like outside. Will they be able to make their way to the front doors, move some debris out of the way, and come and get us? I am thinking, *Get to Chief Prunty!* Just pull the debris up, and get to him. And I am still thinking that we can work our way down the stairway.

But then I start getting my Mayday message out, and the first solid hit I get is Deputy Chief Tom

Haring. He says: 'Okay, Ladder 6, I got you. I have you recorded. I have exactly where you are; you're in the B stairway.' So that was good. I said all right, they know where we are.

But then Battalion Chief John Salka gets on the radio, and he calls for me specifically. 'This is Battalion 18 to Ladder 6. Where are you? We'll come to look for you.' So I gave him the pinpoint location, and then my neighbor Chris Dabner of Rescue 3 radios me. Chief Sal Visconti contacts me. He is an old friend, and I feel really good that he is on the other side of the airwave. I also let him know what to my knowledge is our location, and it is all very comforting to know they are on to us.

Then I hear Chief Billy Blaich, who used to be the captain of Engine 9, which was the engine in our firehouse. His son is in Engine 9 right now, so there is a definite connection. The chief and I are friends, and he tells me that he can approximate exactly where we are, the exact location. He's such a squared-away guy, precise, a colonel in the Marine Corps Reserves.

He tells me, 'I have everybody off duty from Ladder 6 and Ladder 11, and we're coming to get you.' It is also good knowing not just anybody is coming to get me, but guys who have a personal stake for my safety.

There are a whole bunch of guys who are zeroing in on my transmissions because my transmissions are the only ones getting out.

I'm thinking, *Nobody in front of me was bleeding to death, and we were safe.* I feel as long as we are safe in the stairway we just have to figure out what's the best way to go. We don't want to make a move just to make a move. We want to make a move that we know is going to be productive, and is going to get us out of the situation. So we are evaluating everything.

Before we gave out the Maydays, Rich Picciotto yelled out, 'Everybody turn off your radios.' So even I turned off my radio. Now Richie is a floor below me, he's on the third floor. But I'm thinking to myself, after he gave that order. *You know what? He's on the command channel.* They ordered the chiefs to go on the command channel. I'm thinking to myself, *I'm going to turn my radio back on, and I'm going to give my Maydays out on the tactical channel.*

[Three days later, Chief Picciotto is sitting in my kitchen. We're having a cup of coffee, and I tell him, 'I was giving a lot of Mayday messages.'

He says 'What?'

I say, 'Yes. The whole world was coming for us.'

He went pale. He says, 'Oh my God, I thought I had the only line out.' He had been talking to Chief Mark Ferran, who had Ladder 43 with him. But, he thought that his messages were the only ones getting out. He says, 'Oh, man.'

We were both in the same stairway, but my experience, even though we were only a floor apart, was a little different from his.]

We now go into this survival mode kind of thing.

I don't want anybody to use their masks. I say, 'We are breathing the dust, and it is irritating, but it is breathable. We are hearing on the radio that guys are coming, and they are asking to have hose stretched. We may have a fire to deal with in a little while. I don't know. So don't use your masks if you don't have to. Don't use your flashlights 'cause we may be here tonight.'

There's a little bit of light, but not enough for us to realize that even in a collapse, we are in good shape relative to what could be. When I was roaming around the stairway before, I broke into the fourth floor. I found a toilet there, and there are eleven of us in this stairway. So it won't flush, but at least we can cover it up afterwards, and we won't be smelling human waste while we are trapped in this small area. The floor area is not more than fifteen by fifteen feet.

Matt Komoroski broke into the third floor, and they found some sprinkler pipes. They thought that down the road we can break into these pipes and maybe get some water.

We find an elevator shaft, and the door is warped a little bit. I shine my flashlight down and I could see other floors below us. I say, 'We have the lifesaving rope, and could rappel down the elevator shaft and get in the lower floors.' But then I think, *There is no guarantee we could get in on the lower floor. And then how do we get back up?* So maybe that is not a good idea. That goes on the back burner for Day 2 or 3, if we get desperate. We all seem to be patient. We're looking, we're transmitting Mayday messages, trying to get a grasp of what our situation really is.

And then, about three to three and a half hours later, all of a sudden a ray of sunshine breaks through the smoke and the dust. Once we see that, I say, 'Okay, guys, there is supposed to be 106 floors above us, but now I see sunshine. There is nothing above us.'

Rich Picciotto says, 'That's it, Jay, this is our way out, we're out of here.'

I say, 'Rich, it probably is.'

It takes another twenty minutes, and it clears up enough again, and then off in the distance we see a fireman from Ladder 43, about one hundred feet away from us.

Never until this moment did we realize that there was sky above us, that we were virtually at the top of the collapsed building. [In the end] we were able to pull ourselves out. Just seeing Ladder 43 was enough for us to realize that this was our way out. We decide that Rich Picciotto will go. We tie him off on the life-saving rope, and we put a special rescue knot on it. I had Billy Butler tie off with a Munter Hitch, which acts like a seatbelt. It plays out fine, but if there is a sudden jolt it will lock off. I want that knot used because if Rich falls, we will be able to retrieve him.

Richie goes out, and it had to be about a hundred feet before he meets the guy from 43 Truck. Richie ties off his rope on a beam that is sticking up out there. We tie it off on our end, and now I start sending the guys out. [First to go is] Mike Meldrum, who

174

has been beaten up pretty bad. In the meantime, half of our people are out before 43 Truck gets this far in to the stairway.

The guys have to walk on a ledge to get out of the stairway, hold on to this rope, and drop down. There is a lot of sharp metal all around us, and it was good to have that rope up for stability. I'm glad we didn't leave it behind.

The firefighters from Ladder 43 started coming in, and I meet Lieutenant Glenn Rohan. To me he is the handsomest man I ever saw, and I tell him that when things settle down, I'm going to stop by with a box of canolis. I then give him a briefing, and tell him about Chief Prunty.

When I get out and onto a clear pile, I see that 7 World Trade Center and the Customs House have serious fire. Almost every window has fire. It is an amazing sight.

The Customs House is immediately adjacent to this area, and the Secret Service ammunition bunker is there. Just as I get out, that starts exploding. So there's smoke from that, and the smoke from 7 World Trade Center is now coming across this debris pile. And there are large two-story-deep craters that we have to crawl down into and climb up out of, using ropes to pull ourselves up. It is an ordeal, and it takes us a half hour just to make that trek from the stairway to West Street. From there we have to go through the World Financial Center to get where the ambulances were. All in all there were fourteen of us in there, eleven of us, Lieutenant McGlynn, and two firefighters from Engine 39.

Now we're on the other side of the World Financial Center, and I keep saying, 'Where's the command post, where's the command post?'

The guy says, 'Don't worry about the command post. Go get treated, go to the ambulance.'

I say, 'No, there was a lot of guys coming for us.' I know I have to check in. I don't want anybody getting hurt looking for me, especially when I'm out. So I make my way to the command post, and Peter Hayden and Jim DeDominico are standing on top of a fire truck in front of the fire so they can see across the debris. I get to the base of the truck and, naturally, there's hundreds of guys and all kinds of noise, and I yell up, 'Chief Hayden, Chief Hayden!'

I finally get his attention, and it is a very touching moment for me, and I guess for him, because I love working for him, and he loves having me. We have a long relationship, very positive, and he looks down, and I can see him tearing up a little bit. He says, 'Jay, it's good to see you.'

I just hold up my hand, saying, 'It's good to be here.'

I think he says something like, 'Well, now you're going to get promoted to battalion chief.' I think that it will be good to be around for that.

Looking at the tremendous devastation before me, I cannot believe that we have survived it. It wasn't until later that I realized that if we didn't stop to help

Josephine, we would have probably been in the lobby or just outside the building where everyone perished. Engine 28, who had gone past us, were running at top speed to get free of the collapse, and they just made it. And then, to think that all of these men I have talked about, Terry Hatton, Paddy Brown, Dave Weiss, Jerry Nevins, Faustino Apostol, Mike Warchola, and Richard Prunty, are gone — it is just too terrible to speak about.

I get some medical treatment, and they want to take me to a hospital. All I want to do is get back to the firehouse. I can't find any transportation. Our truck is completely flattened. On West Street I see a couple of sector cars parked with one patrolman there.

I grab him and say, 'You have got to give me a ride.'

He says, 'Oh, man, I'm not supposed to leave here.'

'Can you give me a ride at least halfway?'

'I'll give you a ride to Canal Street.'

I say, 'Well, that's something.' So he drives me to Canal and he's all nervous — he's a young officer.

I get out, and I start walking. I get to Broadway and I see there's no traffic. I have worked this neighborhood for years and I've never seen Canal Street with no traffic. But there are a lot of pedestrians, and they begin following me like I'm the Pied Piper. I still have all my gear on, but I'm dirty — like Pig Pen in Charles Schulz's Peanuts cartoon, with a big cloud of dust [trailing behind] me. About thirty people are following, giving me water. Someone says, 'Oh, my God, are you okay?' These are all Chinese people, and they usually keep to themselves.

I say, 'I'm better than most.'

Meantime, I am trying to get a word out to my wife, Judy, and I can't get through. I try one more time, on a cell that belongs to someone from Engine 9. I get a phone call out, and I am talking to Judy. She just keeps crying on the phone, repeating over and over, 'You're alive, you're alive, you're alive.'

And I just keep on saying, 'I love you, I love you, I love you.' ∎

Judy Jonas, Wife of Captain Jay Jonas
Ladder 6

∎ An installer from the cable company came in to change our cable, and he said he was listening to the radio and heard a plane just crashed into the World Trade Center. My first reaction was to call Jay and ask him about it because he works down there, in Chinatown. So, naïvely, I called the firehouse, and the line was busy. And a few minutes later my nephew, Jeremy Cassel, called – he's a firefighter with Squad 61 – and said, 'Are you watching this thing?'

I said yes, and was pretty calm. 'Is Jay there?' I asked. I thought, *The line was busy,* and maybe he wasn't.

'Yeah, he's there, but it's a fire, and Jay is a pretty good firefighter.'

I had a Cub Scout leaders' meeting scheduled, for a den mothers' conference, and Donna McLoughlin and Lynne Bachman came. We are all den leaders in Cub Scout Pack 63 in Goshen, New York. Donna's husband, John, is a Port Authority police sergeant, and he had actually set up the evacuation plan for the twin towers after the bombing in 1993. John was at work as well, in the twin towers. We sat in my kitchen with the television on, and Donna said, 'I know John's not there. If he's there, he's working on the outside.'

A neighbor called, who's in Rescue 3, and assured me that Jay was okay. It was just a fire. And I was okay until the first building came down. When I watched that, all of a sudden it wasn't just a fire anymore. I think I fell apart when that building came down, for I knew that my whole life could be falling apart with it.

Jeremy has asked me if I wanted him to go down, that he would go right away, and I thought, *He has a wife and two children;* I didn't want him to go. But the department had a recall, and he called again and gave me his cell number.

Donna and Lynne kept saying, 'Oh, they evacuated that building. Everybody must be out.' Jeremy and three other firefighters came into my kitchen on their way down, and he said, 'Don't worry, we are going down to get him.' Jeremy was a lifeline for me, for if Jay was in there, they would know what to do to get him out. They did calm me down a little bit before they left. In the next hour I had at least twenty-five calls, from family and friends, and then looking out of my window I see that a police car and the local fire chief's car are pulling into my driveway at the same time. This made me panic for a second. How could they know? But it was two of our friends stopping by to give me their numbers, and to say they are available for anything.

Then the second building collapsed, and I don't know why, but it wasn't affecting me the way it did when the first building did. I was upset, but it wasn't changing anything.

We have three phones, and they were all going at once. Someone called and said that Car 6 had called over the airwaves, and Jay used to be the volunteer chief here and he had Car 6, and I thought he was sending a message that he was okay. I said, *Oh my gosh, this is great, reassuring.* I called all three schools to find out what was happening with our children. I was concerned with Jennifer, who is 15, because they would have the television on in her school. I called the nurse and said just find her and tell her that her father is okay. And I was going to call John's school, he's 9, and Jane, who is 5, but the phone rang again, and I was told that it wasn't him, it was someone else. That made me feel very upset. Because now we're back to not knowing.

We were looking at the TV closely, looking to see if we could see John McLoughlin on the outside, or fire trucks, or anyone I know.

Then at about quarter to eleven, I get a call from Billy Butler's wife, Diane, and she tells me that Jay is trapped, but that they are trying to get him out. So now I am thinking that Jay is trapped, but that Billy is there trying to get him, and Billy is like a football field, so I know he will get to him. Billy can get to him with just his strength alone, so that's a good thing. He's on the outside working in.

But when Jay worked in rescue, he would tell me about certain jobs he had where people were trapped, so I know what the word 'trapped' means. Many people who are trapped don't make it. So I am not very comforted. But it is good to know he is alive and they are trying to get to him.

I am watching the television, and they have only one thing to show, the building going down again and again, a hundred times. Every time I watch it go down I say, 'How can anybody survive this? How can anybody be in there?' It is very stressful, and the wheels start to turn.

I am thinking that Jay was probably thinking about the Yankee game in there. And then I'm thinking, *Oh, my God, how do I tell three kids that their dad is dead?*

My emotions would go up and down depending on who called. When my brother called, I was hysterical because we lost both our mom and our dad in the last five years. I'm one of five, and we have been through those deaths, and I couldn't bear doing it again.

181

Jeremy calls me when he gets to Ground Zero. Every time I talk to him I break down, and I am putting like a thousand pounds on this young man because I said, 'Jeremy, he's in there, go get him.'

Later, when I look at pictures of the site, I can see how it wasn't so easy to just go in there and get him.

At about 2:30, Jeremy called me back and said, 'I can hear his voice – he's out. It's going to take me an hour to get over there, Judy, but I hear him talking on the department radio; he's out, and they are taking him to the hospital.'

We are all under such stress that I say, 'Well, he's not brain dead. He's talking, anyway. I'm still thinking that he was buried up to his eyes.'

In the meantime, when school got out at a quarter of three, Donna McLoughlin went home. Back in '93 when the bombing happened, her husband never called until late at night, and he had not yet called. But later that night, John's brother went to Donna's house and told her that John was missing. He was last seen walking from tower 1 to tower 2 when tower 2 collapsed. I went over there for a little while to be with her.

About fifteen minutes after Donna left my kitchen, Chris Staubner from Rescue 3 called and said, 'I just kissed your husband twice.'

I was so relieved. 'Well,' I said, 'don't get used to that.'

Chris said, 'He's walking over to get his eyes washed out at the ambulance.'

He's walking, I thought. Every piece of information helps. He's in better shape than I expected.

Jay finally got a phone that worked, and he called me. All he kept saying was 'I love you,' and he must have said that a hundred times. 'I love you, I'm coming home.'

I was so happy, I said to Jay, 'I'm going to give Billy Butler the biggest kiss he ever had,' thinking that Billy was one of those who got him out.

And, Jay said, 'Billy? What about me?'

It was then I learned that Billy was inside there with Jay all the time.

The next morning, Wednesday, they found John McLoughlin. He was forty-five feet down in the debris. It was a miracle. He was the last person taken out alive. But he was severely injured, and they rushed him to Bellevue Hospital.

So they had gotten Jay out, and they got John out, too, the next morning, and I thought that was unbelievable, so why wouldn't I believe they would find lots of people and rescue them? We just had that hope. Then, seeing these guys going down there every day, and coming back, talking about desperately digging down for their buddies into these pockets, and then hearing that there was nobody there . . . It is very overwhelming. Every wake I go to I have to say, this could have been me talking to all these firefighters tonight. This could have been me. Thank God, we were very lucky.

I don't know why. ■

Lieutenant Glenn Rohan
Ladder 43

■ It is all devastation. We are walking the piles, searching, and I hear this strange sound, muffled, coming from below us somewhere. It is Chief Picciotto's megaphone. It is the first time in twenty-eight years that he has carried a megaphone into a job, and now he has it in the alarm position. It goes on and off like a police siren. I follow the sound until I come to a sort of a void, and I look into it. There is an open space in there, and I see Chief Picciotto. What a great feeling to see him coming out. I go in and speak with Captain Jonas, and then leave some men to deal with helping those firefighters there. Now I look around and find a little space between a fallen beam and something like a wall. There are two triangular holes, about four square inches, by a stand-pipe riser, and I begin to pick at them. There are two layers of Sheetrock, and so I know there is a wall of some kind. I can only make an opening of about eight inches, and I hear two men beneath me. They are Firefighters James Efthimiades and Jeff Coniglio from Engine 39. So we drop a rope to them. They are only eight feet down, and we take Efthimiades first. We pull him up so that we can grab his hands, and then wiggle him through the eight-inch opening. I ask if the other guy is smaller or bigger than him, and he says bigger. Well, he comes up, and we just have to wiggle him a bit more, but we get him out. They tell us about Chief Prunty below.

I now drop into the hole and onto a slanted landing there. I slide down to the bottom, about twenty feet, and find an open door – I think this door is what saved the lives of these two firefighters, for the debris was packed so tightly into this door frame that I can't pull any of it away. But I could climb to the top and make an opening there. So with Mark Carpinello I climb over that, and we find our way down what might have been another two floors to search for Chief Prunty. We search through the dark, whatever openings we could find.

Mark finds him. The chief is in a bent position. I check closely, and he is gone. We try to give him CPR, and work on him for a while. He is a big man, and we are fifty feet below the hole we dropped into. It is very dangerous down here, I realize, and I have no idea how stabilized anything is. Now Jimmy Lanza and Jerry Suden have come down, and are helping as well. But we cannot dislodge the chief, and so we say a prayer for him there, and I say, 'We'll come back to get you tomorrow, Chief.' ■

Firefighter T. J. Mundy
Engine 36

■ I was voting. The girls had a little television, and they said, 'TJ, look at the helicopter that just hit the

185

Trade Centers.' Broad Channel is a small community [in Queens], and we all know each other.

I was only two blocks from my house at that time. I packed a bag. My wife was still home. 'There's a fire,' I said, 'at the WTC, and I'm going to go in and see if I can help.' I jumped into my van and headed up across Bay Boulevard towards the city until I got up to about 63rd Drive, where it was stop–dead traffic.

I just pulled over and dialed my brother's number. He lives two blocks away from me. I say to him, 'How's the boat? You gotta get me in there.'

He says, 'All right. I thought you were going to ask me that.'

He needed gas and some oil, so I jumped back in the van and grabbed some jugs, but the bridge to Rockaway was closed at this time. This was probably the forty-five minutes that delayed me, so that instead of getting there at 10:15 I arrived at 11:00.

When I get to his house he says to me, 'Tommy, this is bad. I think the towers fell.' He would know. He's a retired captain from Ladder 173.

I say, 'No, they can't fall, you know?'

We got into my brother's boat with about four other guys, two from Engine 285. My original thought was to get to Engine 36, but when I saw how bad it was, I wanted to go straight there. I knew we'd have a clear shot to get there with the boat. So we're in a brand-new, twenty-five-foot sports craft with a Mercury 25 outboard motor, probably capable of forty miles an hour.

When we got under the Verrazano Bridge, we began meeting cops and Coast Guards boats. We were stopped three times but just waved our gear and badges and helmets, and they waved us through. All the way in we see smoke rising from where the towers used to be.

I'm familiar with the pier down at Battery Park City, and when we pull in we see a fireboat already tied up, the *Harvey*. This is a retired fire boat, but they have put it back in service. There was a lot of activity now, and waves and all kinds of choppy water from all the tugboats and the Army Corps of Engineers boat arriving, so it took us a good ten minutes to unload our stuff and try to hold my brother's boat off. We had to make a makeshift ladder to get up onto the *Harvey* and from there to land.

All I could see was ground-up dust and dirt and paper; it was just all over the fireboat, all over the ground there. I climbed up, and you could see guys just covered with the stuff. I'm going, *Oh, my God.*

These other guys are all truckees, so I said, 'I'm looking for engine work, nozzle work, you guys do your own thing.' I said to myself, *I'm going to have more fire than I can handle with this one, but that's my job.* Now I see the devastation. I see the rigs are covered. We were right underneath the southernmost walkway that crosses West Street, Liberty, and West. I still was a little confused about where I was because you couldn't see any street signs. The smoke was right down in the street there. I ran into two or three guys

I knew from 58 and 26, and they had just lost Bobby Nagel, the lieutenant of Engine 58. Someone said, 'Bobby's still in there,' and I put my stuff aside and said 'You tell me where.' They were all visually shook up, but they were in good shape: Mike Fitzpatrick from 58 Engine, John Wilson — we call him Mooky — and a couple of the newer guys were in there in the lobby of the Marriott Hotel on West Street, right by the elevator, when the first tower started coming down. Bobby Nagel got stuck, trapped, but he could [still] talk. He told them to go out, to get some tools to dig him out.

Mike and the rest had [got the tools and] just cut a piece of pipe or something, trying to free Bobby, when all of a sudden another terrible rumble comes down. And that must have been the other tower. Mike says, 'I don't know how we got out. I don't know how we're here, I don't know . . .' the whole nine yards. Then he said again to me, 'Bobby's still in there.'

That's when we got a little concerned because there was not too much to do at this point, and they were stopping us at that overpass. Chief Hayden was on top of one of the wrecked rigs trying to organize teams. I found some Scot packs [air masks], which were all bent up, but managed to put a mask together for myself and milled around, waiting, and waiting. They wouldn't let us do anything, but then gave us a little project to search through the World Financial Center, on the west side of West Street.

When we came outside again, a couple of guys

pointed out a helmet and a knee pad, and we started moving what we could. It was sort of a helpless feeling because it was mostly heavy steel, and we weren't making any progress. Me, Brady and one of the other guys drifted towards Vesey Street until the chief said, 'Hey, get out of there.' The other building, #7, was fully involved, and he was worried about the next collapse.

So I'm looking for obvious things, thinking I'm going to see an arm move, or [hear a] voice. I'm looking in voids. There are guys around, but not a lot, just handfuls.

I had no idea that we had lost so many firefighters. In fact, when I came back to the firehouse that night, I had my head down and I looked at the guys and I said, 'I think we're gonna loose about twenty-five, thirty guys.' They didn't know what to say to me, and they all looked up at me. I said, 'What's the matter?'

They said, 'TJ, there's hundreds of guys missing.'

I was walking back to the main entrance of the World Financial Center when a guy comes up to me, a civilian, [walking] a golden retriever. He says, 'Hey, I can help. My dog can find bodies. We used to do this.'

I didn't know whether he was a kook or whatever, so I asked him a few questions, where he was from. He said Connecticut, but I'm in the marina. And I said, the metal is bad, and there are no boots for the dog.

But he said, 'The dog can take a cut or two, don't worry about that.'

And then I started to believe in him. I worked my way back over to where the guys had found the knee pad and the helmet. He let me help, and [we] carried the dog out. It's a golden retriever, and you know how sweet those dogs are, everybody is their friend, and here he's our best friend now. When we got to the area, he says to the dog, in an excited way, 'Bubba, go find the baby. Go find the baby. There, find the baby. Go ahead, go ahead . . .' The dog's name is Bear, but he calls it Bubba.

It's all metal, and I'm moving the dog around. The dog is sniffing everywhere. And he says, 'What's he got there?'

We all looked down. 'Oh,' he says, 'it's a little blood on a piece of paper.'

With that, I'm looking around now. I say, 'Oh, blood, look for paper with blood on it.'

He [then] says, 'Okay, this is it. This is it.' So I looked up at him and I ask if he's sure. And he says something like, yes, it is a strong indication, and he just held his hands out, [meaning] in a big area. So with that, twenty or thirty guys come over and we just started moving what we could move.

We spend fifteen minutes, a half an hour digging. Guys started drifting, and at this point he said, 'Give me the dog back, let me go, let me keep looking.'

I say okay, but I look at him one more time and

ask if he's sure about this spot. He says yes, yes. And then he left with the dog.

Probably forty-five minutes go by, and there was hardly anybody else in this area now because you couldn't move anything else. And I just kept digging. I stayed probably another twenty minutes to half an hour. I'm on my hands and knees; I'm just looking through rubble. Some guy would pass me a flashlight once in a while. All of a sudden, I was looking underneath this little thing, and as I was brushing the dust away – every time you moved the dust you had to close your eyes, look the other way, because the dust would rise – I see what looks like [part of] a pair of pants, a small spot, maybe six inches in diameter. So I move something else, and I reach in with my hand and grabbed it. It felt like a leg, or an ass, soft.

I don't say anything because I don't want to go alarming people. And I'm just looking around and start to move a few more pieces of tin, when one of the guys from Ladder 14 says to me, 'Hey, TJ, you on to something?'

I say, 'I don't know, I'm just looking. I'm staying here, I'm looking. I'm looking.' And a few seconds later I moved a couple of things, and I could see the stripes of a coat, a firefighter's coat. They were bright, and brand new. My first impression was it might have been an EMT or EMS worker because the coat was new. Now I called out, 'Hey, I need a fireman over there. I need somebody else over here!'

With that, a guy from Ladder 14, John Paul,

comes over and asks, 'What do you have, TJ?'

I say, 'Come here, look down in there.' I shined a light on it. 'That's a little slack spot there. I think we got somebody, just grab that. I see a coat here.'

So he reaches in and says, 'Oh, God, it is; I think it is.' So I say, 'See if you can find something.' He's a much thinner guy than I am. In the meantime, I started to brush the coat off. I see the back of the coat, it says GA, and I'm [still] thinking it's an EMT worker, and all of a sudden I get the end and the CI, and I go, 'Holy Christ, it's [Chief of Department] Ganci.' So I told John, 'Wait a minute, wait a minute; just go in, see what you can find.' He pulls out his wallet and reaches over, but I can't read because I didn't have my glasses.

Now I announced that we have Chief Ganci here. I say to John Paul, 'Make sure this wallet gets to somebody important.' Later on he tells me he gave it to a captain, who eventually gave it to Chief Visconti. Now the guys came back over; everybody came back over, and they gave me a saw. He was under a big piece of steel and a lot of light, thin metal and a small tree, which was only about 6 to 8 inches in diameter, but its branches were pressing hard on his back. One guy asked, 'Is he alive?' I said no, he absolutely is not alive. I still couldn't see much. Now guys are going crazy, so I say, 'Everybody stop. Regroup. He's not coming out my way. This is what you guys are going to do.'

I told them to start digging at the lower part, and

they uncovered the bottom part of his legs. And the big piece of steel was at an angle with the tree, and I say, 'What we're going to do is two or three guys are going to lift up on the metal; two or three guys are going to lift up on the tree. [I tell the others,] you guys are going to pull on his legs.' I say, 'Listen, whatever you do, he's in bad shape, [so] let's not make it worse.' There was no need to do it real fast. But I said, 'At the count of three, lift on the steel, lift on the tree; I'm going to push from this end, and you guys are going to pull.' And I was pushing on his head and his shoulders, and he went under the tree and it worked. We got him out.

We sat there as they moved the body. I had a blanket that I had found that I was carrying around with me, and I just told somebody to grab the blanket. And they covered him up [with it, and], put him on a stretcher. I wish I had stayed with the body. As a marine I should have stayed with him all the way to where he was going, not for any reason other than he was the chief of the department.

But, I was exhausted. I couldn't understand why he was down that far on the sidewalk. And then, at his wake, somebody announced that he had just made a decision to move the one command post farther away from where they were, and he said, 'I'm going to walk down to the other command post and tell them to do the same.' So he must have gotten caught between the Liberty and the Vesey Street command posts when, I guess, the second building came down.

These guys were brave. Oh, God. Whatever made them stay there even way after the first collapse? Boy, they are so brave. ■

Officer Will Jimeno
Port Authority Police Department

■ I am working at the north terminal entrance of the Port Authority Bus Terminal on 42nd Street at 8:48, and my good friend Dominick Pezzulo is working right across from me at the south entrance. Both of us graduated from police school just nine months ago, and we were assigned here together. We are tight, me and Dominick. I can see him just across from me as Sergeant John McLoughlin comes to ask for volunteers to go to the World Trade Center. Sergeant McLoughlin had spent ten years in the emergency service unit, and he knows every inch of the WTC, and so Dominick and I are glad to be able to go with him. Don't forget we are still young on the job, though Dominick is 35 and I am 34.

Our inspector commandeered a bus, and we sped down – the sector cars opened the way for us. We go to the north tower, and then we go under the buildings to go to the south tower, to get to the lobby there. We are one floor under the main concourse area, where all the stores are, and pushing a cart filled with equipment, air masks, helmets, axes, tools, and so on. On the back of the cart, pushing, is Antonio Rodrigues, and just to his left is Christopher Amoroso. [Suddenly I hear a loud noise and] look over to the sarge and say, 'Hey, Sarge, is there a second plane coming?' And, just then, it is like an earthquake when the plane hits the south building. We are just about in the middle of the concourse, between the two buildings, just below and a little south of the big golden globe, when huge parts of the tower and shock waves come down into the plaza area, cracking all the cement. The whole concourse above us collapses. There are a lot of civilians all around, and I don't know what happens to them, but I think it has to be bad. I can see Liberty Street before me as I feel a ball of debris hit us. Now, I see a huge fireball coming at us, and I yell, 'Run! Run towards the freight elevator!' [The fire has come from the fuel that has poured down the elevator shafts.]

Dominick runs first, I am behind, and the sarge is behind me. Antonio is behind the sarge, and Chris is bringing up the rear. But Chris never makes it because the shock wave pushes him back into the main concourse area, and he takes the worst of it. Dominick and I and the sarge just make it around the corner, but Antonio doesn't. Everything just starts hitting us, and then the wall comes down on top of me. I am flabbergasted. My friend Dominick is crushed down in the push-up position, and my legs are pinned completely by heavy concrete. Sergeant McLoughlin sees the walls breaking apart, and they are falling on him. And the ceiling falls on him, [pinning him] twenty feet away from me. I can't see him, but I can hear him. I keep calling out for Amoroso and Rodrigues, calling and calling for two minutes straight. But there is no response.

The lights are flickering, but they don't go out. Dominick begins to wiggle himself out. Sergeant McLoughlin, being from ESU, does everything by the book, and so we are talking about what we have to do. I have an old pair of handcuffs and I begin to scratch at everything around me, trying to free up some of the concrete. The sarge is thinking Dominick will get free and work to get me out first, and then together we will work to dig Sergeant McLoughlin out. The sarge is hurt bad, and he has a few thousand pounds crushing down on him. But he keeps talking to us, to steady us, keep us calm.

Dominick is a weight lifter, and he finally pushes everything off him. He gets free, and he is in an area about three feet wide, and he begins to work on getting me out. My left leg is completely stuck under immovable concrete. He is bending over just a little when we hear the collapse beginning. I didn't know it was the tower. I just hear the most horrifying noise I've ever heard. It was like a huge train coming at me with the roar of the devil. I don't think Hollywood could ever duplicate that sound, and it is right above me, coming down. Everything is shaking. I said, 'Dominick, something big is happening,' and I put my arms over my head. I am lying on my back, like I am

196

in bed, and I am looking upward. I try to get fetal. Dominick stands up, and backs up about four feet. The tower comes down. Nothing moves me or John, but a huge cinderblock the size of a dining room table comes down, and it hits Dominick right across his legs and it slams him down. Then all the debris falls all around us. I can see that Dominick is hurt pretty bad. I keep saying, 'Dominick, are you okay?'

And I can hear him gurgling. He says, 'Willy, I'm hurt bad.' I can see him, and it looks like he is sitting down, but he has all this stuff on him. We keep talking, and I say, 'Dominick, keep awake.'

But he says, 'Willy, you know I love you.'

I say, 'I love you, too, Dominick.' I think about all the things we did together, everything we did in school, the good times at work, the emergencies, the sitting around after work. I know now he is leaving us. He's dying.

'Just remember me,' he says. 'I died trying to save you guys.'

'We'll never forget that, Dominick. Just hold on.' I start to yell. 'Dominick, just hold on!'

'I'm going,' he says. I see him now putting up his arm. He has a gun in his hand, and he fires a shot with his gun, off in the air. A last-ditch effort to say, 'Hey, we're here,' and then he slumps over, dead.

This is real tough. He's a friend of mine and just a few feet away from me, and I can't go to him; I can't help him. He's a family man, and I know how much he cared about his wife and two children. He was a schoolteacher, too. He could have done that instead of being a cop.

I cannot budge an inch. I keep talking to John McLoughlin now, to keep him awake. The radio is dead, but I keep saying, 'Sarge, I know you're in pain, but you have to get on the radio, you know, it's our lifeline.' And through the night, I have to get nasty with him. I say, 'Sarge, keep alive! You can't die on me because I'll have nobody, and I won't make it. I'm dead.'

The whole time we were both awake, and a couple of times he began to fall asleep, but I yelled to him, 'Sarge, stay awake!'

So we were all alone now for hours. It is dark. I worry about fire because I can see flames every once in a while, and then they go away. I can see some light, off a way. I don't know what it is, but I hope it is a void someone will come into. The sarge says, 'Look, they are going to go by the book. They won't come until morning because everything is unstable, and they will need daylight.'

I say, 'Hey, Sarge, I don't know if we can make it overnight.' I am thinking of my wife, Allison, and my daughter, Bianca. She's just 4, and I want to see them again. And my wife is having a baby, a girl. We're going to call her Olivia. I ask God to let me see my little unborn Olivia, and somehow, in the future, to let me touch the baby.

Suddenly, now I hear a voice. 'This is the United States Marine Corps. Is anybody here; can anybody hear us?' This is Staff Sergeant David Karnes and a Sergeant Thomas. I start wailing, 'PAPD *Officers down. 8–13.*'

Before I know it, he is on the pile above us, and I ask him, I say, 'Please don't leave us. This is Officer Jimeno, who has a little girl and another on the way, and Sergeant McLoughlin is down here; he has four kids. Please don't leave us!'

And he says, 'Buddy, I am not leaving you.'

And I believed him, he just stayed. He got on the cell phone and made some calls to his wife, and his sister. His wife is in Manhattan, and his sister is in Allegheny, Pennsylvania, and he told them where he is so they can send people to us. Rosemary, his wife, couldn't get through on the phone lines, but luckily Joy, his sister, got through from Allegheny. That was so smart to use both family members to call for help. He's an accountant, and he put on his Marine Corps clothes and came down from Wilton, Connecticut, to help. About a half hour later, the cavalry came. We are forty or fifty feet south of the golden globe in the middle of the plaza. There is a raging fire on top of us, and because everything is so sharp, the hose gets cut and they begin sending buckets of water up. There is a firefighter, Tommy Asher, who's at the front of the fire, and he gets so mad when one of his water cans, the CO_2 cans, gets empty, he throws the can at the flames.

It takes them almost three hours to dig me out. I think my body just shut the pain off, but once they got the concrete off me then I really started to feel it. I had severe compartment syndrome, a crushing injury where the body swells up and the blood has nowhere to go. When they touch my leg, I am in such pain. The wall had fallen on my left side. My left leg is severely crushed, and my right foot has a very bad sprain, and is still swollen.

It takes about eight hours to dig Sergeant McLoughlin out. He's about fifteen feet back from me, but I keep talking to him all the while. He was completely pancaked. The ceiling came straight down on him. He wants more than anything for them to take the weight off. I hear him saying again and again, 'Can you please relieve the pressure?' When I was on the Stokes basket and going up the hole, I said, 'John, just hold on, they're getting you out.' About a hundred fire-fighters and cops passed me out from group to group.

They take me to Bellevue, and I am in intensive care. They start doing tests, and connecting me to machines. Then they bring in Firefighter Tommy Asher. He was right there in the middle of all that smoke, and I guess he must have collapsed himself. I find out he's in Engine 75. But Asher checks himself out the next morning and goes back to fight the fires. I don't see John McLoughlin until two or three days later, and then only briefly, the back of his head, because they were taking him to the operating room. It was a week before I saw his face, and we really didn't talk for weeks. He is hurt bad, and it is all hard work for him.

The way I personally look at it, I've been to calls with the New York Police Department and the New York Fire Department. To me we are all public servants, and that day at Ground Zero it showed. We all went in there, and we were all wearing the same color made out of the same cloth. We only have a twelve-hundred-man police force in Port Authority, but we all wear shields, we all wear uniforms. We all had a job to do. ∎

Firefighter Phil McArdle
Hazmat [Hazardous Materials] Unit 1S

∎ I get off early because I am going to a meeting with Chief Fanning. We are developing a new training program for hazmat, and so we are going to meet at the Rock [the training schools on Randall's Island]. Chief Fanning worked in Engine 82 and Ladder 31 for a few years, and he is well known in the job as an innovative fire officer.

I am bringing a new system to show Fanning, and I am just going onto the highway when I hear on the radio about the planes. So I race out to Randall's Island but miss Chief Fanning by about five minutes.

Fortunately, he leaves Jeff Polkowski behind for me, so Jeff and I put some breathing apparatus in the suburban and also some Stokes and skids – transport [equipment] for the injured. We head down the FDR

Drive, go under Battery Park, and come up on West Street. Just as we pull out of the tunnel we see a whole bunch of people running towards us and can't figure out why. Jeff stops the suburban because he doesn't want to run anyone over. We don't see it at first, but the building, the south tower, is actually coming down. Steel and debris from the building now covers the car and showers us with debris. We can hear the sounds of the people hitting the pavement.

It seems like a clear path for the people running south on West Street, but then all the debris comes down, and visibility is [near] zero. I can hear the sound of people running into parked cars and into obstructions.

I was going to meet Jack Fanning at the command post, and now I begin to worry about him.

We leave the suburban right there in the middle of the street, and start to walk [uptown] right into the incident. Because of the high concentration of dirt particles and debris in the air, people can't breathe, so we take our masks off and give them a shot or two. They appreciate that, and start to walk out.

At the transverse that crosses the roadway, which has not collapsed yet, we find a photographer who is lying on the ground. His leg is badly damaged, and we tell him that we have to move him. Because he is under the transverse he figures he is under cover, but we say, no, you aren't safe here. He insists that he wants to stay. He is in a lot of pain, but there is no way to move him safely without hurting him. So we pick him up, and carry him.

We put him inside of a deli right behind the World Financial Center, out of harm's way, and tell him that somebody eventually will get to him. Somebody had apparently broken the deli's window [and because] there is so much debris and stuff in everyone's eyes, they were taking water out of the deli and just pouring it over their heads, and into their eyes.

We return to where we had been at West Street and are starting towards the north tower when, suddenly, we look up and we see that the second building is starting to come apart. It looks like sparkles – I guess from the reflection of the sunlight hitting the glass and metal that was starting to come down.

I say to myself, *Thank God this building is so tall. If it was lower to the ground, we would not have this time to run.* I am not very fast, but I am at least fast enough to beat this rubble pile.

We run across the street back towards the World Financial Center and get to those big floor-level windows. Luckily for us the glass had been knocked out when the first building came down, so we [were able to] jump into the [lobby] and hide behind a large column just in the interior. But with all the windows out, it is like a large, open garage space. Then, hiding behind this column, my arms crossed like a mummy, I begin to feel the power of all the debris and rubble rushing past. And there is a great deafening noise. It feels like the world is falling apart. If you've ever been at a demolition site, where you hear a building coming down, if you magnify that by a hundred or a thousand, it might have given you a good idea [of the sound].

So I stay pressed against the column, and first I feel a positive wave of energy go by, and then a negative pressure because it is creating a vacuum, and I am hoping that it doesn't suck me out. And then I can't breathe. I try to put my mask on, and then there is a tremendous amount of heat, but it only lasts for a minute before everything goes black. I don't know how long this blackness lasts. It is very hard to keep track of time, to keep track of everything. I know that I have to remember my bearings, which way was the back of the building. I don't want to get disoriented and not know where I am. I know I have to head in the direction of the river, because that is the way to safety.

All the time I am hoping that this will end. And then, after a few minutes, the noise stops just as suddenly, and I think then that I might be okay. We are still alive, anyway, and Polkowski and I begin to crawl out the back of the building, for it is now filled with debris all over, and we can't see for two feet. I never thought about myself. I never thought about my family. It was really odd. I just kept saying, 'Jeff, come on, we gotta get in there. Our guys are in there, and we gotta get to them.'

We try to find people in the department, at least somebody in charge, and the first person that we see is Chief Cruthers. He looks pretty stunned, and is trying to get everybody to regroup north of Vesey Street. He says, 'If you find anybody, send them up this way.' So we begin running into people. It is strange: I always used to say to people, 'How you doing?' But now people are saying, 'It's nice to see you, it's good to see you.' I guess they are saying that because we are not seeing anybody alive. Usually we find [survivors] in collapses. But here, it is just nice to see someone, because you know that person is alive.

There is so much debris, just so much. It is almost impossible to comprehend. I don't know in a situation like this what's right or what's wrong. Do we start digging immediately, or do we try and get everybody together and regroup because we still have a fire situation? Who's in command? What do they want us to do? What do they want us to do first? What are the best things to do? I don't think anybody knows, really.

We search a few of the vehicles on West Street, right near the pedestrian overpass, the first one that collapsed, to see if there are any people trapped. Then, also, we search the rigs to grab some forcible-entry tools. Everyone is doing that – cops and EMS people are grabbing Stokes baskets, first aid kits. We throw everything into a Stokes and just yank it out of the rubble so that we could set it up somewhere to be available.

We then start to regroup, and one of the first

things that I want to do is a survey of the area, for radiation, and for some of the hazards that we all trained for, the weapons of terrorism – secondary devices and secondary agents. Our job in hazardous materials is 'force protection': to make sure it's safe for our operating force at the site from an environmental point of view. It is a paramount concern. We set up two teams, and each starts at one point and works its way around the perimeter searching for radiation, nerve, and blister agents. We don't worry about incendiary devices, just the chemical agents and radiation, and we aren't worried about the biologicals at this point.

[After completing the sweep] we want to at least tell somebody who is in charge that radiation and chemical agents aren't a problem. At some point, somebody is going to ask that question.

Afterward, there is plenty to do. Because I had [been involved with this type of] work in Oklahoma City, I know what had to be done with the bucket brigades. I help a few of these guys get in line and show them how to take all the big pieces of metal that could be moved and start to put them in piles. Then we [begin to gather] all of the soft stuff into five-gallon pails. Once it is searched, we dump it in one pile. Then we start to set up different piles for metal, soft debris, concrete, and rebar. We make a number of inroads into the site doing this, and we continue to make these piles until after midnight.

During this time I begin to think about my experiences in Oklahoma City. There was a team from New York, [including] Ray Downey and Jack Fanning, who went out there for six days. Here at the World Trade Center, the intense heat from the fires is still below us. Oklahoma City has 13 floors; here we're talking 110 floors. The compression difference between the two is so tremendous. At Oklahoma City we had large chunks of concrete, steel, and debris; at the World Trade Center, it is pulverized concrete, a very fine powder. We are not moving large chunks of concrete, but very, very small pieces. This, in some respects, makes some of the searching easier, but it also tells us what we are going to find when we start to find bodies.

All this time I want to call my wife. It took me about five hours to find a landline. She was fine, she said, up until the point where people started calling her and saying, 'Did you hear from Phil?' I had two brothers who were at the site, too. One's a police officer, an emergency service unit lieutenant, and the other's a fireman in one of the volunteer companies that responded to the site. She hadn't heard anything from anyone, and so that relieved her to hear from me.

At 12:30 A.M. I am told that I am to leave the site and report back the next day at 9 o'clock. I now go to one of the eyewash stations because my left eye is extremely irritated. They wash it out fourteen times, but there is so much debris that it really doesn't do much good.

I leave the site and return to the firehouse. But my eye begins to hurt even more, so I go to Elmhurst Hospital, where they put a catheter in it and rinse it out until 2:30 in the morning.

We are missing nineteen men in my firehouse. This has taught me, and I think it's taught a lot of the guys, how fragile life is. After this is over, I will spend more time with my kids and my family. Also, here in our department and in my firehouse, in spite of everything that we are suffering, I think today a lot of the guys are closer because of it.

We don't really have time now not to be strong. We have to be strong for the people who survived, and we have to be strong for the victims' families. When all of this [happened] I started to get a lot of e-mail because I teach on the outside. Somebody would send me something; I'd answer, 'Thanks,' and move on to the next one. One word was all I had time for. Finally I sat down and composed a letter, which I sent to everybody asking them please not to write me, call me, or anything because I have to stay strong for the families. We have nineteen families to deal with. That is what is important. With all of this stuff that people were sending me, I was afraid that I was going to crack. ■

Battalion Chief Tom Vallebuona
Battalion 21

■ We are at a false alarm, and a firefighter receives a
cell call from his wife about a plane hitting the WTC.
I got out the binoculars and can see that the north
tower had been hit. We are near the quarter of
Marine 9, and we had a pretty good view of the twin
towers from this part of Staten Island. The fire didn't
look that bad. *We can put this* [out], I thought. The
next thing I know a plane is flying almost over our
heads and goes across the harbor and into the other
tower.

The dispatcher ordered me to Brooklyn, and
when I got there, I was told to take another chief into
the city. We loaded up the rig with a bunch of air
bottles, and as we were leaving, the Brooklyn
dispatcher said please come back if you can. They did
not have any chiefs in Brooklyn, [and I thought], if
they don't need me [at the WTC] I will certainly
come back.

I stop right on West Street because I can see the
rigs are starting to bunch up. I don't want to get
blocked in, in case they didn't want me. I start walking up West Street until I get under the overpass, the
pedestrian walk at Liberty Street.

I see a couple of guys setting up a command post
there. There were three chiefs. I looked up at the
burning building, and I said, you know, this is ridiculous, of course they need me. I went back to get my

gear. When I get to the rig, about half a block down West Street, the building came down. It is the most amazing thing I have ever seen in my life.

At first I look up and I hear a boom and a strange noise. I can't really describe it. And it starts to bellow out. It looked like a fountain, like one of those fireworks fountains. I said, boy, that really looks pretty, and I must be back at the firehouse sleeping. This can't be happening, it just happened so fast. And then I suddenly realize that I am not in the firehouse, that I am here, and I am not far enough away from this thing.

Chiefs and their aides don't normally hold hands, but I grabbed Stevie's hand, and we have never held hands better. We hold each other, and then we get engulfed in that cloud right away. And we don't know what it is. It is such a weird color, and we don't know if it is going to light up or not. It is very hot.

We are in front of 90 West Street, just across the street from the south tower, and we really try to outrun it, but the cloud hits us so fast that we cannot run anymore. And I am thinking, *Firemen don't run in emergencies.* So we start crawling, me and Steve, as fast as we can.

I am now doing the Hail Marys. I have done a few through the years, but now, between this and the Waldbaum's fire I was in, I thought I really had it all when it came to being next to death. As we are crawling, trying to work our way south, we find a wall along the street, follow it in the dark, and eventually work our way into a store that is open. There are some civilians there, and other people are coming in. It is phenomenal just how dark it is, how little I can see of anything. You know the old saying – I could not see my hand in front of my face, because of the smoke condition. And this is at 10 o'clock in the morning of a clear day. Imagine. And breathing this stuff is like eating it.

When it starts to lift, after about twenty minutes, we go back to the rig. I grab my mask and give Stevie his mask, and say, 'Put this on. It's only going to get worse.'

Then I saw Chief Frank Cruthers and went to talk to him, to ask what we were going to do now. I know we have to go up the street and do something in the collapse. [But after] a minute or so goes by, it happens again.

I am on Liberty Street and West Street again. As it starts to come, I say to myself, *I cannot run this fast.* Mainly, it was dust, but I can hear stuff coming down, hitting the ground around me. The dust is worse now, because dust from the first collapse [is mixing with] this one. I could not see a thing and I had a hard time finding my way around. I had the mask, but the mask face piece was all full of dust, so it pounded my eyes when I put it on. I did not want to breathe that stuff again because it was so bad the first time. I go right to the gutter and try to find a wall, because you figure if you get down next to a wall, it'll go over you. I hear the falling of bodies, pieces of people all over the place. This is terrible. It's beyond any kind of fire I have ever been to.

Now I do not have any gear, unfortunately, and there is no going back to get gear, so I head up the street. On the other side of Liberty Street, I see Charlie Blaich, though I do not recognize him at first because he is covered with dust. He is on top of a pile directing some guys. I tell him I am going to try to get some water on the buildings down by Liberty, where the fires are pretty heavy. The buildings are on fire, the cars are on fire, and the rigs were burning, too – we had it all. I figure it must be the jet fuel.

Ninety West Street has several floors of fire. There is also fire in the cellar, and the roof began to burn. This is an old and very beautiful building in front of us and has about forty floors of scaffolding on the side that we are afraid will fall down.

After the first collapse we had already hooked up some of the rigs, and the firefighters were starting to stretch the lines, so I figured, okay, when we get water we will knock down the first floors anyway and start putting out the car fires. I still can't see much, and the smoke is banking down on the street. When they opened up the lines, though, they only had half a stream. The mains were shot, and that was a great disappointment. I let them knock down the first floor [of 90 West] from the street, but I will not let them go in because when we open the front door the air is sucking in from the basement fire. It is very dangerous without adequate water. I look at the building burning and think this would normally be a fifth-alarm fire, and here I am with a couple of companies doing what we can.

Then guys started showing up from Staten Island. I sent them down to the waterfront to see if they could find a fire boat. They found Marine 1, and they stretched [hose] from the fire boat. So we knocked down the cars and the first floor, but still we had no masks, no tools or equipment. We also hooked up to the standpipe in the Marriott Hotel, and we hit the fire from behind as well. It was like they used to fight the old loft fires, before we had tower ladders, by hitting them from the windows of buildings across the street. I never thought as a battalion chief that I would be feeding seawater into a standpipe system of a forty-story building.

Someone gives me a report that there are remains on the roof of the Marriott Hotel, and that is almost two blocks away from the World Trade Center. The human loss is just horrible. Then as I am standing there and guys are coming in and I'm directing them where to go, a fireman comes over to me and says, 'Chief, you're standing on a torso.'

I went over to the hospital about 7:30 that night because I couldn't see anymore. The doctor tried without much success to get my eyes opened. I sat there thinking about how fast those buildings came down, the power and the force of the falls, and in all my years as a fireman I never had such a disparaging feeling. I just said, 'Everybody's dead.'

I think this is why I was just happy to take a building on fire and put it out. I felt sorry for myself for a couple of days, knowing that I am all shook up, and that I really have been through a lot. And then I hear some of the other stories, and though I was down saying Hail Marys and stuff was hitting [all around me], I really didn't get buried.

When I was in the middle of the Waldbaum's fire, and the roof caved, I got caught in the cockloft [the area of a building between the ceiling and the roof], and the fire just lit up. I was pretty well burned, but I thought at that moment of my wife and my children. And then I fell out and down into the store, next to a whole storeful of fire. I also broke my shoulder and went unconscious before they carried me out. But I again had that moment to think of my family. Here, though, because my family is now grown and the kids are out of the house and don't rely on me as much, instead of thinking about them, I thought only of how ridiculous it [would be] to die like this, that I was being caught like those miserable people of Pompeii . . .

One fire is too many, but I feel like I was with the marines at Iwo Jima or at D-Day, all in those few hours. It really was historical and something that at first I was depressed about being at. Later on, though, I felt privileged to have been there, in a strange kind of way. ■

Police Officer Anita Rosato
Emergency Services Unit, Truck 10

■ There are 6 women in the ESU out of 450. I work midnights, and I was on my way home. I can see the towers burning from the Grand Central Parkway. My husband, Anthony Romano, who is in Truck 7, kept beeping me. He was on the Long Island Expressway, and he said the response in the HOV lane was incredible – all these firemen and cops racing to the city.

I have a background in ropes and used to teach rope rescue. So I make sure all the ropes are loaded into the truck, [along with] every bit of equipment we think might be helpful: extra Scott bottles, extra Scott packs, generators, lights, ventilating equipment. We made a convoy to the Trade Center.

We got there twenty minutes after the tower came down and parked on the other side of City Hall. Someone brings us an ESU officer who was wandering. We put him in our truck for accountability purposes. Behind his eyes you could see he was really shaken up. He would just look at you. We said, 'Guy, you're all right. You're with us now.'

All he answered is, 'They came down.'

I said, 'I know they came down, I know.'

We stage by the Woolworth Building and get a roster of who is missing and who is not. We then go to Stuyvesant High School and from there to Ground Zero, where we see what kind of equipment we can salvage. FEMA was activated, and they showed up with equipment that had to be stacked. We just walked around, and I really don't know where to begin. I've been on this job sixteen years, and doing emergency services ten years, and everything I knew couldn't help anybody. That is a pretty hard feeling, because normally when we get someplace, we can do things. Then, as I meet people, it started setting in how many are missing. Ronnie Kloepfer is missing, and Santos Valentin is missing, Tommy Langone, Rodney Gillis, Paul Talty, Sergeant Curtin are all missing – these are guys that come to barbecues at my house. The guys from Truck 7 are like my family because my husband works there, so for me it's like losing people in two families.

[While] working the site, which is such a vast area, we would find body parts. I have worked train wrecks and gunshot messes, but here I would have to study something to realize that it was a foot, a burned foot.

My husband was on medical leave, [as was] my brother, who is a fireman. If he hadn't been out on medical, my brother would be dead because his whole company was killed – Engine 219, Ladder 105.

Well, at least my husband and I can talk about it, and that is a pretty good relief system. It was hard for us. I don't have to worry about how my spouse is going to feel when I talk about the horrible things I've seen because he was seeing the same things. We would come home at night [and ask,] Did you find anybody on your end? No. Did you get a hit? No. Day after day, we did not find a cop, and after two and three weeks, it hurt not to find anybody. It was hard for the unit. Then I said I am going to go every day because of the families. I have to give them some sort of closure.

It's just not our guys. There are twenty-eight hundred families there. One day we found part of a scalp, and that gave a family something, knowing that a loved one was really in there.

We are angry in my unit. We lost fourteen, and twenty-three in our department. But knowing what we know now, would we have done it again? Yes. If I could have been there a half hour earlier, I would have been. ∎

Lieutenant Michael Penna
Rescue 1

Michael Penna was born into the New York Fire Department. His dad, Manny, had been a longtime firefighter in Ladder 48 when it was one of the busiest companies in New York. I worked a whole summer with Ladder 48 and knew Manny Penna well. He was a low-key kind of guy, but he was recognized by all of us as one of the stalwart members of the company. It is a good legacy to have a father with such experience, especially if you are a young firefighter and new on the job. Men in the busier companies learn to be very straightforward, and they quickly learn the ropes of firefighting. It is a good thing to have a dad who can set you straight and safe.

Michael knew from his first days in proby school twenty years ago that he wanted to one day end up in a rescue company. That is where the fire duty is and also the other interesting duties. Every job was an adventure, and each jumper, auto wreck, air crash, or building collapse brought a new set of challenges. It took Michael five years before he got the assignment he wished for at Rescue 2, in Brooklyn, which for a decade had been one of the busiest fire companies in the world. Five years ago, he was promoted to lieutenant and was assigned to Rescue 1 in midtown. Although most firefighters disdain elitism, they recognize that to be an officer in Rescue 1 is certainly one of the most prestigious jobs the fire department has to offer.

On September 11, Michael was sitting at home on Long Island when he received a call from his brother-in-law, just after the first plane had hit. He flipped on the TV, where the crash was being covered on almost every channel.

■ As soon as I saw the fire in the one tower and the heavy smoke, I got dressed. I figured it would not be easy to get into Manhattan with a car, so I just hopped on the Long Island Railroad. On the train I meet a paramedic with a portable radio, and together we heard that another plane just hit the south tower.

At Flatbush Avenue and the Brooklyn Bridge,

218

there was a staging area. They had a caravan of rigs going into Manhattan, so I just jumped onto the back of one of the pumpers. When we got down there, blocks away from the actual scene, the cloud of dust was [so thick that] you couldn't even see anything. We went into one of the command posts right across from City Hall, and I wound up hooking up with some other members of rescue with Chief John Norman. We were notified through our command that we were also part of the urban search and rescue task force through the Federal Emergency Management Agency, which was bringing equipment and tools to the scene by boat. John Norman sent me off to coordinate getting the tools into the area, because we knew all the rescue companies were there already.

Along with a couple of firemen I climbed onto the piles. Coming onto the scene as an individual, you more or less have to do what I did – grab a couple of guys together and have them work as a team. It wasn't that the command structure was not complete; there was no command control of the scene at all. No one issued orders, but no one needed to be told what to do. Engine companies were stretching lines to put fires out while everybody else was on the pile taking victims down, treating victims on the street.

I've been to plane crashes, building collapses, train wrecks, everything. Here is a scene with two towers down, two planes into the towers, but it wasn't your typical collapse. You had to actually search for the people or parts of people. Body parts weren't just lying around; it wasn't the gore that you would expect from that type of devastation.

We came across a lot of people deceased, under the rubble, and I told the guys to just mark the spot and keep going. We couldn't stop and start extricating the people that were deceased. We had to worry about the live victims.

And as the day went on, the magnitude of the list of names of guys who were missing, from the staff chiefs down to the firemen, was incredible. Just word of mouth on the pile gave us all the information.

We worked through the day, and that night we were still on the pile. Later that night we located a Port Authority police officer in tower 2, the south tower. He had a radio on him, and he was able to key his mike. So we were able to locate him in the middle of the pile. He was buried almost from his neck down in a hole. We were able to crawl into the hole, which was just a few feet wide, and get an IV started. We had a doctor with us, and he began administering morphine as we started digging.

I rotated the guys around for about six hours, and he was conscious, the whole time talking to us. We had him uncovered to about his midthighs by the time we were relieved by Squad 270. They came in and freed the rest of him.

While we were working with the Port Authority police officer I saw that the golden globe [that stood] in the middle [of the plaza] was still intact. There was a crack in it, a hole in the middle, but it was still standing. It was amazing. But a lot of the area behind the globe, including buildings 4 and 5, was still standing. They were completely burning, but still up. Everything in front of the globe on the West Street side was totally gone.

What hit me was the amount of debris that *wasn't* around. In a typical wood-frame or tenement collapse the bricks break up and spread like water all over the place. It is just one big pile. Here it was like everything just came straight down. To a civilian it may seem huge, and it is nine square city blocks. But to us it isn't like a large debris field at all, considering that these two buildings went up 110 stories. Most of the steel that came out was from the outside of the building. It just fell, but the insides of the building came down right on top of each other, like a pancake collapse.

I've heard some talk about how we were going to approach high-rise fires after this great tragedy of the World Trade Center. If I have my two cents to put in, we're going to approach them the same way we always did, only we're going to do a little better sizeup. What's the [alternative]? To see a plane crash into a building and say, 'Sorry, folks, we're not going in'? That's not an option; that's not what we're about. Maybe there could've been some things done differently here, maybe fewer people could've gone in, but that's all hindsight. We can use [the experience] to plan for the future, but we certainly can't change our operation to the point where we're not going to protect people in a high-rise building if a plane crashes into it. ■

Police Officer John Perry
40th Precinct

Patrolman John Perry, 38, has spent eight good and solid years in the New York Police Department. After the academy he worked first in the Central Park precinct, an assignment he loved because it gave him an opportunity to meet New York's foreign visitors and to practice the languages he spoke: Swedish, Russian, French, and Spanish. He became interested in languages during a junior-year trip to Belgium, where he stayed with a multilingual family. Whenever he found the time he would reach into his pocket full of language tapes, and a headset would cross the top of his dark, curly head, which was a full six feet four from the ground. He would say, jokingly, that each new language increased dramatically the number of women he was able to communicate with.

His next assignment brought him to the traffic division, and there he learned how movies were made, for he often had to stop or reroute traffic to accommodate filmmakers in the city. He saw how the extras all seemed to have other careers and that to work bit parts in movies was a reliable way to supplement an income.

And so he took his good looks to the appropriate clipboard-carrying person on a set one day and applied for a job. He then started acting as an extra on his days off, and soon received his Screen Actors Guild card. It was like John to introduce his friends and family to anything he found interesting, and soon his mother, Pat, and his brother, Joel, had SAG cards as well. John loved people and the interaction among extras between takes when making movies. Even when he went to Stony Brook College he chose to live in the foreign students' dorm so he could meet new people. Like his language ability, the social dimension of filmmaking also broadened his chances to meet women, and that suited the smiling, handsome bachelor just fine.

In his most recent job he has worked at police headquarters for five years as a prosecuting officer – a police officer who is qualified to write and present prosecution documents for court cases. John Perry had graduated from New York University Law School in 1989, in the same graduating class as John F. Kennedy, Jr., but unlike the president's son he passed the bar on his second try. He had gone through law school on student loans and then studied for the bar, afterward jumping from law firm to law firm. He needed more excitement in his life, he decided, and so took the exam for the police. It was a good decision and gave him experience that money could not buy.

His position is a prestigious one as he is responsible for advocating the interests of the police department, and so he has control of his own time, a luxury to which most police officers are not accustomed. He uses that time to advantage, studying the objectivist philosophy of Ayn Rand and anything that supports his libertarian view of life – that each person should do as he or she wants as long as it doesn't interfere with the peace of others. He often told friends that he joined the police department to exercise his Second Amendment right to bear arms, and some occasionally believed him. His view of freedom also led him to join and then to be asked to serve on the board of the Nassau County Civil Liberties Union and to work for the failed election campaign of Norman Siegal, New York's ACLU head, for public advocate.

There were some stormy times in those years, as he is telling a civilian clerk on the morning of September 11, a clerk who is making certain he is signing all the forms in the right places, but now he is ready for the sunshine period of his life. He had enough time in to take a vested interest pension, and he had been offered a good position with a medical malpractice firm in Long Island. It was a reasoned decision to retire early for he had had some great experiences in NYPD, but now he would see if all the time, energy, and money invested in attending one of the country's most prestigious law schools would add anything more to his life. That life would get easier, perhaps, and he certainly would be better compensated. He had been waiting for this day with some anticipation, as has every police officer and firefighter who has faced it. John was on the tenth floor of 1 Police Plaza, the department headquarters building in the shadow of the Brooklyn Bridge, and the forms before him were his retirement papers. He was 'putting his paper in,' as the saying goes.

He had just handed the clerk his badge – something of a ritual for all police and firefighters, turning in the 'tin' – when they were interrupted by a thunderous bang, the likes of which they have never heard before. John looks out of the window and down toward the tip of Manhattan, just a few blocks south of the City Hall steeple. What he sees alarms him – a plume of thick, curling black smoke rising from the World Trade Center.

The department radio in the office is cackling with rapid and disturbing communications: Tower 1 had been hit by a plane. He reaches across the desk and retrieves his badge, the badge that will get him through any barrier, and heads for the door. He hasn't yet signed all the papers.

'What kind of plane?' John asks himself, as he turns into the police store on the first floor of 1 Police Plaza to buy a pullover golf shirt for thirty-five dollars. He is in civilian clothes, and at least now, since his shirt is emblazoned with NYPD, he will have something of a uniform. Just two minutes later he reaches the corner of Church and Vesey streets. *This is a big job,* he is thinking. He isn't going to miss a big job his last day as a cop.

Captain Tim Pearson
NYPD, PSA [Police Service Area] 2

Captain Timmy Pearson loves his job. He is just approaching his twenty-first year in the police department and has attained the highest rank a police officer can reach by taking tests and having passed in quick order the sergeant's, lieutenant's, and captain's tests. All of that studying was now behind him and any further advancement in the department will now come only with accomplishment and merit and signs of approval from the chiefs at the top.

September 11 is turning out to be a routine day, he was thinking on his way to work – nothing out of the ordinary had been scheduled, no community meetings, no internal investigations, no fiats from the high floors of 1 Police Plaza that would require a response. He was planning to catch up on some of his paperwork, the forms and evaluations that pile up so quickly in the headquarters of the Brooklyn Housing Police, which he commands.

At this moment he is heading for his office and talking on his cell phone to a fellow staff officer who is just down the hall from the PC's office. The officer interrupts their conversation and says, 'Hold on, Timmy, I think a plane just hit the World Trade Center. . . .'

■ **The first plane that went into the World Trade Center generated a lot of 911 calls, and so we were brought from a mobilization level of 1 to a 4, which** is the highest. I'm out in Brooklyn, in my police car, traveling from my residence to my work, which is in the East New York section of Brooklyn. I'm commander of Police Service Area 2, which is a former housing police unit that has been integrated into the NYPD.

So now I'm hearing all the transmissions about the plane and the different responses. I hear the chief of department, Joe Esposito, [say] that it was a plane that went into the WTC. I turn on the lights and sirens and start heading quickly into my office because I know that it's going to turn into a very serious event. And as I'm traveling, I can see the WTC on fire. At my office I turn on the television, and after twenty years in the job I know to immediately start getting into my uniform. I direct my staff, my command, to get ready for the mobilization: Everybody suit up, because we're probably going to be responding. As I'm putting on my uniform, the second plane runs into the WTC. I hear the chief of department [say on the police radio] that we're under some kind of attack and start yelling orders, clear this, clear that, clear 1 Police Plaza. We talked to the borough command and were told to respond.

So I and my staff – Police Officer Lukas, my driver, Officer Kenny Deere, and a group of one lieutenant, one sergeant, and eight police officers, who were in a separate van – all started traveling towards the city. We mobilized at Church and Vesey at about 9:20, some fifteen minutes after the south tower was hit.

It was around Church and Fulton Street that I ran into John Perry, who had come from 1 Police Plaza and was heading towards the WTC. He sees I'm in uniform and says, 'Hey, Captain.'

I'm looking up at the building, watching, assessing the situation. There's carnage all over the place. There are a lot of bodies. I realize that people are trying to run out through the doors on the other side of the WTC, to cut through the plaza. But there were still people jumping there, and debris still falling. So John says, 'Captain, what can I do?'

I've known John for six or seven years. He was a prosecutor while I was an internal investigator, which is how we became colleagues – close colleagues who would socialize from time to time. So I say, 'Come on, let's go into the building.'

I thought we should first start helping with evacuation, and we also wanted to prevent any more people from getting injured by trying to run across the plaza. We went down through Borders, through the concourse going towards the PATH train escalators, and then down that corridor into the north tower. From there we went to the [lobby of the] south tower, where the fire department had set up their command center, to see what assistance we could provide. We also wanted to inform the fire department about the reports we had gotten from various people running out of the building [about workers] trapped on different floors. They were mustering their firefighters and sending them up in the building, and the security director was also there trying to point out how to get around the building, showing them the different areas, the freight elevators and things of that nature.

John and I and Officer Lukas then went from the south tower back to the north tower. Lukas has received a call on his cell phone from his sister, who is trapped up on the 103rd floor. He is frantic. I have to hold on to him physically for he wants to get into an elevator, and then he wants to go charging up the stairs. It is now after 9:30. We go up the escalator to the second floor. A human chain was set up between the people coming out of the north tower, going down into the concourse, and then back up by Borders bookstore. If they had tried to go directly out through the second floor, through the swinging doors, they would not have been able to get through. There was carnage all over the place, and there was debris and stuff falling through the roof, and there were people jumping. When [people did] try to run out, [the jumpers] would land on top of them, so to prevent that from recurring, we sent them underground.

After we'd been there for approximately half an hour, it'd gone from a flood of people coming out to a trickle. But there are still people, and you can hear them on the staircases as they're coming down. And we're trying to tell them, 'Let's go, let's go, let's go let's go let's go.'

From the second level, I could see that a concert

had apparently been scheduled to take place on the plaza for there were a lot of chairs set up there. I thought it was pretty interesting that if that concert had been going on, there would have been a hell of a lot more people hurt in that area, which was right around the big golden globe in the middle of the plaza.

There were body parts in those chairs. There were so many bodies there, I don't know if the people were – well, they looked like they were victims from the plane crash itself, and then it could have been people who were just knocked out of their offices. But there were still a lot of people coming down, jumping. And I could see plane parts, luggage – it was just a mess there. As we're standing there a woman came down the stairs, a female Hispanic, I would say middle aged, and she was having a hard time. She had another young co-worker with her, and when they got down the stairs, this woman started to have an asthma attack, with tightness of the chest and difficulty breathing. The Port Authority police department had Scott packs [air tanks] and started administering clean air to her.

At this moment I hear some heavy debris fall, heavier debris than had been falling previously. After a while, we became immune to the different [sounds of] things dropping. We realized they were bodies, people dropping, and they sounded like an M-80 fire-cracker, that's how loud they were. The first couple really scared me, but then after a while, it was how the people would just fall. But now, something heavy fell and set the stage on fire.

I say to the firemen, 'Hey, fellows, look, I think we better start getting out of here, because heavier debris is starting to fall, and the place is on fire over there.'

So a fireman says to me, more or less, 'Ah, you cops. So there's a little something falling. That's over there, that's not going to bother us. We're all right over here.'

John Perry, myself, Lukas, and another guy named Jimmy, or something like that, from Port Authority began to lift this woman up. We initially start to look for a chair, so we can just carry her out. But then we realize, forget a chair, because we don't have the time. So I grab her by an arm, John grabs her by an arm, as does Lukas, when a cop comes up to me and says, 'No, Captain, let us do it. Don't you carry the woman, we'll carry the woman. You take her personal effects.'

As I'm starting to pick up her personal effects, we hear this rumbling. *Boom boom boom boom*. It's coming, this roaring sound. The lights flicker, but as I look outside to see what's going on, I start to see what initially looks like a storm, but it was the dust cloud starting to form from way up on top of the building. All of a sudden, I could see the façade of the south tower falling against Battery Park City. And then all hell broke loose, and *boom boom boom*. Everything came crashing in. A tremendous wind comes through. You know, you can hear the glass, the steel, the concrete, dirt, it's just a loud sound crashing down all

231

around us, and all in this tremendous wind. I'm sitting there, saying to myself that I'm dead, because there's no way I'm living [through this], through just so much debris coming down around us. I'm waiting for the big brick to hit me in the head. It shook your brain, it was that powerful, the stuff coming down. You could hear it and feel it.

When the south tower fell, it ripped apart the north tower. You really couldn't see it because of that big cloud, but I think it compromised the building itself. On the second level [there is] nothing but soot smoke and the acrid smoke that is coming from the fire. All the windows got blasted out, and I could hear the glass breaking and I can hear steel, dirt, and everything falling. Now we're in total blackness, darkness. We can't see anything, I can't breathe, I am really disoriented.

In the confusion, John, who had been right next to me along with Lukas, and Deere all became separated. I yell out in the dark for Port Authority, because I'm figuring, I don't know the building. I haven't been here in months, and when I am here, it's just passing through. It's not like I'm familiar with the corridors.

I have no flashlight, and I can't see or hear John. I knew he was just with me a moment ago. I began to crawl along the floor and I realize I'm still alive because somebody touches me. It went completely silent at first; nothing could be heard. I don't know if that's because the sound was so loud it deafened me, but after a short period of time, I started to hear the moans and groans, and people yelling for help. I'm still yelling for the Port Authority, and then I begin to yell for everybody to stay together because for everybody to make it out of there we will have to stay together.

I keep crawling along the floor, trying to get my bearings, trying to make my way towards the escalator, so I could get back downstairs. That's the only way I know to get out of the place. So as I'm feeling my way along, we yell for a light, and suddenly a light appears, though I don't know who has this light. There's a human chain that begins to take form. People are holding on to me as I work my way down. There's one person in front of me, because I'm holding on to that person and guiding that person as to which way we're going to go. We walk down the escalator, holding on and inching along, because the light doesn't really provide enough [illumination] to show us where we're going. We knew we had [reached] the escalator because we felt the rubber – that's the only way we knew. We really couldn't see each other; nobody could make out a face.

As we're going down, I can see the floor had collapsed. The south tower had collapsed the south side of the north tower. And I see nothing but fire all along there. It's all fire down in the basement con-course, too, where we went with the swinging doors, where they'll take you to the plaza. I'm looking, and I'm saying, God, how are we going to get out of here now? Our way is totally blocked, there's nothing but debris and fire before us. And a lot of the doors are crashed in, so that you cannot see beyond them. Eventually we make a right and go through the middle of the elevator banks, but on the other side of the banks is a big, open area filled with debris and fire. Now I can see, and obviously smell, the jet fuel that had come down the elevator shaft and that was all over the floor. The fire helped us in a way, because it illuminated the area for us, and we knew which way *not* to go. We then started to work our way towards West Street. It seemed like an eternity, but it must have taken about ten minutes to get out of the building after the initial collapse. I couldn't tell you who came out with me because I never saw their faces. It was too dark, and when we got towards the street, there was so much soot all over everything and everybody, and so much in the air still that you really couldn't [distinguish] anybody. But I knew there was a lot of people hurt.

I headed towards Battery Park City and then down West Street towards Warren and back to Broadway. I went to the command center on Vesey between Broadway and Church to report the fact that my people were trapped in there, because I thought Officer Lukas, Officer Perry, and the others were still in there.

As I'm reporting that I have people trapped, tower 1, the north building, came down. I began running again and got caught by the big storm, the second go-around. At this point, I'm totally wasted. I am also totally disoriented, and I'm choking, and I'm thinking how it appeared that the building was falling down on top of me, right there by Trinity Church. Thank God it imploded, as opposed to falling over.

Lukas eventually made it out of the building. He said that he heard my voice in the dark. He said it was like *The Poseidon Adventure* in there. Two or three groups formed: My group went one way, another group went another way. You didn't have much time to try to figure out how to get out of the building because the rest of it was going to come down. But Lukas's sister didn't make it out; nobody I know of working that high up made it out.

Once I found that Lukas had made it out, then I assumed that John had, too. I called around looking for John all afternoon, but I didn't realize that 1 Police Plaza's phones had been knocked out. I knew that John had been working in the advocate's office as a prosecutor, so I tried to call that office, too. But as night began to fall and we were still working on the emergency, I [stopped worrying] about him, because I said, 'You know, John must have made it out.'

The very next day I called a friend who knows John and said, 'When you see John, you tell him I'm mad at him, because we had a near-death experience together, and he never even bothered to give me a call. I'm worried about him!'

So he tells me, no, John had been working out of the 40th Precinct. I say, 'You call up there and tell

235

John what I said.' When the guy doesn't get back to me and the day goes on, something keeps bothering me about John. I decide to call the fortieth myself; I want to talk to him and see how he is making out. So I phone up there, and I speak with the commanding officer. He says, 'It's funny that you are calling now. John's mother just called me and said he's been missing.'

I said, 'Missing? He was with me in the WTC!'

He says, 'In the WTC? What are you talking about?'

I say, 'We were in the WTC, together.' They didn't even know that he had responded from 1 Police Plaza to the WTC. I was the first person to alert the department that John had had anything to do with the WTC. They had no idea that he had anything to do with this emergency. They just thought he had gone down to put in his retirement papers.

I went back down there and looked at the area. There's no way. At first I had held out hope, because if those concourses were still intact, he might have made it, but I knew before I left the building that the concourse was compromised.

These losses are so . . . I keep holding my composure together. But as I get quiet and as I sit in the corner, believe you me, there is many a day I want to break down about it. The more I talk about it, it kind of relieves me. But you might hear me sound a little excited, because it really was a very traumatic experience.

Sometimes I sit there and think about my feelings at the time all this happened. I was so petrified and so afraid. And for a minute I thought I was the only one alive. I think of what I was confronted with, and how I struggled to survive, and live. It really scared the hell out of me. It makes me want to cry, because I know that I really fought to get out of that building. We were battling and pushing debris out of the way just to get out of there. I don't know how I managed to do it as quickly as I did, and pick the right way, but God guided us out that door. And we didn't have the time to spare to try to figure out an alternate way. We had to get it right the first time or I was going to get trapped when the rest of the building came down.

I go down there and see that site, and the smell. There's something that always will stay with you. My knees buckle every time I go near Ground Zero, because I look and I'm almost – I was originally afraid to go near it. I really was. I was petrified to go near the site, for what it represented to me.

I know what it means for others, too. I was up there, I know what they went through, I know the initial experience of feeling trapped, and feeling hopeless – thinking that I was going to die and it was just a matter of time before it all comes to an end. I was there, trapped, and I know I didn't hear anything more from John. It's strange because there was some yelling and screaming back and forth, but I never heard his voice again. It's like one of those things, you go one way . . . Oh, my God, I wish he was here.

I feel blessed and fortunate, but it scares me sometimes. Last night I woke up. It really is hard some nights, I just think about it, that's all. My God. Just to look at the mangled building like that. Just to think that I would have been in there, just to think of the people who are in there, it just sinks you. ∎

Joel Perry
Attorney

∎ I had a trial scheduled and was driving to the courthouse in the Bronx, the one by Yankee Stadium. I heard on the radio that a plane had gone into the twin towers. I couldn't wait until I got to the Throgs Neck Bridge, because I knew I would get a good view of the towers from there. The twin towers looked like two smoldering cigarettes from that distance, and I imagined the amount of fire that was causing so much smoke. I called my mother to get my brother's cell phone number, but then I couldn't reach him. I tried and tried and tried.

When I got to the courthouse, and it came to be 10 o'clock and the south tower collapsed, the judge told us to go home, he was closing up shop. I then drove to the 40th Precinct, where I fully expected to find John because I did not know that he had gone downtown to put his retirement papers in. We are only two years apart, so we grew up pretty close, and when I didn't find him there I began to worry. Now my mother called me and said she left messages for John all over the place, but that she had not heard from him. I told her, 'Don't worry, he'll call you back.'

But I knew my mother was concerned, and we both started to have that sinking feeling. I went home, and then I met friends, and then I went home again, but I couldn't sleep. There was a missing persons number announced on the television, and my mother called it at 2:30 A.M.

The following morning *Newsday* suggested that people call their local police department if people are missing, and so we called the Nassau County Police. An officer, Tom Savino, came over, but we had a problem about John's Social Security number, and I just didn't want to deal with more problems. So I decided to go into the city to John's apartment with two friends. John had a very nice apartment in public housing on 67th Street and Amsterdam Avenue. He got it through a police department program that tries to place police officers so that they will become part of a city neighborhood and not move out to the suburbs.

My heart sank when I saw two police sergeants waiting in front of John's building. They asked me if I was John Perry. Then I knew that John was missing. I told them who I was, and then we went up to John's apartment on seven, and we brought a locksmith in.

John had the place so burglarproof that it took a lock-smith forty-five minutes to get the lock off the door. The apartment was empty except for John's kitten. He loved that little animal.

So now we go down to 1 Police Plaza, and they have a family reception area set up in the auditorium, with tables of food and coffee, tea. There is a list of twenty-two police officers printed on a sheet that is pasted on a wall. I notice that another name is added to the list in ink. It is John's name, and it brings the list of missing police officers up to twenty-three. But we still didn't give up hope, at least not until we were invited to meet President Bush and Mayor Giuliani at the Javits Center. I knew they wouldn't ask us to that meeting unless they were certain that John was lost, and so then all hope evaporated. ■

About Captain Terry Hatton
Rescue 1
(As Told by Lieutenant Michael Penna)

■ Terry Hatton had been captain of Rescue 1 for approximately three years, [having succeeded] John Norman, who is now a chief and leads the special operations command, the rescues, and the squads.

Terry and I were firemen together in Rescue 2 when I first arrived there in 1985, at which point he had been there for about a year, having come over

240

from 105 Truck. I remember Terry as a fireman so well. He was always on the rig, he always had a tool at the workbench, taking it apart, checking it out to make sure it was operating. Any questions about any tool that I or anyone else had, we knew to go right to Terry. When he came in to relieve me or any officer, we would laugh, because after we would go over what we had to do in the office, he would be right downstairs, and you could hear the rig doors opening and then closing, every one of them. He was a nonstop worker, never one to sit in front of a TV or read a magazine. His dedication never [wavered,] no matter what rank he had risen to. When he transferred to Rescue 4 as a lieutenant, he went over to their rig and learned every tool, in every compartment, and became known for that.

We went, typically, to fire department functions together – Memorial Day services, Medal Day, St. Patrick's Day, parties, and everything else. We were very close and had a good relationship and were on the same page with a lot of things. As his wife, Beth, always put it, we shared the same brain. He would come to work with a new idea, that we needed to do this or we needed to send in a request for that, and I would just reach over to my box, where it would already be typed up. Without realizing it, we were working and thinking for each other, always moving in the right direction, always on the same road. We would crack up whenever we realized this.

It wasn't Terry's thing to go out and look for a

date – he just wasn't that type of person. When we went out, it was just as a group of guys.

In that respect Terry was like Paddy Brown. If you met them off duty, asked what they did for a living, and they told you they were firemen, you might first have to question it, because they never really had that fireman image type of thing, and they never really bragged. They weren't loud and boisterous and pounding-their-chest type of people. They remained very low key, even when talking to other firemen.

I have known both of them for so long that I have always been able to see their differences. Paddy was the [image of the] fireman jumping through the flames, coming out with the baby in his hands, making the spectacular rescues. He never did it for the sake of publicity, but just happened to be at the right place at the right time, and somehow a photographer was usually there, too, and the next morning there was Paddy on the front page. Terry, on the other hand, would go around to the rear end of a building where no one would see him, [perform] the same rescues, and nobody but the rescued ever saw him.

In fact, either of these men would risk anything to save someone. Paddy just did things that, if you analyze it, no normal person would be able to [bring off,] and he was always at the right place at the right time. Terry went by the assignment, whatever it was, and made sure it was done. [If faced with a terrible] amount of fire on a fire floor, a typical firefighter would be on the fire escape or a ladder. But Terry was inside the fire, searching. There was never a time where he'd say he couldn't make it, or he couldn't do this or that. He always found a way to do it.

Terry Hatton had a reputation on the job as being a hard guy, a hard boss, a guy who would hook you up in a second, give you charges for not doing something correctly. But he just wanted the work done right, and his way usually was the correct way. When he was named captain of Rescue 1, I would just sit back in the kitchen and laugh at what the guys [were saying about him –] Terry was going to come in and cancel the twenty-four hour tours, cancel the mutuals, cancel any side jobs people had. And for the first week that Terry was there, everyone was walking on eggshells: Nobody wanted to approach him, nobody wanted to say anything. Only three months later, if you made any comment about Terry to the guys in his group they would practically kill you. They had tuned in to him so quickly that they were suddenly explaining what Terry wanted done, and why. This was true for everybody in the company.

When Terry came to Rescue 1 we had a certain way of operating. All the officers in Rescue 1 had been in Rescue 2, in Brooklyn, together, and had an established way of fighting fires. You put the fire out, and the problem was over – that was our experience, a very aggressive way of firefighting, and we more or less molded Rescue 1 [along those lines]. Then Terry came in and modified the assignments for each man on the apparatus to protect the firefighters operating on the fire floor and the engine and ladder companies. The rescue mandate was now not just to make sure the fire got extinguished by any means possible, but also to support the firefighters on the fire floor to make sure their jobs got done. He changed the Rescue 1 operation to what it still is today. The firefighters cover every aspect of the fire building when they pull in. Everyone knows his assignment, and you do not hear, 'I don't know what type of building this is.' Every position is documented. Each guy, as he comes in to a fire, is trained to know where he has to go and what tools he has to carry.

Rescue 1 responded to a lot of [false] alarms in midtown offices, hotels, and skyscrapers, [often because these so-called class 3 alarms were triggered by faulty alarm systems within the buildings]. Because we were close to Times Square, we had all the areas on Broadway, 6th Avenue, and 8th Avenue, and we had been doing a lot of running around, not getting anything accomplished. Terry came in with the belief that a rescue company should not be used on alarms like this, so he created a pilot program, changing many of the boxes we used to respond to down on 6th Avenue and bringing us up to 9th Avenue instead, [which effectively put an end to our involvement in] the majority of class 3 alarms in the high-rises.

This left us available to attend to more actual jobs. We had already been going to every all-hands fire in Manhattan, and Terry then modified [our duties so that we would be responsible for responding] to any type of trauma runs coded through the EMS system that involved people pinned in truck accidents, trapped in machinery, and such things. Before Terry arrived, maybe one quarter of our runs actually involved helping workers, and after this new pilot program came out we brought that number up to one half. This brought about a total change in company morale, one that was incredible, phenomenal.

Terry was in the process of changing the [way the] fire department [operated, and leading it in the direction it would have to go] in the future. Fire duty now is not much like it was years ago. Today, we are getting more into the hazardous materials end of it, bioterrorism, more emergency work. We are getting a lot more industrial accidents and a lot more construction accidents, because of all the construction going on around the city. We are still doing fire duty, but since that was dropping off a bit, we were making up the work by dealing with all these emergencies around Manhattan. And Terry loved New York. The guys called him Captain Man-hatton.

No one knows what Terry was doing as the alarm for the World Trade Center came in. He could have been in the office working, he could have been going over the rig, or he could have been having a roll in the kitchen. We will never know, because all eleven firefighters of Rescue 1 who were in the firehouse got on the rig, and none of them came back. They left sixteen children and seven wives behind.

We do know that he met his friend Tim Brown

in the lobby of the north tower, kissed him on the cheek, and said, 'I love you, brother. I don't know if I'll see you again.'

There is one blessing that has grown out of all of this, and that is on the day before Terry's funeral, his beautiful wife Beth was informed that she was pregnant.

We are devastated in the rescue companies, but so are all the engine and truck companies in Manhattan or Brooklyn. Rescue companies are especially hurting because we have a lot more senior men and have been around a little bit longer. But you know, it's not any different from the proby in an engine company. It's the same loss all over. They have families; they have children.

The talent we lost from the fire department that day – from Pete Ganci to Ray Downey, Terry Hatton, Paddy Brown – these men were the future. While a lot of training is done on paper and books and everything else, 90 percent of the job is learned from the senior men on the job. We relied on these guys for the experience of fire duty – the proper use of tools, everything else. The leadership – all the guys that came through the ranks, that became chief officers, company officers – was more or less molding the younger men from what they learned, and that's gone now. It's gone.

It wasn't until a couple of days after 9/11, when I

came home for the first time, and I was with my wife and children, that we started realizing the loss of our personal friends. There are a lot of firemen that I had worked with – Peter Martin in Rescue 2, Timmy Higgins in Squad 252, Timmy Stackpole in Division 11 that day, Terry Hatton and Dennis Mojica. So my wife, besides mourning the members of Rescue 1, those eleven men who were her friends, too, also has to deal with the even larger number of our friends who are missing. These are the families and wives who went to all the parties in Rescue 2, the picnics, the annual dances, and everything else.

My wife Dianne is very supportive. She knew that, as I was the only remaining assigned officer in Rescue 1, that I had a large responsibility to take care of the company. Whenever I get to the point when I feel I can't do it any longer, when I hit the wall, she will say, 'You have to do it. You have to do it for the guys, you know. These guys depend upon you.' ∎

Firefighter Tom Schoales
Engine 4

Tommy Schoales is working in Engine 4, in a modern firehouse it shares with Ladder 15. The firehouse stands just along the East River on South Street, in the long shadow of the Brooklyn Bridge. Tom has almost two years on the job, but because he had three years working for the New York Police Department, he is nearing the five-year mandatory for full pay. As a probationary fire-fighter he is still on rotation, going from one company to another for experience, but that period is now coming to an end, and Tom is looking forward to return-ing to his assigned company, Engine 83, on 138th Street in the South Bronx. The South Bronx is where he wants to be. It is where his father, Battalion Chief Ed Schoales, worked when he first became a firefighter thirty-two years ago, his first stop before a long line of promotions.

Around the firehouse, Tommy is known as a step-up guy, always smiling, and this morning he is sitting at the kitchen table up on the mezzanine floor, flipping through the pages of the newspaper, wondering if Michael Bloomberg has a shot at the election today. He is a quiet sort of man. Unless you knew his dad person-ally, you would never know he had a father who was a chief. He never bragged about the fact that there was 'weight' in the family.

The company chauffeur of Engine 4, Bob Humphrey, is downstairs at the housewatch desk, mak-ing entries of all those who have reported for duty. Frank Vaskis, the chauffeur of Ladder 15, is rushing to check out the rig. He is handsome, single, Lithuanian, and lives out in the Sheepshead Bay section of Brooklyn. He arrived a little late this morning for he was up late last night studying for the lieutenant's test. The doors of the firehouse, looking out toward the underpass of the FDR Drive and the East River, are wide open. A few of the firefighters are standing there greeting the dozens of people who are passing by on their way to work in the high towers of the financial district. And then, out of the clear morning from the south end of Manhattan, a man comes running up the street. The firefighters are alerted for the man is yelling, 'A plane hit. I saw a plane hit.'

Battalion Chief Ed Schoales
15th Battalion

At the same time, up on 233rd Street in the Bronx, Battalion Chief Ed Schoales is filling out the TPRs (time and payroll sheets) in the office of the 15th Battalion, where he has been working for six years. He is an older chief and has spent a good many of his thirty-two years with the department in the busier sections of its heavy-response areas. Ed has the casual style of a country gentleman, and you can sense when you talk to him that it would take something out of the ordinary to ruffle the maple leaves on his collar.

Every now and again thoughts of one or more of his six children will occupy his attention, and as he works his mind turns to the youngest of his four sons, Tom. He spoke with Tom at the firehouse just last night, and so he knows his son will be down at Engine 4 today, working the day tour.

After finishing the reports, Ed takes a leather brief-case, 'the bag,' and begins to make it ready with the paperwork that has to go to the division chief before it is sent out to the Brooklyn headquarters of the department. The quiet of the battalion's morning is suddenly broken by a firefighter who rushes into the office, saying, 'Quick, Chief. Turn on the television.'

■ I looked at the television and couldn't believe my eyes. The second plane hit the building, and we were watching it on television. I know Tom is down there for Engine 4 must be second due on the box at the World Trade Center, maybe third due. I watched the buildings come down and I still couldn't believe it.

The total recall came in, and I finally said, I just can't sit here and watch this any longer. I called the division and told them that I couldn't stay in the office, that I had to go down there. The deputy sent a chief over to relieve me, and another deputy who was going down there gave me a lift.

At first I was hoping that he wasn't there, but I knew he had to be there. I assumed. The first thing I did was try to find out if anyone knew anything about Engine 4. I didn't make any calls because I wasn't going to get anybody, anyway – they would be all out. I went to the command post to find out what they had. There were so many people there, and I could pick up from their looks that Engine 4 had been mentioned. Then I went to victim tracking and asked if they knew of anyone who had gone to hospitals. Mike Currid called the hospitals in New Jersey, and they had firefighters there but that was all [the information] we had.

My son Ed came down with the police detail that had been sent by Stoney Point, the Rockland County task force. Someone told me, 'Your son the cop is here looking for you.' We both went all around, looking here and there. We went over to the firehouse of Engine 4, by the river, and nobody knew anything there, either. Most of them there were volunteers from the island, and they had volley fire trucks there also.

We went back to the site. When you see the mess that is there, you have your doubts, but you want to keep a positive attitude, which I tried to do. There was a possibility they would rescue.

The company chauffeur tells me that they had gone up I don't know how far, but when they were told to evacuate, they came down to the twenty-ninth floor, and for some reason went from stairway C to stairway B. He heard this in the handy-talkie location. Engine 28 was with them, and then Truck 7, and then Engine 4. Engine 28 all [got] out, for when things began to fall they made a mad dash. One of the guys in 28 heard someone say, 'Hold up, hold up,' thinking that the falling debris would stop, I guess.

I stayed until midnight. I had a final talk with the chiefs at the command center, [but] there was no point in staying any longer at that time.

They all knew. My daughters and sons knew where he worked, and they knew he was probably there. We made numerous calls letting them know where we were and what was going on. And when my son Ed and I got home, probably about one in the morning, everybody was at my daughter's house. The three sons and their wives, my daughter and her husband, were all sitting in the kitchen. They knew. Just the look. I had no good news, anyway. My other daughter was in New Orleans, and she calls her sister every day anyway, and today she is calling quite a bit. Her husband was up here, it happened, because he works for Mercedes, and he was looking to get another job with [the company] up here, and so he was interviewing. She was going to drive up.

[Tom] lost his mom at nine years old, and his older brothers and sisters looked after him. I worked at nights, but I was never afraid to leave my kids at home. They were never going to get into any trouble. I always knew that about them.

The next day my son Ed Junior and I went down again together, and we hear on the radio that they have rescued five people, and supposedly two of them had even walked away. And my heart stops beating, and I'm thinking, *Oh, God, maybe it's Tommy.* But then we get down there and we find that the guy who walked out of the rubble was one of the rescue workers who had fallen into one of the voids of the rubble. Then the heart sinks again.

After a few days, I see things as they are. But of course, I don't want to say anything to the family because I want to keep their hopes up. After four days, I stopped going down to the site for a while. Everybody really knew it was hopeless. The department or the mayor didn't want to say it. It was still a rescue operation according to them. By that time I knew it was not a rescue operation. It was a recovery. I think the mayor was doing the right thing, but anyone with any time on the job knew it, but nobody wanted to say it.

I decided I was not going to put the memorial service off, and we had it. Afterward, I began to go down to the site again, to work.

There is always a chance that you will lose your life in firefighting. Even me – I think there were six battalion chiefs who [died]. But it is something you don't think about. The rewards are enough for me, to keep me going, to keep me interested. When you do a good job, when you have a good fire, you save a few people, it's something to be proud of, not that you'll brag about it, but it gives you a good feeling. I figured Tom is a firefighter, too, and we're proud of him. You never think something like that is going to happen. You talk to anyone about Tommy. Of course, no one is going to say anything bad about him. But they all loved him.

A female police sergeant who worked with Tom in the three years he was a cop came by. She told me, 'He was so pleasant, and he never fell into the trap of following the crowd. He always had a sense of doing what was right.'

Tom loved being a firefighter, and I know he always remembered what I taught him, and that is that the uniform is supposed to say something about you. You get it for nothing, but it comes with a history, so do the right thing when you're in it.

Bob Humphrey
Engine 4

Bob Humphrey dropped the company off at the north tower so they could report to the command center. He watched for just a moment as Tommy Schoales, Calixto Anaya, Paul Tegtmeier, and James Riches hopped off the rig and then pulled away to find a hydrant at Vesey Street and West Broadway.

■ I watched about twenty people jump. I heard them land, but I couldn't see them. It was good I couldn't see them. They landed between the north tower and building 5. As I was watching the second plane came in, and at the last second, I saw the flash, and then the boom came. I jumped into the rig to escape everything that was flying around. I didn't see anyone jump out of the south tower, but I couldn't see the south side of that building.

I heard the company on the radio. They got to the thirtieth floor, and then the south tower collapsed. I ran into the delivery area for the post office, and it was completely black. I couldn't see. I had no mask and no light, and it was hard breathing. I thought maybe there would be enough air. [I was in a] cellar and had to feel around the walls for twenty minutes, slowly, because I didn't want to fall into anything. I had this thick crap all over my face, and I tried just to breathe through my nose. And then I found myself on Barclay Street, where you could barely see anything, and there are fires all around. I am thinking about the guys in the company, and I hear them on the radio. They are evacuating, but they are far up. They meet Engine 28 and Engine 7, and when they get to the lobby they say they are going to make a break for it. Engine 28 and Engine 7 made it.

I saw a guy from Engine 28 on the street, and then the north tower collapsed. When everything hit, the force lifted the cars off the ground and set them at all angles. I jumped down to the street and covered my head. It was terrible. Too much. I tried to call the guy from Engine 28. I screamed for him, but there was no answer.

He made it, though.

One thing about Tommy Schoales is that he was a quiet guy. He could be there without you knowing he was there. He didn't try to be pushy, but he always did what he had to do. He loved having the nozzle. He was one of the nicest guys to fit in here, just a better kind of guy. He always came up with a good one-liner – the kind of guy I want in my firehouse.

I just took my son to the firehouse last Sunday, to show him around. This is something to think about. These things can happen anytime, and I don't know what I would have done if my son was with me when this happened. I'm so glad it didn't happen when he was with me just two days before. ■

Firefighter Frank Vaskis
Ladder 15

The voice alarm had announced that the plane went into tower 2, but the department radio was saying tower 1, the north tower, when Frank Vaskis rolled the big rig of Ladder 15 out to the street. They are the second due truck company on the first alarm. Because Lieutenant Leavy did not want to take the traffic tunnel at the end of the FDR Drive to the World Trade Center, because he was afraid of getting caught in the tunnel in a traffic nightmare, they stayed on the streets and went up West Street. It was slow going, and at one point the lieutenant orders Frank to drive up a road divider near the entrance to the Brooklyn tunnel to circumnavigate the traffic. Frank does, and when he comes off the divider he hears the air rushing out of his brake. He has snapped a valve somewhere. He pulls the truck up until he is in front of the Marriott Financial Center Hotel at Albany Street, and the brakes lock, leaving them stuck.

Still five or six blocks from the north tower, the lieutenant and the firefighters, tools in hand, rush up the street. Frank is disappointed for he should be with his company, but he is left behind to fix the blown relief valve of the air brakes and then to take the rig to the base of tower 1 to the command center. He looks through a small tool box on the apparatus to find a replacement valve. He finds a valve bolt, but it is too large, and he shoves some plastic into it, hoping to force it onto the broken valve. *It might last a few hours,* he thought as he slid himself under the rig. Just then, at 9:03 A.M., a plane approached in a semicircle overhead and crashed into the eighty-first floor of the south tower. Hearing the powerful explosion, Frank jumped out from under the rig.

■ I know now it's not an accident, that this is a second plane. I suit up in my bunker clothes, put a mask on, my helmet, grab my tools, and head down West Street. There are body parts all around, there is a foot in the middle of the street, and what can only be described as intestines. I go under the overpass at Liberty Street. I heard that a command center was set up just past the overpass, but it wasn't there. I am looking for my company. I notice my guys are not there. I go to the next overpass, over at Vesey Street. I see Timmy Stackpole, who's going in the other direction, but I don't see anyone in my company. So I head into tower 1. The windows between the big columns are gone, and I walk straight into the lobby of the building to where the fire command center is supposed to be. There is nobody there, no chiefs, and so I walk [toward] where the staircase is. I figure there are no elevators, and I'll have to head up the stairs to find Ladder 15. When I am three quarters of the way to the staircase, [near] the middle of the lobby, all of a sudden this great wind comes, chops me to my stomach, and basically pushes me all the way back through the same window I came through. That was building 2 coming down.

It is now pitch black. I'm on West Street. I crawl back into the building [to] where I could feel where the glass frame was. I figured that was probably the safest, with everything coming down. I fumbled to get my face piece on, to get some air. It is pitch black. I wait. Pitch, pitch black. And then it gets gray; now it's just cloudy. In about five minutes I can see: I can see where my helmet was blown off my head; I can see where my tools were blown out of my hands. I go out [to retrieve them] and come back in to where I was, when people come out of the stairwell, and they're screaming, 'We're in the stairs. We're in the lobby. Oh, my God, now what do we do?' They began heading out the main entrance.

When I had been outside, I looked up, and it looked like the top of the south tower, above where the plane had hit, had collapsed and fallen on top of the Vista Hotel. So I said to them, 'Listen, there's a lot of debris falling out there. You probably can't get out that way. Come out this way. It's safer this way.' They come out by the window that I went through, and I said, 'Stay close to the building, just work yourself around, go up West Street, and just keep on going.' I go back in and head towards the stairwell when I hear on my radio an urgent message: 'All firefighters out of the building. We're going to regroup on Murray Street and West Street.'

I turn around and when I get to Murray Street, the north tower comes down. And I just ran – I guess I got to Chambers Street, by Stuyvesant High School, just as the dust cloud got to us. We ducked into the building.

I ran into a few guys from Ladder 1 and Ladder 10, [and about forty-five minutes later] guys from Engine 4. I started asking any member I met what they knew about Ladder 15. I met Bobby Humphrey, the chauffeur of Engine 4, and we talked there on Chambers Street. He pretty much told me that both of the companies were still in there, they never came out, as far as he knew. People were walking around looking for faces that they [recognized]. It was really a horrible situation. And then we were waiting for a roll call [by] chiefs who were just taking notes. We gave [our] names and the company designations to three or four of these chiefs, who were taking these notes, and still we both ended up on the first list as missing.

I didn't take any medical leave. I felt it was necessary to be down there. It was adrenaline driven. When I was thrown I landed on my elbow, and so I had a good swelling. When I was thrown outside, I had lost my helmet for a time, but I was pretty much covered with my bunker gear [so] I had a cut and some bumps and bruises on my head, but miraculously I was pretty much unscathed. ■

Fire Patrolman Mike Angelini
Fire Patrol 3

■ The fire patrol is a department of people employed by the New York Board of Fire Underwriters, the central organization of the city's insurance industry, to protect property and inventory during fires and emergencies. It was the first paid fire department in America, started in 1839 when the board hired four fire patrolmen at $250 a year. We are not a part of the FDNY, but we have been associated with it since the mid-nineteenth century and were actually in existence before the FDNY became a paid department. The fire patrol has about a hundred patrolmen working out of three firehouses, two in Manhattan and one in Brooklyn. These companies are connected to the FDNY alarm system and normally respond automatically on all third alarms, or by special call from the FDNY dispatcher. The company officer can also make a decision to respond to any alarm [that might involve] salvage operations.

I have worked in Fire Patrol 3 for two and a half years, and my firehouse is on Dean Street in Brooklyn. We saw it on the news, and I said, 'This will be a long pack, we're going to be carrying stuff up the stairs.' We could see it [on the way] across the bridge, and I knew right there. I fly airplanes and have been licensed for more than two years. I could see this wasn't a small airplane, but a big plane, and I knew there was no way a big airplane could make a mistake like this on a clear, sunny day. And then when we got there, the second one hit.

I knew my father was working. He has forty years on the job, starting with Engine 7 and then going to Ladder 1. For the last twenty years he has been assigned to Rescue 1 in Manhattan. They call him a legend in that firehouse, and he told the guys that he wants to do 'fifty and out.' He was on medical leave and was just assigned light duty at the special operations command. I thought for sure he would be down at headquarters in Brooklyn. He knew I was working that Tuesday and had said the night before, because I still live at home, 'Maybe if something big comes in I'll jump on the TAC unit, and I'll see you there.'

So I said to him, 'Ah, Pop, you're working, so nothing big is going to happen.'

My brother, also Joseph [named after our father], had worked a twenty-four hour tour the night before, so I figured he was out of it, in his car going home. My father was probably stuck at SOC answering phones or something. This was only his second tour over there, and they didn't know what they were going to do with him just yet. He was going to be mad that he didn't make it over. And my brother [might have] heard it on the radio on his way back home, and [he might have been] trying to get back to his firehouse in all the traffic. [I thought I was] going to be the only one there.

So we are in the lobby of the north tower, just

waiting, and I figure they know better than me what these buildings can take. I see Father Judge there and he's praying. I say hello to him, because he is such a good friend of my father. It is a nice thing. When the south tower came down, I went into the lobby of the Customs Building and then up the escalator. It was a matter of twelve seconds and the whole building was down. I had gotten up there, and then it got quiet. And all I knew is that I am by myself, I can't see much, and I'm okay. Let me go back down and see where my guys are. Everybody just took off when the collapse came, going, getting out. So we regrouped, and everybody was pretty much all right. Nothing had fallen in where we were. There were a couple of chiefs there, guys in white helmets, who said, 'Everybody grab someone and let's get out of here.' So we go through the muck to the escalator stairs and at the bottom of the escalator stairs we find Father Judge – I almost trip over him. He appears to be dead. I open up his jacket, and a chief undoes his collar and tries CPR, I think Chief Pfeiffer, but that doesn't help. So one of the chiefs said, 'He's dead, and we have to get him out of here. C'mon, let's go. We all have to go.' So I found a board and we put him on it and carried him out.

But a couple of our guys had gotten separated as we were going out. They found another stairwell, on the mezzanine level, and they went out directly to West and Vesey. We didn't know that stairway was there, so we went all around the building, which was good, because the building had an overhang. People were still jumping, and it was dangerous. It was me and a fireman and Chief Hayden who were carrying Father Judge, and I think an FBI guy. I put him down to get another board, and when I came back they had put him in a chair and carried him off. So my officer said, if we're not carrying him, let's go this way. [A little later] the second building came down.

After all is said and done, after the two towers came down, we ended up by the river, and they said a gas line there was going to go, so we went to Chambers and West Street. [On the way] I meet Bill Mirro [the recording secretary of the Uniformed Firefighters Association], whose son is on the fire patrol. 'Have you seen my son?' he asks, but I didn't see him all day. Things start to slow down then, and my first thought is that my father and my brother are going to be mad they missed this one. I didn't find out until that afternoon that they were down there, when I called home to tell my mother that I was all right. My mother said, 'What are you doing there? You're in Brooklyn.'

She told me that my father and my brother were [working at the towers] and that she hadn't heard from them, so she knew they were [still] there.

We heard all sorts of rumors. 'Oh, yeah, I saw your father, he's okay.' But I was very worried. My father would make it a point to call us or to let us know he was okay. They sent me home later that afternoon because my eyes were bad – I had gotten a lot of stuff in them. I wasn't sure if they sent me home because of what was going on. I came back in on Thursday morning and stayed until Monday morning, back and forth between Brooklyn and there. We were doing eight-hour shifts. I was off duty, but I just kept going over with them.

I went home on Monday because my sister-in-law, Donna, told the kids, finally. Joseph John the third, who is 3, Jacqueline, 5, and Jennifer, 7, were wondering what was going on. 'Where's Daddy?' they were asking.

'There's a big fire, and he's got to work,' Donna kept telling them. But even that young, they knew something was wrong. There were so many people at the house. Joe was 38. Everybody in his company, Ladder 4, who responded was lost, and there were seventeen in his firehouse who never returned. When my nieces and nephews were [finally] told their father was missing, they said, 'Oh, Pop will find him.'

I thought, *You know, my father, he's going to walk out of there. He's in there, but he just won't talk on his radio.* That was always the joke, he never uses his radio. So he's in a void somewhere, and he's just getting mad because he's running out of cigarettes. When he gets out of cigarettes, he'll come out. And I really expected him to come out of there, until I heard otherwise.

And Donna told the kids, 'No, Pop is missing, too.'

And with my brother, I just kept hoping and hoping, until after a certain point, [after] three weeks.

The next conclusion they jumped to was that, being [my father and brother] were missing and they hadn't seen me, that something had happened to me, too. So I said I'll go home, I'll step in the door and let them see me, let me pick them up, and I'll go back. So I did that and came home, and figured I would take a nap, sleep in my own bed, and go in again in a few hours. But I slept through the alarm, and the next thing I know the fire department is at the door, saying they had found my father. One of the retired guys from the rescue company comes, with a chaplain, [to tell us] they found him two hundred feet from where I was working. Had I been there that day . . .

My mother understands everything. She's been married to my father for forty years, so she knows how things are in the fire department. Most guys with eighteen or twenty years on the job start looking to go to slower fire companies, or retire, but my father went to rescue for another twenty.

My mother isn't alone because I am there, and my sister comes in regularly. Donna has a group of wives from the firehouse on 48th Street and 8th Avenue, and they support each other and go to things together.

I still don't get it. I dug through those piles for days, and I still don't realize what has happened. I go home and see my mother, and I remember that when he was home, Pop always had a cup of tea with her before they went to bed, and if he wasn't home, he would call. So I have to catch myself from saying,

'You waiting for Pop to call, Mom?' And with Donna, too. Whenever I talk to her I have to catch myself before I say, 'What's Joe up to today? Where is he? How come he's not home?'

I like my job, but I have always had the dream of being a New York firefighter. I scored pretty high on the last exam, but they had a special five-point gift for paramedics coming on the job, and so some of those people went ahead of me, and I got one hundred names away from being hired when the list expired.

Maybe. Someday. Now, we just have to think of the family. ■

Assistant Chief of Department Harry Meyers

I have known Harry Meyers for thirty years, since we worked together in the same South Bronx firehouse, on Intervale Avenue. La Casa Grande, the firehouse was called, the big house, because four companies were located there in the late sixties and early seventies. Harry Meyers's first assignment in the department was Engine 85, a very busy engine company and a twin to the fire engine that sat right next to it in the firehouse, Engine 82. Harry was always a likable sort. He joked like everyone else, but he had that serious part to his nature that is required of students in the fire department, and I knew even then he would rise up the ranks in the job.

There are four assistant chief of department badges

in the FDNY, and Harry has one of them. There are only two uniformed men in the department who outrank him: Pete Ganci, the five-star chief of department, and Dan Nigro, the chief of operations.

Since golf has become the preoccupation of many firefighters, there are numerous golfing trips organized by members of the department every month. Every now and then when Harry would see one of these outings advertised, he would call a few pals and associates and they would make their way to a golf course.

On the morning of September 11, he is playing golf with his good friend Pete Rice, a battalion chief, at a course as far away from the city as they had ever gone without taking a plane.

■ We were upstate in Margaretville, and luckily the people there knew we were firefighters. We had an 8:00 start time, and so we had been playing for almost an hour when a little before 9:00 someone comes running out over those beautiful green grass mounds of the course, waving his arms, and yelling about a plane that has crashed. My first thought was that it was a small plane. Maybe it was just down the road, in a cornfield . . .

We got back into the clubhouse, where they had a big-screen TV. There was an announcement that the roads and bridges had been closed, that no one could get through. Shortly after that, the first tower collapsed. When that happened I knew I had to get down there somehow. I had to pack and check out of the hotel, and by the time we left there, the second tower had collapsed. It's a long way from Margaretville to the [New York State] Thruway, so I had the lights on, going pretty fast, probably a lot faster than I should've. I got on the thruway and started to create the sizeup. My radios weren't working. My cell phones weren't working. Even my beeper wasn't working. I pulled over at the first rest stop and got my 800 megahertz radio out of the trunk. I thought to get some gas, not knowing what I would be facing when I got down to Manhattan or how much traffic I was going to be sitting in. I got down here about 11:30 and went to Engine 6, [because] I wanted to get some fire gear on.

There were some guys back in the firehouse, from No. 6 Engine, who were shell-shocked. Everyone was shell-shocked. I changed and drove the car up to Park Row as far as I could get. I had to do a damage assessment, and see what had to be done. I walked from Park Row to Church, and the first thing that struck me when I got to Church Street was — it jumped into my mind — *The Planet of the Apes*. I saw the façade of WTC 4 standing there, eerie-looking fingers stretching up in the air. It reminded me of that movie.

And 4 and 5 were burning pretty much throughout. I continued to go as far as I could, to Liberty Street. There were a lot of people working in the piles.

I was trying to stay as detached as I could at this

point. In a way it was an advantage for me to be coming in later. Because I had not been involved with the original collapses, I could view [the situation] with a fresh pair of eyes. This helped in doing a risk analysis of the surrounding buildings and the piles.

As I went around the site I tried to reposition people, move them based on what I saw in the surrounding structures. One of the biggest problems that you have in an incident like this is secondary collapse, where you can end up getting the rescuers killed. There were all kinds of people working on the pile, a lot of them working on rubble in the street, but some were in places too dangerous, where they shouldn't have been. I couldn't get all the way down to West Street on Liberty, so I went back to Cedar, up on Cedar Street to West Street.

Ninety West Street took a pretty good hit. There were fires there. Scaffolding on the north side of the building was ripped away on the lower part and hanging from above. The building didn't look like it was in any danger of collapse, but the scaffolding certainly looked insecure, so I had to move some people away from there.

I worked my way around West Street as far as [possible, but] I couldn't get by the pedestrian bridges that had collapsed. So then I went back around to Park Row to [pick up] the car and parked it on Vesey and West. The field com unit told me the command post was at Chambers and West. I knew we had a lot of people that were missing before that, but I didn't know who at this point. So [it was there] I found out that Pete Ganci was gone and Bill Feehan was gone.

They had a field com board set up and there was a captain there, and he was the first one I saw. Joe Callan was looking kind of out of it, kind of shell-shocked. At that point he was trying to make telephone calls, though the cell phones were all out.

I had been trying to call on the way down to let them know at home that I was going there and also to find out if my sons were involved in the collapse. I didn't know where they were at that point. My son Michael, a lieutenant, started out from his home. I didn't know it but he was working in the pile. And the younger guy, Gregory, who is a police sergeant, was there when the north tower collapsed. He started out from Queens when the first plane hit the first tower. My third son, Dennis, was in Engine 222, and when Danny Suhr of Engine 216 was hit by a falling body and killed, the dispatcher moved Dennis's company into 216 to cover. So Dennis stayed in the firehouse. My fourth son, by the way, is a fighter pilot, now stationed in England and waiting for his role in Afghanistan.

Since I felt I had the freshest perspective of what was going on up at the site, I wanted to have a strategy meeting.

Earlier I saw three chiefs at what seemed to be a command center, standing in mud in the middle of Vesey Street. And then they moved it all back, three blocks north, where field com set up their status board. We managed to get the staff together and went into the auditorium of Manhattan Community College, because that was the only quiet place we could find. There were a few guys resting there, and we told them to leave so we could talk.

My biggest concern at that time was #7. It was burning, and I felt that was a danger to the rescuers and to any trapped survivors there might be. We decided to keep the command post back on Chambers and to keep all but our necessary people out. At this time the recall was in, and people were staging out in Queens and in the outer boroughs. They were calling, looking to come down. We held a whole bunch of companies at the command post up at Chambers because we just didn't want them down at the site.

We were all pretty much on board that tower 7 was going to fall. We just didn't know when and in which direction. Everybody forgets about 7, but that building was forty-seven stories, and that in itself would have been a very major event.

We couldn't fight the fire in tower 7 because we didn't have enough water. All the water mains were down. What water we had was being supplied by the boats, through large-diameter hose. It would have been a chaotic situation, in any event, trying to get in there under those circumstances and fight that fire . . .

The building did go down eventually, all forty-seven floors, about 5 o'clock.

That was the first, for lack of another term,

stampede that I have ever experienced. In my thirty-three years on the job I have never seen anything like it. We were all the way up at Chambers and West, and this building started to go. We were clearly out of the collapse zone. Professional people, police, fire, a few military people, were stampeding like they were a herd of cattle.

After it collapsed I told them that I was going to make another assessment because I thought the collapse was a good thing [because] we could now begin our operations. I walked up Greenwich Street to Barclay Street, one north of Vesey, and it was an amazing scene. I was the only one on the street, the first one after the collapse. Cars were angled in their parking spaces for the force had lifted them. When the building came down it was completely involved in fire, all forty-seven stories. The rush of heat and burning papers landed under the cars and [set thirty or forty of them on fire].

The moonscape that I talked about, *The Planet of the Apes* [landscape], was the first thing that struck me. It was a light, light talcum powder dust, and there were papers everywhere. Documents were shin deep. There was no wind, which was a good thing, because if there was a wind stirring this up, I think operations would have ceased.

On Greenwich, I saw a few battalion chiefs coming up on the side streets. I told them to get some lines stretched in and to start to put out some of the car fires before the tanks started going. I told them to keep the guys back and not to treat them like car fires [but] like bombs.

I worked my way back up to Church Street where buildings 4 and 5 were still burning, each a nine-story building a block long. The pile that was the rubble of building 7 was still burning. But that was not a concern. That fire was not going anywhere.

There [was a lot of ground to be covered at] the site: nine square blocks, and then the surrounding blocks, which we had to be aware of. That night I went to the aid station for some bandages because my heels were bleeding from being in those fire boots.

I called Chief Callan on the radio to ask that the command post be moved to West and Vesey, to officially move it, because the collapse was no longer a concern. So we did that. It was around 7:30, just getting dark.

Battalion Chief Pete Rice showed up about seven. Pete had gone to the disability board and on 9/11 he was already out of the job on his leaves and heading towards retirement. But he came down and stayed the whole time. He has bad knees, and one is completely shot. So I knew he was hurting, and hurting a lot worse than I. Pete was the former captain of Ladder 3, which took a big hit. They lost ten men in that fire company. These were his guys, and he just had to be there.

We managed to sector it out that night – a lot of walking, continually walking around, pulling guys off the piles in places where we felt it was too dangerous to work, especially under the scaffolding. There was also some big steel hanging off the roof of building 6 that looked very dangerous. There was a huge piece of steel hanging down off the Bankers Trust Building on Liberty Street that was also very dangerous.

The rest of the night we just tried to get some organization to the chaos. It was organized chaos, but it was still chaos. We broke it down so that a particular chief was in charge of everything in his sector. We let them know that they were in charge no matter who was working with them – whether it was police or steelworkers or whoever. That command policy carried through probably for about two weeks.

We had a problem with the volunteer firefighting force. There was a call for mutual aid to Westchester and Nassau counties. The original plan was for them to come in and man our outlying firehouses near the city lines, which would free our firefighters to come into the operation. But that was not what the volunteers had in mind. They wanted to come in to the scene, and no one seemed to be in charge of this. The first we found out about it was at the command post. It was a good thing, but it took probably a week and a half to two weeks to get it under control.

Through that first night my focus was to keep the living alive: We would make any obvious rescues, we would put out any easily extinguishable fires, and we would do limited searches without adding to the body count. I was concerned about the safety of everyone at the site. We were very, very fortunate. I don't know how we got through this the first couple of days without adding to the body count. And credit for that goes to our guys. You know, our guys are the bravest guys in the world. You have to be very careful what you ask them to do, because they will do anything to try to do it.

My relief came the following morning, at 5 o'clock or so. He got there pretty much on time. But there was so much going on that it was hours before I got out of it. Then I got home, got a couple hours' sleep, and came back in that night.

On night two I did a limited [assessment]. I call this strange walk through Ground Zero a walkabout. We had a lot of radio communication, because [of the successful] sectoring of the site. Heavy machinery was in by now, and the steelworkers. We never called them. Local 40 of the steelworkers just showed up.

They were some really, really great guys and good workers. The key to the search and rescue operation was all that heavy machinery. When I say 'heavy machinery,' I don't mean the stuff that's there now, those huge cranes, but heavy for that point of the operation: the Cats, front loaders, and things like that. And the key to the heavy machinery was the dump truck. They were key for the first two weeks, certainly, and they still are key because all those big machines can move all that stuff but don't have a place to put it. If the dump trucks are not making good time, it just gluts the operation, stops you from moving anything. We had a constant battle early on to get the police to let the dump trucks move through the streets, to let them go in and out of traffic, to make sure the lane in front of them was open.

On that second night I had great hopefulness about finding trapped employees, living, firefighters, cops. Absolutely.

But [the extent of the disaster] is just totally amazing. Even considering the size of it, there are things that don't strike you right away. For instance, there were always documents, papers, and files [among the rubble, but] I never saw a file cabinet, never saw a desk, a chair, never saw a telephone, never saw any type of office furniture. There is no glass. It just disappeared and had become part of this fluffy white or gray dust.

On the third night, I think we really started to get a grip on it. Chief Fellini and I both worked that night. Fellini took the Vesey Street side, and I took the Church Street side, and we started to make an effort to coordinate what was going on on the top of the piles. We got some honest control. We had a battalion chief on the edge of the pit, with about six or eight men, and we had another six or eight on reserve that he could either use as relief or use them if any of his first six or eight got in trouble. They spent a long time out there on the piles. Some of those battalion chiefs were phenomenal.

We came to an agreement with the steelworkers,

because they were really good. They knew what they were doing, but when I grabbed a few of them and asked who was in charge, they told me nobody. If nobody is in charge you can't be here, [I told them,] so you'd better figure out who your spokesman is going to be and have him deal with me. So they did that. We [organized] where they would go, what they would do, how they would keep track of their people. I wanted a list of who went in and who came out so we could start to account. We had people searching voids, and we didn't know who they were.

On the third night we had a guy arrested who followed one of the urban search and rescue teams. He had an orange jump suit on with a backpack, and he was wearing a helmet that said RESCUE across the back. He had a hook of some kind and followed one team in and broke off from them after they were in. He did a little shopping in the stores down on the concourse. Someone else tried to accompany another USAR team down, but its leader was a pretty sharp guy from Massachusetts. 'This guy is not one of ours,' he said. 'You may want to check him out.' We called one of the cops over and found a ton of watches and cell phones in his backpack. They arrested him. Then we had another guy arrested, a photographer. 'I gotta take these pictures,' he said. I asked the cops to put the handcuffs on him right in the middle of the plaza and walk him out very visibly through the crowd on the other side of Church Street. So between him and the guy in the orange jump suit, that slowed down all the stealing and picture taking for a while.

The third night, they also started getting more engineers coming in, and one thing I learned from this whole experience is that engineering is not an exact science. Numbers don't always add up the same, because you can get one engineer, like an expert witness in court, to tell you one thing and [another to] tell you another thing. And so I had to be careful about which advice I could take from whom.

One night, an engineer from the Department of Design and Construction came to the command post. He said, 'Get everybody out of here, out of that debris pile, right now.'

I looked at him and said, 'You are here to advise, not to order.'

He said, 'Well, it's very dangerous.'

Everything I do is very dangerous, and he didn't get it. I asked him to leave the command post, but he didn't want to do it. I told him that, shortly, I was going to have him arrested. So he left.

And then there were the snake oil salesmen. Everybody that is working on some new type of technology shows up at the scene of an event like this, like characters that say I have this machine that will pinpoint a cell phone signal, or even will detect a heart beat through thirty feet of concrete. People trying to push their products and push their ideas. The Department of Justice said they were already aware of some of them and knew that some did not work. So I asked them to consider in the future, if we get a catastrophic event like this, we should have someone from the Department of Justice on the scene just to talk with the inventors that come around. They should talk to them, screen them, and then evaluate the ones that have some merit.

One of the early things that we wanted to get done was to get West Street open. To do that we had to get that pedestrian bridge down, under which we had so many fire apparatus, crushed. That was impeding the whole operation, and it was right in front of us on West and Vesey.

And then in the pit up on Church Street, that space between buildings 4 and 5, it became obvious that we were searching the same places over and over again and going through the same risks. That was when we started using spray cans. We developed a policy in that any search you did, any round you did, anywhere you went, you had to spray it, so people would know that the area was searched. It was red spray paint. The same thing was done with the PATH tube down below.

When we got down to the PATH tubes we had information that there was a train there. We did get somebody to look at it, but it turned out it was at a dead end, an abandoned train. Structurally, it was very bad down there, so we had firefighters spray-paint [warnings] on the walls: DANGEROUS STRUCTURE and INSTABILITY IN THIS AREA.

But not everything below was dangerous. As it turned out, the [underground] arcades would've been a relatively safe place to be because they were pretty much intact. The middle was down but all the side walls, the walls underneath the sidewalks on Church Street looked like they were untouched.

We have very good interaction with the cops. They knew we were in charge, and they never pushed that issue. There was always somebody there from the police commissioner's office or the chief of the department office. Every tour I worked someone would come up and hand me a card and say, 'I'm the guy in charge of my side tonight. If you have any problems, call me.' The emergency service unit guys were also superb.

One police chief did come up to me early on, upset like our guys were upset. Why aren't the cops being used, he wanted to know, when on television they see volunteers of every kind giving interviews? He felt the police officers were being held back, but the [truth] was that they were being held back because someone was in control of them. There was no control organization for the volunteers.

[A number of] surgeons also came up to me, pointed to where they had been standing, and said, 'I'm going to be right over there, and I am going to be there all night. If you come across anyone that is trapped in the rubble, and they can be removed but there is an amputation that has to be done, I'm a trauma surgeon and I would do it.' So that was good news.

Later on we had other decisions to make [in that

respect]. This is now getting into Day 7 or 8 or 9, and we start finding bodies that are pinned in the steel. We know that we are not going to be able to get them out without using the heavy cranes. So, do we do an amputation? We talked about it and thought about it as if we were in a family. I said, 'I would rather know for sure if that was my loved one.' So we have to think the things differently. We have to take whatever we can get. [If you had to] take a finger and do a fingerprint, or do a DNA on other body parts, [then we went ahead and did it], because you have to have some closure. And then this was wartime. It got to the point that when we needed a surgeon to do that, it required the approval of the command. Each time. And that was a good thing.

We had good management of our people, but we were still finding we needed to get the dogs to do something, to concentrate on a particular area. Even as late as Friday we were still going with the dog work. The dogs can tell you what's in there before you know it. There are different kinds of dogs. Cadaver dogs are trained to find nonbreathing persons, and people dogs are trained to find breathing persons. And then there are also trauma intervention dogs. We had some people showing up with house pets: 'Here's my dog.' There were a few light spots, a few funny things.

As time went by, we went from expectation to hope. You know, in the beginning you expected to find a survivor, and then you hoped to find someone breathing, and then you hoped to find the body.

On the second day, there was a woman under a piece of steel, and she had her hand sticking out, and then held our hands. The rescuers were so up to get her. It was a great moment.

I got home one night and saw on TV that two firefighters had fallen into a hole in a pile, and they worked for hours to retrieve them. I was shocked and surprised that they got them out of that hole. It was a good thing in the middle of so many disappointments. I can't tell you how many reports we had that someone was in the rubble and their cell phone was activated, or that a call had been made from a cell phone from someone missing in the piles. They were all false alarms.

John Vigiano is a retired captain out of Ladder 176. A very, very intense, dedicated into-the-job guy. Drill, drill, drill – a former rescue guy. And his company is noted for being a good truck company and for how much drilling they do.

John has one son in the fire department and another in the police department. John II is in Ladder Co. 132, and Joseph is in Police Emergency Truck 3. Both brothers were working, and both responded, I guess neither knowing the other was there. And both died in the line of duty. From what I understand, though I don't know it to be a fact, these were his only two children. We haven't recovered either body, so John has been on the scene every day, searching for his sons. The police department has a driver assigned to him.

John Vigiano is such a gentleman. He stays out of the way and doesn't interfere with anything. He just wants to know what's going on. He's that kind of guy. He's a very knowledgeable guy and knows what should be being done. He also wants to know that everything that could be done *is* being done. When I came in one morning I walked out of my way to ask how he was doing. He asked me if I really wanted to know, and I said, 'Yes, that's why I asked.' He told me he had some problems with the way things were being handled. I understood, and as soon as I took over I did correct what was a problem, and he was right.

The bucket brigades, as time went on, got to be called conga lines. It was a very good description. We talked about how many there were and where they were. Although [they did not serve] too much useful purpose as far as a recovery, it was a very [beneficial] thing for the guys at the scene to be doing something, not just sitting around and doing debris examination.

John saw that that wasn't going on and that everybody had been taken off the pile. There should be some work going on the pile, he said, within the limits of safety. He was right; there were areas that we could safely do it.

I can only [imagine] the loss John is suffering as I think about my own children. As I told my wife, Cecelia, 'You know,' I said, 'you never have to worry about hitting the lottery – you already did.' What are the odds here? There are three boys and myself, and for whatever reason, we were not working or not called to respond, and so we were not there in that crucial first ninety minutes.

I saw Eddie Schoales down here the first night, and he broke my heart. We all worked together in that Intervale Avenue firehouse in the blaze days. Eddie was standing around outside the command post, and Eddie being Eddie, he would never be in your face for he's a very reserved guy. I talked to him, and he says, 'My son is in there, and my son is missing. Have you heard anything?'

I said, 'No, we haven't. But there's no point in your being here, Ed. You should be home with the family.'

And he says, 'I can't.'

'Well,' I said, 'I'll make you a personal promise. If we recover your son, I will call you, and you will be here when he's taken out. No matter what the condition.'

And at that point he gave me his phone number and left to be with his family. ■

Dennis Smith

From where I live on 77th Street, the closest firehouse is Engine Co. 44 on 75th. But Ladder 16 and Engine 39 is one subway stop away on East 67th and has a police station next door. I thought I could always get a ride with the police if the firehouse was empty.

It is 9:59 when I turn onto Lexington Avenue. The streets are filled with people walking quickly, or trying to get on buses, or looking for cabs. They aren't panicky, but a pained look on every face seems to say, 'We are involved in something very great and very terrible.'

I try the subway and learn why so many are on the streets: The subway system, at least the Lexington Avenue line, has been shut down. I think to take my car, but I might not be able to get close enough to the area before I get hung up in traffic. A bus comes down Lexington and stops, but it is crammed with people and the driver does not open the front door. The back door opens to let someone out, and the waiting crowd rushes toward it, one or two managing to squeeze themselves in. Someone in the crowd has a radio to her ear, and yells, *'The building has collapsed!'*

I close my eyes momentarily and say an ejaculation, something I learned to do when I was in grammar school at St. John the Evangelist's on East 56th Street: 'Christ, help them.'

I am thinking of the firefighters. They may have successfully evacuated the building, and there may not be any civilians in it at all. But the firefighters will still be in there. Without a doubt they will be there, searching, stretching hose, trying to get to the fire. How many firefighters will be there? How many companies?

I see someone getting out of a cab on 78th Street, and I run to it before anyone can beat me to it. In the backseat, I think for a moment. Should I take this cab as far south as I can and walk to the site? No, I decide. I can be put to better use if I go with the firefighters, so I asked the cabbie to stop at the 67th Street firehouse.

The men there will have responded to this emergency the way they would respond to any fire alarm, I know. When the alarm is pulled, the firefighters leave the station, not knowing what is awaiting them or what they will be called upon to do. Will it be a fire, a drug overdose, a knifing, a car wreck, a suicide, a burst water pipe, an elevator stuck between floors with four screaming children within, a forgotten pot roast in a stove, a false alarm, an explosion, a building collapse? The not knowing is part of the excitement of the job. But here, in the collapse of a 110-story burning building, I know at least part of what will be awaiting the firefighters, and the police officers, and whoever else is an emergency responder: Danger and death will be found there, as inevitable as the earth's going around the sun.

But maybe there has been a miracle. Maybe they themselves evacuated to let the fire burn out by itself. Maybe . . .

A group of firefighters is milling about in front of the firehouse quarters of Engine 39 and Ladder 16, holding boots and helmets in their hands, their bunker jackets thrown over their arms. They have the look of concern all over their faces. It is an anxious concern.

The firefighters have managed to commandeer a crowded 67th Street crosstown bus. A lieutenant tells the passengers, 'You can get off at this time, right here, or you can go with us to Amsterdam Avenue and 67th

Street.' About half the passengers get off. This was the first indication they have had, I think, that life is going to change for everyone. We go without stopping from Lexington Avenue to the staging center on Amsterdam. We don't talk much on the bus, and not a single passenger complains about missing his or her stop.

At Amsterdam we board another bus, along with fifteen or so additional men from the staging area at Ladder 35. Again, everyone is silent, until the quiet is finally broken by Lieutenant Wick, who says, 'We'll see things today we shouldn't have to see, and there will be things we might think we should attend to, but listen up. Whatever we decide to do we'll do it together. We'll be together, and we'll all come back together, boys.' He opens a box of dust masks and gives two to each of us. I put them around my neck, as the others do, and then we sit within our thoughts as the bus speeds down the West Side Drive, past the luxury liner piers, the air and space museum on the aircraft carrier *Intrepid,* and the Chelsea Piers sports complex. At 14th Street we can see that the quality of the air has changed radically, and as we go farther downtown it becomes almost viscous. The bus passes fire engines, police cars, ambulances, and chiefs' cars parked for blocks on West Street up from the remains of the north tower; there could easily be a hundred vehicles. It discharges us by Stuyvesant High School, and I leave the firefighters from Engine 39 and Ladder 16. They have to report in as a unit to the chief in charge, and as I look around, I see several chiefs huddled beneath the overpass by the school. I recognize

Chief Visconti and Chief Nigro, who is the chief of operations. They are having what looks to be a spontaneous meeting at a command post unlike anything I have ever seen. There are no aides taking notes, no staff of field communications firefighters, no command post desk, no onlookers.

It is like approaching a beach as I walk south down West Street. First there is a little concrete dust, like powdered soft sand, and then suddenly every step kicks up a cloud. Paper debris is everywhere, strewn between window casings, air-conditioner grates, and large chunks of what had once been the tallest structures in the world.

As I near the site I meet the chief fire marshal, Louis Garcia. I have known Lou for many years, since his firefighting days in Rescue 1, and have watched him rise through the ranks of the Division of Fire Investigation. He is followed by a staff of fire marshals, and they are all covered with grime.

'How bad is it?' I ask him as I fall into step with the group.

'Could be three or four hundred,' he answers, and I know immediately that he is talking about firefighters in the buildings. 'And we can't find Ronnie Bucca. You know, Ronnie came out of intelligence in the special forces of the army, and he was so certain there would be a major attack in these buildings that he had a set of plans of the World Trade Center in his locker. He knows them like the palm of his hand.'

'How did you make out, Lou?' I ask.

'I got caught by the first collapse, and in the second

288

The lobby of the north tower, where Chief Pfeifer and then Chief Hayden set up a command post.

JOHN LABRIOLA/GAMMA

A firefighter escorts a group of people down to the lobby from the mezzanine of the north tower.

AP/WIDE WORLD PHOTOS

Firefighter Mike Kehoe makes his way up the stairway of the north tower shortly after the first crash. Fortunately, Mike and his brothers from Engine 28 escaped after a "mad dash."

AP/WIDE WORLD PHOTOS

Rich Ratazzi's photo of Lieutenant Ray Murphy, Ladder 16, heading in to make a search after the first collapse, just minutes before he perished in the second collapse. RICH RATAZZI

In the immediate aftermath of the two collapses on September 11, there were hundreds of ground fires, car fires, and major conflagrations in some of the surrounding buildings.

JAMES NACHTWEY / VII

Chief Hayden, Fire Patrolman Angelini, and others carried the body of Father Mychal Judge from the lobby of the north tower and delivered him to these rescuers.

SHANNON STAPLETON/
REUTERS/GETTY IMAGES

After rinsing his eyes, a firefighter returns to the site.

DAN CALLISTER/SPLASH NEWS

Triage on the corner of Church and Dey streets. Paper and ash fell to the ground in a blizzard of destruction. JUSTIN LANE/NYT PICTURES

A Port Authority police officer assists an injured policeman to an ambulance. Thirty-seven Port Authority police officers lost their lives in the course of duty. BLACK STAR/LISA QUINONES

Firefighter Dan Potter was forced to flee from the collapse of both towers as he searched for his wife, Jean. Here, after the second collapse, he wonders what his next step should be.

AP/WIDE WORLD PHOTOS

First Deputy Police Commissioner Joe Dunne's sunglasses, found in his car at the Staten Island dump. In the heat of the fire, the lenses were reduced to the size of small pebbles. NYPD

Josephine Harris is surrounded by her rescuers (*left to right*): Matt Komorowski, Tom Falco, Sal D'Agostino, Captain (now Chief) Jay Jonas, Bill Butler, and Mike Meldrum. All were trapped in the second collapse.

DATELINE NBC/LYNN HUGHES

The first night on the piles in front of what remains of the Marriott Hotel on West Street. The single mission: search and rescue.

TIMOTHY FADEK/GAMMA

The bucket brigades of Ground Zero.

MARIO TAMA/GETTY IMAGES

The fallen pedestrian walkway connecting the World Financial Center to the World Trade Center that was used by Chief Pfeifer to escape the north tower. The pumper is one of sixty-three fire trucks that were destroyed at Ground Zero.

MICHAEL J. LEDDY II

Chief Fire Marshal Lou Garcia confers with his men as Chief Callan (in white helmet) discusses strategy with Chief Fellini (in white hat) at one of the many command posts that were established on the first day. MICHAEL J. LEDDY II

The officers of NYPD's emergency service unit (ESU) go to Ground Zero to assist in search and rescue. Of the twenty-three NYPD officers killed in the line of duty on 9/11, fourteen were from ESU. AP/WIDE WORLD PHOTOS

Tents line every street around Ground Zero. At the right is the temporary home of NYPD's organized crime control bureau, at the corner of West and Liberty streets. AP/WIDE WORLD PHOTOS

An out-of-town police rescue officer searches the site with his dog. Canine units from NYPD, the Port Authority Police Department, and other departments that volunteered their help were very effective in locating victims. STEVE WOOD/REX USA LTD.

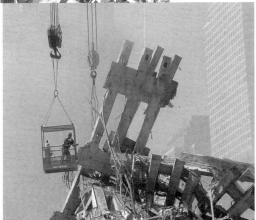

When there were no buckets available, pieces of steel and debris were brought out from the wreckage hand over hand. MICHAEL J. LEDDY II

An ironworker with an acetylene torch goes to the top of the steel façade of the south tower. MICHAEL J. LEDDY II

At his promotion to lieutenant, Kevin Pfeifer poses with Chief of Department Pete Ganci. All the firefighters of Engine 33 who responded to the attack, including Lieutenant Pfeifer and Firefighter Michael Boyle (*right*), were lost.

Battalion Chief Joe Pfeifer is given a special greeting by the president of the United States.

Searching at the site every day were (*left to right*): Captain Bill Butler, whose son, Tom, was lost; Lee Ielpi, whose son, Jonathan, was lost; Firefighter Pete Bondy and Chief Marty McTigue; and Lieutenant Dennis O'Berg, whose son, Dennis, was lost. These men, along with Captain John Vigiano, set the standard of dedication and resolve at Ground Zero. MICHAEL J. LEDDY II

Transporting a victim in a Stokes basket. DAVID TURNLEY/CORBIS SYGMA

Time, again, for prayer at Ground Zero.
AP/WIDE WORLD PHOTOS

The cross that was pulled from the pile on West Street provided a comfort for rescue workers of all religions. AP/WIDE WORLD PHOTOS

The mayor and thousands of other mourners bid Chief of Department Pete Ganci farewell at his funeral service.

The funeral of Lieutenant Dennis Mojica of Rescue 1.

A shrine built by neighbors in front of the quarters of Ladder 4, Engine 54, and Battalion 9. Seventeen men were lost in this firehouse, including Michael Angelini's brother, Joe.

we were just going up Vesey Street to a bagel shop to use a land line. When it came down we ran into the shop. Two firefighters came out of the cloud carrying Pete Hayden under his shoulders, and they brought him into the bagel shop and poured water into his eyes. He took a big impact. And then they brought in Brian O'Flaherty, who has a broken and shattered shoulder and was in a lot of pain. We just came from there, and we are going now down to where Chief Turi is trying to marshal people together.'

When I reach the corner of West Street and Vesey, I am frozen in my step. I am startled by the scope of the utter destruction I see before me, thoroughly shaken in a way I have never before been. Everything is gray. There is no color anywhere. On the corner is building 6, the Customs House, an eight-story structure still fully involved in fire. One extended ladder truck with a stream is going into the middle of the building, but it seems almost a prop in the midst of this ruin. All around it is rubble, overturned burning cars, and buried fire apparatus, ambulances, and emergency vehicles. Incredibly, the streetlights above the street divider on West Street are still standing. But just south are the remains of the crosswalk that linked the Winter Garden to the north tower and that seem as if they are rising out of the debris. Farther south, at Liberty Street, I can see several tall buildings still in flames, and one, 90 West Street, is almost entirely engulfed. And then there are the piles, the many mounds, mostly sixty and seventy feet high, of fallen steel, broken concrete, shards of cement and rebar, pulverized powder, and paper like large confetti spread over everything. Imagine six or eight city blocks filled with hot, smoldering, and smoking debris, lying in drifts and hillocks and windrows, as if they had been laid down by some titanic wind. In the street itself are large sections of bare and bent steel. Firefighters are walking carefully through the devastation in small groups, searching, calling.

At the intersection of West and Vesey, ankle deep in the mud of the pulverized material that has mixed with the water of the broken mains and fire hose, are two fire chiefs, their ears to their radios, directing groups of firefighters. One chief calls to move the hose going into the ladder truck in front of the Customs House Building, and I join a few of the men who bend down to pick up the charged line and carry it about forty feet north. I feel unprepared to work in this disaster and wish I had gloves and my fire helmet. Still, even with inadequate safety equipment I know I can be of some help somewhere.

I want to go to Church Street, which is the thoroughfare directly to the east of the WTC, but Vesey Street is completely blocked off by rubble, so I decide to walk around to the south area along the western edge of the site. I overhear a few firefighters on Vesey talking about Chief Ganci, saying that he has been killed, along with Bill Feehan, the first deputy commissioner, and Father Mychal Judge. I had seen Pete Ganci and Bill Feehan not long ago, when the fire commissioner promoted me to honorary assistant chief of department, and they came to the small ceremony outside of the commissioner's office. They are both wonderful men and have been friends for many years, and I don't want to believe that they are lost. Pete always had that huge but incongruously shy smile every time we met, always connected to something kind to say, always something positive and upbeat. And how often would I meet Bill Feehan in various places around the country – at the firehouse convention in Baltimore or at the annual meeting of the Congressional Fire Services Institute in Washington? He always had an intelligent take on things, mixed with a great amount of New York irony and humor. And Father Judge, the spiritual fixture of the fire department, a priest for all seasons and all emergencies. I know this is a dark day, and I have known it from the first moment the plane hit the north tower. But this, now, is a darkness so profound it almost blinds me, and I have to rest for a minute.

After a while I make my way to the World Financial Center. I know these buildings well and go into the lobby shared by American Express and Lehman Brothers, stepping in through the broken-out windows along West Street. A small group is working right outside the window, a chief and two firefighters. I stop to see if I can help, but stand few feet away as they work. They have found the middle section of a body, and the chief directs them to pick it up as best they can and lay it in the body bag he has spread on the ground. He is whispering encouragement as they do, and seeing there is no way I can assist, I continue through the building and up the stairs.

The area is completely abandoned, and there is a one- to two-inch layer of concrete dust on the floor, like an even and pristine snowfall, and mine are the only footsteps in the floor's dust as I walk down the corridor. I pass by the monumental oil paintings of New York Harbor and the waterside of Rio at the entrance to American Express and think of all the times I have admired these paintings, and now it is a small but real satisfaction to see they are not damaged.

At the Winter Garden, a beautiful atrium with restaurants and stores covered by a glass dome, I am confronted with a large pile of debris that I am guessing is part of the north tower and part of the east exterior wall of the Winter Garden itself. The rubble spreads halfway down the large and majestic stairway that leads to the dining area, and I have to climb around it. The glass above is mostly broken and large threatening shards are hanging from their bent frames. I go through the Merrill Lynch Building, and still my footprints are the only ones on the floor. It is a haunting feeling to realize that these buildings, which have had people in them every second since they opened, are now abandoned. All the commerce and business of thousands have been stopped by these terrorists.

I go down the stalled escalator stairs to Liberty Street and continue around to Cedar. There is still a hanging cloud of gray and brown all around, and I can see clearly for perhaps half a block at most.

Only a few firefighters are in the streets here. Covered with dust and debris, they seem to move like statuary. They seem indolent and shaken, and they are walking slowly and aimlessly or sitting on curbsides. Many are dazed, but they do not seem injured. But living through what they have been through, they have suffered in unimaginable ways.

I pass many cars that have either been overturned or angled in a way that makes them appear scattered. I cannot read the street signs on the lampposts, which are completely covered in dust. Many personal items are scattered along the street – a shoe here, a cap there. On Liberty Street, where Ladder 10 and Engine 10 share a firehouse, I find a woman's pocketbook. I contemplate opening it to find the name of its owner, and to bring the reality of this huge and complex disaster down to the simple, reassuring act of establishing one person's identity. But I decide to leave it, remembering Lieutenant Wick's counsel that we might see things we want to do but that we should keep with the group effort, and the group effort here is to search for any survivors. A fire truck from a Brooklyn engine company is nearby, and I tell a firefighter there what I have found and throw it into the front seat of the pumper. I know that he will give it to his officer to take care of. Maybe, I hope, some documents or papers in it will help a family come to grips with a terrible loss.

From Liberty Street I can see large, gaping holes in two sides of the Verizon Building all the way over on Vesey Street, and I think of my friend Paul Crotty, who is group president of the Verizon company. He must be confronting the complete outage of telephone service in the area and trying even now to patch it up as soon as he can. The building seems to be formidably strong, but it will require a historic job to repair it.

At the corner of Liberty and West a corner of the south tower still stands, rising sixty or seventy feet into the air, and on the Liberty Street side are twelve fifty-foot Gothic arches that only hours earlier held immense windows. Those windows nearer the east side of this row seem to be less wide than all the others, which conveys the illusion of balance from the side angle. It is a compelling image, this churchlike structure, serving as a reminder of all the art and all the engineering that support our culture, the human intelligence that has brought us to where we are in today's modernity. To me, at this moment, this progress seems as if nothing. What do the power, the statement, and the utility of this architecture matter, now that it has been leveled?

Yet this particular piece of devastation before me has a haunting aesthetic, in that all that has been preserved of the south tower is reminiscent of a church, and I come to think of this south wall as a cathedral wall. Like all cathedral walls, it separates the outside world from the inner, the secular from the holy. On the other side of this wall is a sanctuary of human spirit. Still, I can see nothing beautiful here in this consequence of ugly evil; the only beauty it is capable of is the sound of the voices of survivors.

The photographs and videotape I see in days to come cannot render the awesome reality of the destruction present all around the site. It is the size in relation to the human scale that makes me gape. It is the absolute absence of color, for the landscape looks as if it has been painted in grisaille. It is the smell, the peculiarly clean smell of plaster and concrete mixed with the dank odor of blood and death.

There is some activity now on the West Street side, where firefighters are gathering. A few dozen construction pails appear — large, white, seven-gallon plastic pails. A line forms, and these pails are passed from person to person. There are mostly firefighters in the line, but I notice a few police officers and ironworkers as well. I take a place far back in this line and have no idea what they have found or what they are digging for. I just know that there is something to do here, and I will do it until my arms get tired. I put the dust mask over my mouth now because I have begun to breathe with effort and more deeply, and it helps. After an hour, this line suddenly ends as mysteriously as it began.

I see on these piles, and hear while crawling over them, the most extraordinary things. These are men determined to find anyone who can be helped in any way. There is no small talk, just words of cooperation. The firefighters went into these buildings as teams, and their brothers are now working piles of what is left of them as teams.

A new and unique human seed has fallen to the ground. Because of this brazen, daylight attack, we will grow into different people, not so much changed people as charged ones — charged with becoming more aware of the perils of the world we and our children inhabit.

There are people in our country who are trying to kill us, and kill those we love, and we must be vigilant as we travel our streets. Where we had been open and trusting we must now be circumspect and acutely perceptive of our surroundings. But I myself am still the same American I was yesterday, a child of immigrants, and I resist the impulse to distrust every person who is not like me.

I return to the Vesey side of the site and see as I make my way over that a bulldozer is heading for Liberty Street, where it will try to clear a lane between Church and West streets, along the south side of the site. Several tables have been set up offering water and small safety equipment, and I take a pair of black hard rubber gloves. A chief is directing a group of firefighters to carry a large-diameter hose, six and a half inches, that has been stretched the two and a half blocks from the river to the north tower, and I grab a section of hose and pull. It is the heaviest hose I have ever handled, for it comes out of a fire boat lashed to the seawall at the end of the street, and the fire boat typically uses it to feed inland pumpers where there is a water shortage at big fires. Here I meet Mike Carter, the vice president of the Firefighters' Union. We are good friends, and Mike is fond of telling people that he became a firefighter because of reading *Report from Engine Co. 82* when he was in the military. Just two weeks ago I brought mayoral candidate Michael Bloomberg down to meet the union officials, Mike and Tom LaMacchia, and just afterward we ran into a friend of mine, Dave Weiss of Rescue 1, whose rig just happened to be parked on 23rd Street. I remember how pleased Carter and I were that Bloomberg had had an opportunity to meet the men of Rescue 1, and get to know the serious work of the line firefighters. But today Mike and I hardly speak besides a warm hello. We both are focused on what we are doing – heavy manual labor that takes all of our effort for the moment. This job takes about a half hour, and soon the hose is lined up near the water. I learn then that the chief has been clearing the area in anticipation of the collapse of the forty-seven stories of 7 WTC.

Suddenly, several hundred firefighters are milling about, and when the order is given we all are moved back another few hundred feet from the corner of Vesey and West. We just mingle there, talking quietly. I meet a police captain, and I ask if he has seen Joe Dunne, the first deputy police commissioner. The captain shakes his head and says, 'It's not good, what I hear.'

The Dunne family has been friends for thirty years, and I question the captain, who won't say more. I know that word spreads quickly at multiple-alarm fires, but I am hoping that, like all emergency situations, this one can be filled with false information, too.

All of a sudden the forty-seven-story building before us begins to shake. 'It's coming down,' someone yells, and everyone gets poised to run in the opposite direction. I head with a small group into the bagel shop at the uptown side of the street, but I am watching the build- ing as it begins to crumble. It falls almost as if in slow motion, sinking straight down, and then leans to the

297

south, leaving most of its rubble against the north wall of building 6. At any other time it would be a major catastrophe to lose a building like this, but now it is just an afterthought. At any other time it might be called an awesome, even a beautiful sight, but now it is just another event in a terrible day. The regality of this high building is transformed in a few seconds to mere rubble. And now I fear that it has fallen on those we seek.

I wait for the dark cloud of concrete dust to come hurtling toward us, but it rolls in the other direction, toward the northeast.

Now, it is as if we were in the eye of a storm. The stagnant quietude is unlike all the emergencies I have ever been to. No sirens. No helicopters. It only lasts a minute, for the firefighters are stunned that this sky-scraper has come down in front of them. Then, finally, the area of the north tower collapse becomes safe again, at least safe relative to what it had been moments before.

No one wants to estimate the number of firefighters missing. As I speak with the men around, I learn that the department's elite squads – Rescue 1, Rescue 2, Rescue 3, Rescue 4, Rescue 5 – have not been heard from. Ray Downey is also missing, Ray who was one of the first at the scene in the Oklahoma City bombing and who is well known as the embodiment and spirit of rescue efforts throughout the country. All firefighters are proud of their companies, but no one could have been prouder than Ray of his rescue companies. These firefighters have met difficult and rigid prerequisites to get into the rescue companies, including endorsements from other company commanders and tests of mechanical and engineering skills. I remember thinking then that these were truly unusual men, smart and thoughtful, the kind of men into whose arms I would put the lives of my children. I know the captain of Rescue 1, Terry Hatton. He is married to the mayor's assistant Beth Patrone and is one of the universally loved and respected men in the department. Someone else tells me that Paddy Brown is also missing. Paddy is one of a small cadre of the most decorated firefighters in the history of the department, and in the nineties he was on the front page of all the city's newspapers when he lowered one of his men on a rope to pick up a victim in a Times Square fire. He is an old friend, and I recollect many a time that we elbowed a bar together.

Now I think of Brian Hickey, the captain of Rescue 4, who just last month survived the blast of the Astoria fire that killed three firefighters, John Downing, Brian Fahey, and Harry Ford, two of whom were his own men. He was blown out of the building along with Captain Dennis Murphy of Squad 288. The funerals were terribly sad for there were children up front wearing their fathers' caps and helmets. I remember particularly the sadness in Brian's eyes.

As we all begin to go back to work, I see Kevin Gallagher, the union president, who is looking for his son, Kevin, who is unaccounted for. Kevin cares for nothing but the safety and the security of his men and has managed to make enough of a contribution within the politics of union activity to get himself elected twice. I can see Kevin is biting his lower lip, and I hope he finds his son soon.

Someone calls to me. It is Jimmy Boyle, the retired president of the union, the man who provided such inspired leadership during my time in the job. 'I can't find Michael,' he tells me. Michael Boyle, his son, was with Engine 33, and I have already been told that the whole company is missing. I look into the bright blue of his eyes and am reminded of the sense of optimism and good will I feel whenever I meet Jimmy. He has helped me innumerable times, and I am always filled with affection for this man. It distresses me, wrecks me, to see him like this. He is benumbed but not yet in shock. He wants an answer; he wants someone to say, 'Oh, yeah, Jimmy, I just saw Michael around the corner.' I can't say that to Jimmy, but just throw my arms around him.

With 7 WTC down, the army of construction workers, police officers, EMTs, and firefighters begins to work. People who have never met begin working side by side as if they have practiced for months. Cars are lifted, hoses and fire trucks are moved, and the heavy equipment begins to roll in.

I join another bucket brigade, and many more men are involved than before, along with a few policewomen as well, their sleeves rolled up and their forearms covered with dirt. As I watch the steelworkers, I think of how very confident they are in their work, how self-reliant, and as they labor they progress in my mind from admirable to heroic. These men, some of them covered in tattoos, remind me of the seafarers of a hundred years ago, who would go off in the whalers knowing that often there was no return. These men knew no danger, nor do the ironworkers. It is fascinating to see how they carry their center of gravity as they walk on a beam, never giving a thought to a fall.

I find a pay phone that is working in the World Financial Center, and I call Tom Dunne's home. His wife, Nancy, tells me that Joe Dunne is all right, that he has called home. It is nearing midnight. I don't have boots and am wet to the knees. I will 'take up,' as we say, and go home. At the end of this horrific day I think of Shakespeare's line about evil living forever, and I realize how most of the good of everything I know about this world is interred beneath the rubble before me, and in the spirits of those around me.

It will be days before there is a final accounting of who is missing. Lists will have to be generated and checked against one another. Families, friends, and employers will be contacted to determine the last known whereabouts of particular missing people. Representatives from the police, fire, and Port Authority will meet to decide on press releases, and the department chaplains will make their way to the doors of as many spouses as possible.

Before I leave I stand on the corner of Vesey and West streets, looking across the grayness that has now been illuminated by banks of theatrical lights. Fires are still burning all around me, and thick clouds of smoke rise from the piles here and there. The large neo-Gothic building behind me, 90 West Street, is smoldering, as is the Bankers Trust Building up the street from it. In normal times, these would be fifth-alarm fires.

Hours after the collapses the air remains heavy with the dust that is thickly layered everywhere. I cannot see the concrete of the street anywhere. The white-gray dust covers everything – the steel of the piles, the overturned cars, the façades of buildings. I know too well what each particle of dust represents – the dead, the warriors and the victims, many of whom have been integrated with the dust in the air. I turn and begin up West Street, looking for a bus that will go up the east side of New York. A chief offers a ride and warns me that it won't be door-to-door service but that they will get me within walking distance of home. I take one last look around at the site, grateful for the one good memory I will be taking home. Earlier, I saw Kevin Gallagher kissing another firefighter. It was his son.

Twenty or so firefighters are sprawled across the seats of the bus, all filthy, all exhausted. Again, there is little discussion and no heroic stories of pulling multitudes from the wreckage. Finally, a firefighter named O'Hagan, from Engine 50, says, 'I have four brothers on the job.'

I remember having heard earlier that Lieutenant Tom O'Hagan was working with Engine 6, a single firehouse down on Beekman Street, and that the whole company was missing, and I fear the worst.

But this firefighter on the bus continues, 'It took me all day to find them. Everyone is down here, and they're all right.'

I grieve for one O'Hagan family, and I am glad for

the other. But that happiness is short-lived as I remember Captain John Vigiano, whose two sons are missing; Chief Jim Riches, whose son is missing; and my friend Chief Eddie Schoales, whose son, Tom, is missing. . . . And Captain Bill Butler, whose son, Tom, is missing; Lieutenant Dennis O'Berg, whose son, Dennis, is missing; Lieutenant Paul Geidel, whose son, Gary, is missing; Fire Patrolman Mike Angelini, whose father and brother are missing; and Lee Ielpi, whose son . . .

Aftermath

Try to visualize the World Trade Center as it was mere days ago.

On a normal workday, some 50,000 people were at their desks, and 140,000 tourists excitedly rode the elevators to the observation towers or to visit on business. It is said that when the skies were clear one could see as far as 45 miles from the observation deck. For New Yorkers, being here was as natural as being at the Metropolitan Museum or Central Park. It was grand, it was big, and it was expensive, and recently the prize of the largest real estate deal in New York's history, when Larry Silverstein leased it for 99 years from the Port Authority for $3.2 billion. It had a 97 percent occupancy, and its tenants included 430 companies from 28 countries, companies doing business in banking, finance, insurance, import, export, custom brokerage, bond trading, and transportation, as well as a number of trade and professional associations.

The towers were not known for their architectural

beauty, but when you stood at the base of these buildings in the WTC Plaza and looked up, you experienced a definite sort of thrill at their sheer size and what they represented as a human endeavor. Groundbreaking occurred on August 5, 1966, and it wasn't until four years and four months later that the first tenant moved in. Nine city blocks were leveled to make way for the WTC, and before construction could begin the Port Authority of New York & New Jersey had to first build a huge rectangular 'bathtub' in the ground. Three city blocks long on the east and west sides, and two city blocks on the north and south sides, the structure was seventy feet deep.

This was created by building a slurry wall, in which forms were sunk into the earth and filled with a mixture of water and clay. For strength, each slurry-filled form had placed in it a cage made of reinforced steel. Each form was three feet thick by seven feet long and twenty-two feet deep. The slurry stabilized them, and then concrete was poured in, displacing the lighter slurry that would be used in the next twenty-two-foot form to be connected like a patchwork quilt. Gargantuan concrete slabs were made in this way, seventy feet deep, and they were connected all around the four sides of the bathtub. In New York, the rule of thumb for distance is that twenty blocks equals a mile, and this slurry wall would have been half a mile long if it had been stretched out in a straight line, or three thousand linear feet. It descended seven stories from the ground level and was bordered by West Street on the west, Liberty Street on the south,

Church Street on the east, and Vesey Street on the north. The slurry wall took a year to build and was completed in March of 1968.

The waterproof walls that make it up were built as much to keep the wet soil and the Atlantic-fed waters of the Hudson River out as to contain the construction of the six buildings of the World Trade Center.

Because of local geography, this bathtub was an essential component in the building of the towers. In earliest New York history, the eastern shore of the Hudson River ran along what is now Church Street, just east of the WTC site. Over the past two hundred years, the city fathers filled the downtown western shore of the island with garbage, animal carcasses, leather shoes, bottles, cannonballs, oyster shells, timber, concrete, and other debris. The landfill expanded the city westward to what is now West Street. When construction teams began the digging to create the bathtub, they had to first go through this landfill to what is the river bottom, and then through fill that had been left by glaciers that once covered the city, and a level of hardpan clay. Beneath that is the bedrock of mica that lies below most high-rise buildings in New York, called Manhattan schist. This rock is seven hundred million to eight hundred million years old, the base of an ancient mountain range that existed long before the island of Manhattan.

If you look at Manhattan from New Jersey, Brooklyn, or Queens, you will see a configuration of buildings that is shaped like a hammock, with high-rises at the south tip of the island, shorter buildings as you head north, until you reach the midtown area, where they grow tall again. The reason for this configuration is the way the schist folds within the earth beneath the island – it is nearer to the ground surface downtown and in midtown.

When the bathtub walls were complete, the digging started, and long steel rods called tiebacks were drilled through the wall and anchored into the bedrock. The more than 1.2 million cubic yards of earth removed for the bathtub excavation was dumped into the Hudson and expanded the city even farther west, creating 23.5 acres deeded to the city, the blocks on which the World Financial Center and Battery Park City now stand. Once the basement concrete was poured, all openings in the wall were welded over with steel plates, and the tiebacks were cut to avoid any permanent part of the structure from encroaching on the adjoining privately owned property. Then floor by floor, the seven lower levels of the WTC were built, and these floors provided the lateral support needed to diminish any chance that the bathtub walls would cave in. The World Trade Center basement covered 16 acres and ultimately housed seven levels of shopping, parking, and at the bottom, a PATH train station.

The towers were finally opened to the public in the early seventies, a period when New York had been for some years a city in financial difficulty. President Ford faced a notorious public relations problem when the *Daily News* ran a headline that said 'FORD TO CITY: DROP DEAD.' City employees, including police officers and firefighters, were being laid off, and Wall Street firms were downsizing and moving out of the downtown area to more convenient locations in midtown. A new world-class development was the city's best hope to keep the financial district from becoming a ghost town. The Rockefeller brothers, David as chairman of Chase Manhattan Bank, and Nelson as governor met with the officials of the Port Authority of New York & New Jersey, and the idea for a mega, world-class, building was born.

The food critic Gael Greene may have best expressed the promise of the World Trade Center after a visit to Windows on the World, a four-star restaurant that was on the 107th floor of the north tower: 'Suddenly I knew New York would survive. . . . If money and power and ego and a passion for perfection could create this extraordinary pleasure, this instant landmark, Windows on the World, money and power and ego could rescue this city from its ashes.'

New York has historically had its ups and downs, and the years just after the completion of the WTC were not good times for the city. The Port Authority had great difficulty renting space in the towers through the seventies and into the eighties, and it was only in the great bull market in the nineties that the towers would come to be almost fully rented and universally recognized as the symbol of America's success.

At 110 stories high both towers of the WTC were, for one month, the highest buildings in the world – a distinction supplanted by the Sears Tower in Chicago. Curiously, the north tower (tower 1) stood 1368 feet, 6 feet taller than the south tower (tower 2). It was also crowned with a 360-foot antenna that facilitated the transmission of ten television stations in the New York metropolitan area, including all the major networks, numerous auxiliary stations, and a master FM transmission. In addition, six stations transmitted high-definition digital television from this antenna.

The buildings were designed by Minoru Yamasaki, who was selected from a group of a dozen architects who were asked to submit proposals. He was challenged to design twelve million square feet of floor area on the site and to accommodate new stations for the Hudson tubes and subway connections.

The bamboo reed was a powerful icon in Yamasaki's view of the world. In an interview, he talked about holding a straw in his hand – the straw representing a sort of man-made reed – and then making it square. It came to him that if he laid in floors and controlled the bearing weight at the middle and at the edge, he could build a very high building. It was in this concept that a fatal flaw was integrated into the building's design, certainly without his being aware of it.

Yamasaki eventually designed a building with 60 percent of the bearing weight in the center, around a core of elevators and stairwells, with the remaining 40 percent held on the outside walls, with the trussed flooring connected to the outside steel of the skin. As a result of the plan the building had large areas of unobstructed space on every floor, and each floor had a total of about an acre of square footage.

This innovation permitted 75 percent of each floor's area to be used for occupancy, when more conventional methods would have allowed just 50 percent. And what, after all, is a major part of an architect's job but to find ways to create the greatest amount of occupancy space on a given piece of property?

The early construction was a massive effort. At times, about 3500 construction workers were employed in the building effort each day. More than 425,000 cubic yards of concrete were poured – enough to lay a bicycle path from New York to Palm Beach. The buildings were cooled with the world's largest air-conditioning system, with 60,000 tons of cooling capacity. Each building had 97 passenger elevators and 6 freight elevators that could travel up to 27 feet per second, and each car could carry 55 passengers.

Hyman Brown, the supervising project engineer of the World Trade Center, announced that the cost, originally budgeted in 1966 at $540 million, had risen to $800 million by 1973. Yamasaki, who died in 1986, originally projected the cost at $350 million. 'Economy is not in the sparseness of the materials that we use,' he explained, 'but in the advancement of technology, which is the real challenge.'

The World Trade Center was constructed from two hundred thousand tons of steel – more steel than was used in the longest suspension bridge in America, the Verrazano, and four times the amount in the George Washington Bridge. It has been reported, though, that the steel had been allowed to rust while waiting to be used and that the company responsible for applying the fireproofing material might have applied the spray-on fiber-based fireproofing over the rust. An engineer named Ron Hamburger was recently quoted in *The New Yorker* saying of the mineral spray fireproofing: 'If you knock it, that spray-on protection will fall off.'

This distinction regarding fireproofing is an important consideration in the collapse, because it is said that most fire professionals at the scene expected the buildings to withstand a heavy fire load for a much longer period of time than that which elapsed between the attacks and when the towers fell. But if the steel was exposed because the fireproofing came off on impact, or had been subject to years of deterioration, it would stretch as soon as it reached 1200 degrees. (The matter is further complicated by the fact that there had been years of litigation between the company that made the fireproofing and the Port Authority. It was also alleged that Louis DiBono, the head of another company that was contracted to apply the fireproofing, was a member of the crime family of John Gotti. DiBono was gunned down in 1990, allegedly on the orders of Gotti himself, and his body was found filled with a multitude of bullet wounds in the garage of the World Trade Center.)

Even if the steel held, however, it is doubtful that anyone on the floors above where the crashes took place could have survived. The planes came in at an estimated speed of 500 mph, and the people working on the floors that were impacted would not have an instant to react to an explosion. Those on the 100th to 110th floors of the north tower and the 85th to the 110th floors of the south tower probably searched desperately for an escape, but found none. The planes came into the buildings at such high speeds that they crashed through the center cores of the towers where the three stairwells were located, rendering them unusable.

After the bombing of the WTC in 1993, Leslie Robertson, one of the engineers who worked on the towers' structural design in the 1960s, claimed that each had been built to withstand the impact of a fully fueled 707. In a recent interview with *The New Yorker,* he said, 'I'm sort of a methodical person, so I listed all the bad things that could happen to a building and tried to design for them. I thought of the B-25 bomber, lost in the fog, that hit the [seventy-eighth and seventy-ninth floors of the] Empire State Building in 1945 [killing fourteen]. The 707 was the state-of-the-art airplane then, and the Port Authority was quite amenable to considering the effect of an airplane as a design criterion. We studied it and designed for the impact of such an aircraft. The next step would have been to think about the fuel load, and I've been searching my brain, but I don't know what happened there, whether in all our testing we thought about it. Now we know what happens – it explodes. I don't know if we considered the fire damage that would cause. Anyway, the architect, not the engineer, is the one who specifies the fire system.'

Tragically, neither the engineers nor the architects could have anticipated an airplane as advanced as the 767, which carried ten thousand gallons of fuel in wings that spanned 156 feet from wingtip to wingtip. Nor did they anticipate that planes could crash with a force of about twenty-five million pounds. Also, it seems that no one could have anticipated the consequences of the extraordinary fire load created by ten thousand gallons of fuel and the consequence of a fire that burned at about 2200 degrees uniformly and comprehensively on one or more floors of the building.

Recently, Henry Guthard, 70, one of Yamasaki's original partners who also worked as project manager at the site, said, 'To hit the building, to disappear inside the building, to have pieces come out the other side, it was amazing the building stood. To defend against five thousand [sic] gallons of ignited fuel in a building of 1350 feet is not possible.'

Of course, when Yamasaki was designing the buildings he was aware that steel, when it reaches an inherent temperature of 1200 degrees, will stretch at the rate of 9½ inches per 100 feet. He undoubtedly took into account the possibility of a plane's hitting the building and causing the steel to stretch in a resulting fire. There might even be a collapse, but only on the side of the building that was hit. Partial collapses often happen in burning buildings. To a large extent, the World Trade Center is a disaster that could be as much a result of its fundamental design as was the *Titanic*. The plane hit the south tower and distributed the burning fuel throughout. The temperature of the fire was such that steel was stretched throughout. The steel trusses that held the weight of the corrugated steel and the 3 inches of concrete that formed the floor stretched. The bolts connecting the trusses to the exterior wall stretched. The weight of the floor was shifted to the interior columns, all 47 of them around the elevators and stairs, but those stretched and weakened columns could not withstand such a burden. The trusses on the fire floors separated from the exterior walls, almost all of them simultaneously, and the exterior walls buckled. Since the weight of the floor was no longer sustained by the exterior wall connections and no longer held by the interior columns, the floor collapsed. But it did not collapse partially; it collapsed fully in a pancaked layer to the floor below. The floor below could not sustain the dynamic weight of a uniform falling body, and it, too, collapsed, to the next floor, and to the next, and to the next, until, moving at 120 miles per hour, the building fell in 12 seconds. Forty thousand tons of 110 stories of a building project that took eleven years to build, with 4,761,416 square feet of space, came down in twelve seconds. It brought weight loads of 2000 pounds per square foot, plus the dynamic force of the fall, onto floor designed to bear 100 pounds per square foot. Twenty-nine minutes later, the north tower followed.

Day 2, September 12

The breaking light through my bedroom window wakes me, and the first thought in my mind is the families of all those who are missing. We found so few bodies yesterday, so few survivors. I think of Jay Jonas, Tommy Falco, Mickey Kross and the rest, and I wonder what thoughts they are having this morning, and if they slept at all last night. Can anyone sleep with the memories of being trapped under a falling 110-story building? Even I have trouble keeping my thoughts from running in every direction in the face of what has happened and what needs to be done.

I go to my desk, an old piece of furniture with a leather top that came from the Geneva home of François Marie Arouet, Voltaire, a lawyer and a playwright, and a writer through and through. He was a sought-after intellectual in Paris and was exiled several times from France for what he wrote. But Voltaire was also a notorious womanizer and an obsequious favor seeker at the court of Frederick the Great of Prussia. My desk once again reminds me how complicated people and events can be. Nothing ever quite turns out to be the way you first perceive it. The destruction at the WTC is a clear-cut act of terrorism, but in reading the papers this morning I find that people around the world are not convinced that Osama bin Laden is responsible for it. This will be a complicated issue, and I hope the leaders of our country deal with it accordingly.

Early in the morning I receive a call from Geri Fabricant, a friend who writes the entertainment business news for *The New York Times*. She commiserates with me about this terrible loss of firefighters and suggests that I write something for the Op-Ed page of the newspaper.

'You're the natural, Dennis,' Geri tells me. 'If you don't do it, who will?' The idea appeals to me, for having served so long in the fire department and written so much about it, I can serve as a voice for many of the issues it is now confronting. But, I am just leaving to go down to the site, a place newscasters have begun calling Ground Zero, and I haven't much time. There is also a small reserve of doubt in my mind about chronicling this disaster in any way. This is the biggest story of my lifetime, and certainly bigger than anyone could capture in a few short paragraphs. Also, I am hesitant to say anything public about firefighters in the middle of this sudden and profound sadness, when it might be more appropriate to come together as a firefighting family, close off our ranks from the rest of the world, deal with our losses, and support our families in need.

Mayor Giuliani is on television this morning, and seems he has the media under control. He has undoubtedly made a decision to speak for the city and that all information will be supplied by a single voice, his. He promises to appear each day at a 10 o'clock press conference. He had been up on 50th Street and Fifth Avenue when word of the first crash came, and he rushed down to his bunker, at #7. But not long after he arrived, the south tower collapsed and he had to leave through the basement. He went to city offices at 75 Barclay Street, but that area was also determined to be unsafe, so he traveled to the firehouse of Ladder 5 on 6th Avenue and Houston Street, where he worked for a few hours. He was quite a disheveled sight, and I surmised from the photos I had seen of him that he had been on the line, where he should have been.

I have been a supporter of Rudy Giuliani since I first met him more than fifteen years ago. His demeanor, refined during the many years he served as U.S. attorney for the Southern District, assured me then that he would always be an independent-minded politician. If there is any criticism of Rudy that I will abide it is that he may occasionally be too brazen and out front, but that is a small price to pay for his intelligence and honesty. His ability to take absolute and reassuring control during this crisis has made me respect him all the more.

My son, Sean, calls and suggests in his easy way that I reconsider my reluctance to speak to the press. Sean is a news producer at ABC and he had phoned late the night before to ask if I would come to the studio for an interview on a national special on the disaster. I would do anything for my son, anything, but I told him that I didn't want to talk to the media about this matter — it did not seem appropriate while some of my friends were missing, still deep under the rubble. I could tell that Sean did not want to argue the point with me, and I promised him to think the matter through.

Many firefighters are talking to reporters, he now tells me, and says that I should as well because I have a history of being a spokesman for the heroism of fire-fighters. Also, I have been at the scene and can present a clear firsthand account as well as some historical context. And, finally, Sean stated, that if I don't do it, who would? This, of course, was precisely the argument that Geri

Fabricant had made earlier, and the advice of these two people whom I trusted made me think about the responsibility I might have in this situation.

The world was now hungry for information about firefighters, and they would get it one way or another. I have always been exceedingly grateful for having been given the honor of wearing two hats in my life, that of firefighter and writer, and now, given the coincidence of time and place, I was being asked to draw upon my experience to help people understand a momentous and historic event. *I am a firefighter. I am a writer.* I also remember a quote from the Bible: 'Write down the revelation, and make it plain on tablets so that a herald may run with it' [Habakkuk 2:2]. This should be the standard bearer for all journalists. I call Geri and tell her I would be interested in doing an Op-Ed article. *Whatever I do write,* I think, *must not do any damage.* I will not say anything that might infringe on the privacy or sensitivity of the families of the missing thousands, or the many police officers and firefighters who are grieving. I will just write a clear account of what I have seen.

Before I leave for the site I take an hour at my computer to try to put into words what I have been feeling since 8:48 yesterday morning. The writing does not come easily for I have trouble conveying the sheer size of the story, and dealing with how confrontational a subject it is for a writer. The firefighter in me could face the tragedy for I found specific tasks to do at the piles, but the writer faces the challenge of grasping the wholeness of it.

I decide to take my car to the site and go down the East River Drive, but I am stopped at 42nd Street by police officers who are directing everyone but emergency vehicles off the road. I realize that I might not be able to gain access to the site today, that my credentials might not be good enough. I search my pockets for my badge, which I often remember to carry. When I retired from the New York Fire Department, the president of the Uniformed Fire Officers Association made me an honorary deputy chief, and, a year or so later, the department itself made me an honorary deputy. And just less than four months ago the fire commissioner, Tom Von Essen, promoted me to the rank of honorary assistant chief of department and presented me with my third gold badge. Luckily, a department picture ID came with the honor, which I find as I flip open the leather badge holder. The cop examines it and waves me on. Without a photo ID, I realize, no one will be allowed onto the site of this disaster today.

I had been told the previous day that fire marshals and the police had arrested a number of unauthorized people wearing firefighters' coats. The day was hot, and so trauma filled and frenetic that firefighters were leaving their bunker gear on the top of the burnt-out cars so they could get right to the search and rescue work. Some of those arrested had been speaking Arabic into cell phones, and others were looting. One man had filled a pickup truck with equipment he had cleaned out of the fire trucks at the scene – tools, lights, power saws, and generators.

After passing through a series of police checks, I park my car near the entrance to the Brooklyn-Battery Tunnel and walk to another police line. I have my gloves and mask from yesterday, and I begin the four-block walk to Ground Zero. The ground is still muddy, and papers are strewn everywhere. There is no wind. It is a hot day, and the day's news reveals that the worst imaginable has come to be.

Whole companies of the fire department are missing. Everyone who responded from Ladders, 2, 3, 4, 5, 7, 11, 13, 15, 101, 105, 118, and 132 is missing, as are the members of Engine 33, Engine 40, and Engine 235. Every member of every one of the five rescue companies that responded to the first crash is missing. Every single firefighter from the 9th Battalion save one is missing. Before yesterday, in the whole history of the New York Fire Department, since its inception as a paid department in 1865, there were 774 lost in the line of duty. Today, though, it had been reported that, in total, 343 men from the fire department are feared dead: 249 firefighters, 46 lieutenants, 21 captains, and 23 chiefs, including the chief of department and the first deputy fire commissioner, 1 fire marshal, 2 paramedics, and 1 chaplain. There are 23 police officers, including 1 policewoman, and 37 Port Authority police officers. So many. So many family members are now sharing this grief, including the children of these emergency service workers, perhaps over 1000.

Yet it cannot be said that this disaster was unanticipated. After the bombing of the WTC on February 26,

1993, Melvin Schweitzer, a member of the Port Authority Board of Commissioners, told me that he repeatedly inquired of the Port Authority about the possibility of a plane's running into the towers.

Also, our preparedness for dealing with terrorism did not advance significantly after that incident. On March 21, 1998, Ray Downey, the chief of the FDNY special operations command, testified before a committee of the U.S. Congress investigating the federal response to domestic terrorism involving weapons of mass destruction. 'I was on the scene immediately after the bombing. . . .' he said. 'Five years later, we are better prepared [in that we] have more knowledge about weapons of mass destruction, but I still see many shortfalls in the area of first responder capabilities for dealing with and mitigating upon incidents of weapons of mass destruction. The fear of chemical or biological terrorism is foremost in the minds of every firefighter. What we read about and hear about regarding the nation's preparations and training for these incidents does not go far enough. . . . victims were treated for various injuries after the bombing. . . . What would have happened if they had been contaminated by a chemical agent? . . . Can the fire service handle the same potential incident in 1998 after five years of additional preparation and training?

'The answer in most cases is No.

'Why? Lack of sufficient funding and training for weapons of mass destruction.'

Battalion Chief Ray Downey is now among the

missing, lying before me under these piles of debris. More than ten thousand body bags were delivered to Ground Zero today, along with ten refrigerated tractor trailer trucks to house the remains.

About a dozen bucket brigades are working the piles this morning, which doesn't seem much changed from yesterday. In a curious way these mounds of steel and concrete remind me of an ancient fort I visited on the Aran Islands called Dun Aengus. It was an open rock structure, pre-Christian, about half the size of a football field, built in a semicircle against a seventy-foot cliff facing the Atlantic Ocean. Fronting the fort, in a large, erratically lined ring, was a *chevaux de frise,* which was like an abatis, a thirty-foot-wide rampart at most five or six feet in height, made from cut and broken rock placed tightly in haphazard angles so that an invader could not just pass through. These stones had to be climbed over carefully, and if you fell while traversing them you would undoubtedly be mortally injured.

Although these piles at Ground Zero are formed by pieces of concrete, rebar, and steel, some large, some small, every one of the thousands upon thousands of them has a sharp and jagged edge. I recall that every concrete floor of the towers was an acre in square footage and that there are 110 acres of broken pieces of concrete before me, and up nearer to Vesey Street, another 110 acres.

The mayor has given responsibility for clearing the site to the city's Office of Design and Construction, though the fire department has complete charge of the operations. And though this is a fire scene, it is also a crime scene, which means a large unit of crime scene investigators is present, working from a tent at the corner of West Street and Liberty.

Up Liberty Street, in the direction of the Hudson River, other tents have been raised as well – eyewash tents, rest tents, first aid tents, and food and supply tents. In such a massive job, the needs of hundreds of workers have to be served. Under the overpass from one of the World Financial Center buildings to another I see that a group has set up a supply table on which gloves, hard hats, and coveralls are stacked.

Hundreds if not thousands of volunteers have already come to Ground Zero, offering their assistance, including paid and volunteer firefighters, police officers, military people, doctors and nurses. I have heard that one doctor, a prominent brain surgeon, cleaned and prepared his hospital's operating room, and rushed to the site early yesterday, fully expecting that there would be many brain injuries among the survivors of the collapse. But there was no triage, no foxhole bandaging, and no victims to attend to. Finally, he returned to his hospital, with his staff, to his regular duties.

Other volunteers are being put to good use. Some of the downtown firehouses are being covered by volunteer firefighters from Long Island and Westchester County, who have responded to the call of a mutual aid agreement with their fire trucks and equipment. Their suburban fire trucks are desperately needed, for the city's fire department lost twenty-seven fire engines yesterday, thirty ladder trucks, five rescue trucks, seventeen battalion chief SUVs, one division suburban, and twenty-two sedans. Some of these dedicated and generous men and women are also on the piles and I can see shirts that announce companies from Great Neck, Port Jefferson, Greenwich, Massapequa, Valhalla, Larchmont, and so many more.

There is a single focus for all here, and that is search and rescue. Yet I know the world beyond these piles is changing very rapidly. The World Trade Center had housed thirty-two Wall Street brokerage firms, but today both the New York Stock Exchange and NASDAC are closed. Several hundred thousand people have not been allowed to come to their lower Manhattan offices. Billions of dollars are being flowed into our banks by the Federal Reserve to reduce the possibility of economic panic, and Europe and Japan are supporting their own financial systems. Airports are closed, meetings and conventions have been canceled in every state across the country, advertisers are scrapping plans to present new products, movies and television premieres are being postponed. The New York mayoral election, scheduled for yesterday, was canceled, as were Broadway shows and concerts. Bridges and tunnels are closed. New security measures are being put into place in every public building and in every center of public gathering from coast to coast. People are told to be prudent, to be watchful. We have not yet identified our enemy, though we are certain this horrendous act was born in the Middle East.

I notice a space between a firefighter and a police

officer on the bucket brigade line and step into it. Almost everyone is in shirtsleeves or T-shirts, all covered with the pervasive site dust. The police officer has large, developed arm muscles that flex with each bucket that is passed into his hands. The firefighter is from a ladder company in Brooklyn. He tells me that a Port Authority police officer had been pulled from forty feet down in the debris this morning, a man named John McLoughlin. They had taken fourteen men out yesterday afternoon, members of Ladder 6 and Engine 39, a man last night, Will Jimeno, and now another this morning.

I begin to fall into the rhythm of the work. Some of the buckets are light, and carry only small pieces of steel and aluminum of the type you might find around a home air-conditioner while others are very heavy with concrete and rebar. I have to put these heavy ones on my knee and then kick them over to the man next to me. From where I stand I cannot see the end of the bucket brigade.

There is a Captain Metcalf next to me, a good worker, strong. I ask myself if I am taking the spot of someone who might better serve on this line, but I believe I am keeping up. There must be thousands of people who are wishing they could be here, especially the relatives of all the victims, feeling that they are at least doing something.

There is a chief or a fire officer overseeing each of the many digging sites across the landscape of piles. There is so much to do here, and thousands and thousands of tons have to be moved, perhaps in millions of buckets, before these victims will be reached. These chiefs know their business, and when they call for a change in the operation, all is changed quickly. Lines move from one spot to another, the configuration twists and turns like a prowling reptile. A few large Cats have begun work at piles, and clusters of firefighters gather around them. A crane has also been rolled to the edge of the site and is waiting to enter in.

I learn that many probationary firefighters from the three most recent classes at the training school have been lost at the collapse – some say as many as twenty. I hope it is a rumor for these young men have barely been on the job, six or nine months at most [seventeen were lost]. They haven't even had a chance to go to the Christmas parties or the company dances or annual picnics. They haven't yet experienced the fun of the job, the joy of being in the company of a group of men who grow as close as a family.

I think back on the day I was appointed to the New York Fire Department. We still used the term 'fireman' then, before the union came out with a bumper sticker in the early seventies that said FIREMEN STOKE FIRES, FIRE-FIGHTERS FIGHT THEM. My next door neighbor Jack O'Keefe was a fireman, and he had walked me to the subway the day I was taking the exam, prepping me all the way. We passed Engine 8 on East 51st Street, and Jack told me that he could put a good word in for me if I wanted a firehouse close to home, but that I had better pass the test first. He then gave me a piece of advice I have never forgotten about multiple-choice exams, and that is: 'When in doubt, take "C."' I was just 21.

Almost two years would pass before I was summoned to the probationary fireman's school on Welfare Island. I raised my hand and was sworn in to qualify for the silver Maltese cross badge of the fireman. I could not have been more proud and remember writing, 'I had made it, and I had it made.' I was not yet 23 years old.

I trained for three months. There were pushups every morning until my arms felt as if they were not any longer attached to my body. There was hose pulling and ladder carrying, knot tying and rescue techniques, trips through the smoke house and the heat house, on my stomach all the while, coughing and choking – training that would not be permitted today by OSHA [Occupational Safety and Health Administration]. There were hours of building code classes, fire science classes, and department regulations classes. When I graduated I truly felt I was prepared to work side by side with seasoned firefighters, men who made their reputations in the days of 'leather lungs and wooden fire hydrants.' I was assigned to Engine 292 on Queens Boulevard for three years, with Rescue 4. Captain Finnerty taught me how to crawl on my stomach through a burning house to reach a back bedroom. He taught me how to 'take a beating,' which is to breathe in the smoke environment between coughs until you are nearly unconscious. It was the way we fought fires then, before self-contained masks, when it was a matter of pride for every firefighter to make his way through the smoke to the seat of the fire and get the job done.

Today, new firefighters spend eight weeks in school, called 'the rock.' They're then assigned to a ladder company for seven weeks and an engine company for seven weeks, and after that field time they return to probationary school for three weeks to integrate and share their experiences in the trenches. It is a good system, one that gives the new firefighters a wider view of a department that has 8599 firefighters and 2629 fire officers in 203 engine companies, 143 ladder companies, 5 rescue companies, 7 squads, 3 marine companies, and a hazmat company, all stationed in 225 firehouses scattered through the five boroughs. In a cruel irony, the department had only just recently begun to rotate these new firefighters as a way of giving them more hands-on experience during their training period. A year ago, probies would not have been put into the field, but kept in school for three months, and then assigned to one permanent firehouse. The current probationary firefighters had just finished their first eight weeks on September 7, and this was, for some, their first day working in an actual firehouse.

I look around as I pass the buckets, and it strikes me that all the high-rise buildings of the financial district are empty. We have a power blackout covering an area the size of the city of Syracuse, and I know it will be a long time before these buildings are cleaned out and again made habitable.

Later, as it begins to get dark, I walk through the

mud to the Vesey Street side of the site and run into a few firefighters I know from Engine 82. They have volunteered to be down here, like most of the men, I am guessing. Some of the workers, though, are here as part of their defined duties. I see the first squad of the urban search and rescue teams pass by, and I know they are specially prepared, equipped, and paid by the Federal Emergency Management Agency.

I hear someone remark that the GE company has pledged $10 million to a fund created by Rudy, The Twin Towers Fund, as did AOL Time Warner, and that Microsoft had donated $5 million, and another $5 million in services. The Starr Foundation of the AIG Insurance Company is donating $10 million in stock to the New York Police & Fire Widows' & Children's Benefit Fund and Deutsche Bank has given $9 million. *The New York Times* has raised $50 million, the American Red Cross and the United Way have raised several hundred million. This outpouring of generosity is so commendable and reassuring that I can only hope that it is well managed by those organizations who are receiving it.

Standing in front of the Verizon Building on Vesey Street is a crowd of workers who are facing the biggest challenge in their lives to return wire and phone service to the downtown area. The New York Stock Exchange needs its computer terminals and phones to work before the financial nerve center of the world can resume operations. The Verizon building has five subbasements, called vaults, that house switching devices and line connections for 300,000 voice lines, including those for

City Hall, police headquarters, and the Federal Plaza Building, as well as 3.5 million data circuits. The company had been able to continue without interruption its 911 emergency system by transferring the calls to another site, a remedy that was not possible for the other phone services. The WTC collapses caused the water mains to break near the building and to flood the five vaults beneath. In some places, the cables that need repair run beneath thirty-foot mounds of debris from the north tower and from #7, both of which stood directly across the street from the Verizon Building.

Out of the corner of my eye I see my friend Tom Dunne. He seems to have commandeered a bulldozer and is directing the operator to lift a van and eight cars out of the Vesey Street area so that Verizon can get in there to start digging to reach its cables. Normally, 115 million telephone calls are placed each day in Manhattan, but yesterday that number doubled. Verizon expects significantly increased usage today and tomorrow as well, yet more than 200,000 phone lines are out of service south of 14th Street.

A deputy chief stops me and asks how I have been. I haven't seen this man in many years, and I am glad he has risen through the ranks to be a division chief. We talk a little about the job, and then I ask if he heard that Brian Hickey of Rescue 4, with whom I know he was friendly, is missing. The chief suddenly begins to cry and says, 'You know, every name is like a stake that goes deeper into the heart.'

The preoccupation with numbers that a tragedy like

this gives rise to is misleading. To look at the building rubble before me and to know that it once had two hundred elevators and a separate zip code, 10048, is to integrate empirical information. To know that 343 fire-fighters are missing means also taking into account the families and children and the kitchens of firehouses. Only then can we realize how much suffering will have to be endured.

I have more mud on me today than yesterday, if that is possible, and as I arrive home at eleven or so, Katina is waiting for me. She tells me that our friend and neigh-bor Nancy Berry has lost her son in the disaster. I meet Nancy often in the elevator or go to her apartment once in a while for a drink, and now my eyes well up with tears at the news. David Berry, 43, had everything good in life – degrees in philosophy and physics from Yale, London School of Economics; was a well-known and respected financial analyst; had a happy marriage, a great wife, and three boys. Nile, Reed, and Alexander. He was working on the eighty-ninth floor of the south tower when it was struck. I did not know David, and I do not know his wife or his children, but I do know how sensitive Nancy is and how she will feel this pain.

I sit at my desk for a while in my dust- and mud-caked clothes, complete an article about my first day at Ground Zero, and send it to *The New York Times*. After a shower, I pick up a book of Seamus Heaney's and read through that rugged verse of stone and dirt, but sleep cannot come easily in these dark days. I hear a firefighter say today that he used to dream about water, and now all he sees in his sleep are fields of steel. The close of every day should be filled with the anticipation of tomorrow, but I am still preoccupied with our failures of today – why we didn't find anyone after John McLoughlin was taken out early this morning. Still, there must always be a time to say good night, a time to come to terms with the day's transgressions, to resolve that the day wasn't spent entirely in vain, to remember that tomorrow will bring other opportunities. As long as a body draws breath, there is time to contribute something.

Day 3, September 13

It is 2 o'clock, and the sun is beating down on hundreds of firefighters from every neighborhood in every borough. I pause for just a moment in my work and look around the site. I can see police officers from the emergency service unit, the crime scene unit; and, I am certain, from every precinct house in the city. Port Authority police officers have also come in from the airports, railroad terminals, bus stations, and docks, and hundreds of volunteer firefighters from the suburban fire companies. Some members of the eleven volunteer fire companies that are registered within the boundaries of New York City are here as well. Even old-line firefighters are surprised to learn there are so many volunteer fire companies in New York City – three in the Bronx, one in Brooklyn, five in Queens, and two in Staten Island.

The site is expanding, and for every rescue worker there seems to be several volunteer service workers. The level of activity reminds me of a gold rush camp, or perhaps that of the followers of a large medieval army. Every two tents seem to beget another one instantly. The volunteers are extraordinary, and one cannot walk forty feet in any direction within the twelve-block perimeter of Ground Zero without a volunteer's offering water, soda, or an array of energy bars.

This afternoon I have joined a bucket brigade in front of a large crane that has been constructed overnight. Bucket brigades are working to the left and to the right of me, many lines of men and women passing these heavy yellow, white, or blue five-gallon buckets, hand to hand as efficiently as a conveyor belt. The contents of each of the buckets are emptied on a dump pile on West Street, which, as it grows larger, is picked up by a Cat and deposited into the back of a truck, and then hauled away. There are two dump sites: One is in Fort Hamilton, Brooklyn, and the other is Great Kills, Staten Island. At each location police investigations unit detectives and FBI agents are spotting and sifting through every truckload, searching for the flight recorder of the planes and for any remains of victims.

The crane is lifting steel girders, each thirty feet long and weighing several tons, but the critical work is being done by hand. As I look over these fields of hundreds of emergency workers, not one is idle or standing with his hands in his pockets, and I feel certain that we will make good progress and that there will be men and women still breathing beneath these mounds.

One person has mentioned that the airports, which have been closed since the second plane hit the south tower, were reopened today, and that flights were operating. They were scheduled to be reopened last night, but a man with a counterfeit pilot's license was discovered trying to board an aircraft at Kennedy, and all flights were subsequently canceled as a prudent measure.

Generally, the men and women on the piles do not introduce themselves. It's not necessary, for no one is here to make new friends. There is only one reason to have come to the site, and that is to dig through these piles of steel and concrete, find a void, open it, and hope that there is a survivor.

I move up to the front of the line and see an opening at the edge of the area that is being excavated. Some are digging with shovels, others with their hands. I get down on my knees and begin to work at the ground, my black rubber gloves providing good protection. Small pieces of sheet metal lie like stripping, just two inches wide, all around. Those I can pull loose I throw into a bucket while others will have to be moved after a crane lifts and loosens the dig. A firefighter standing next to me puts his hand down into the small void we are enlarging and brings out a foot attached to the lower half of a calf muscle. There is a sock on the foot, and the firefighter hands it to me with instructions to put it in a clean bucket. I do so, thinking that it is a man's sock, but it could very well be a woman's. There is no softness in the skin, no give, and I place it gently in a bucket that will be conveyed to a chief. I want to reflect for a moment on the person whom this was once a part of, perhaps utter a prayer. But the finding of a body part has given everyone a renewed resolve, and the men around me begin working with such concentration that I have to focus carefully on what I am doing. We dig and pull, pull and dig for another half hour around this area, but find nothing. A chief has seen the operation and comes over with a firefighter holding a batch of triage tags. The firefighter marks the location on the tab and attaches it to the bucket.

It is disappointing to not find more remains in this location, and underlying the thoughts of everyone here is the frustration at not having found a trapped person for more than thirty hours. It is the third day, and if there are survivors here they are now beginning to ache for sustenance. The window of time for search and rescue close slowly, but it closes unrelentingly and inevitably.

Several Cats and bulldozers are picking at the piles all along West Street. Part of the pedestrian crosswalk still stands across a section of the street, a large corridor held up by thick concrete stanchions. A construction crew is all over it like ants, for it has to come down. I am certain the command chiefs want to create an unobstructed passageway from Vesey to Liberty here on West Street.

A truckload of heavy timber has been left in the mud in front of the piles that are lying against the Winter Garden. I take a seat there among a few firefighters and learn that about fourteen thousand pounds of gold is stored in the basement vaults of a bank in #7, and also that there are seven safes belonging to the Secret Service that hold vital and valuable government secrets. I suspect someone will be made responsible for watching that area of the debris very closely, but from my point of view, the finding of someone's foot is more important than recovering a safe full of gold.

I take a walk over to the command center on Vesey Street, where a white tent has been put up. I see Denis Oneal, a man I have known for a long time, since the days he studied for a Ph.D. at New York University. He was a deputy chief in the Jersey City Fire Department then and now serves as the director of the National Fire Academy in Emmitsburg, Maryland. I usually see him at jacket and tie dinners in Washington, but today he is in casual clothes, though I know that he would rather be in bunker gear if he could. He is here to help the fire department establish a hierarchy at the scene and to inform the staff of the resources that are available to them through FEMA. Normally, the department would be thoroughly familiar with the FEMA programs, but the authorities on the national urban search and rescue programs – Chief Ray Downey, Commissioner Feehan, and Chief Donald Burns – are all missing.

Today Denis has been working with Chief Nigro, who was just appointed chief of department, Chief Sal Cassano, and Deputy Commissioner Fitzpatrick to inform them of the amount of money, equipment, and manpower that is being sent to help in the disaster. Denis tells me that he has brought in trailers to be used as permanent command posts and that he has alerted all twenty-eight urban search and rescue teams from coast to coast. Some of them have already arrived, for I saw the group from Seattle and another from Texas, all in very identifiable dark-blue coveralls, with the initials USAR in large letters across their backs. Each team includes about fifty men and women specially trained in rescue work.

'We had a guy,' Denis tells me, 'a fire buff from our department in Jersey City, Joe Laero, who came in with a few firefighters in the early minutes of Day 1. He was a blue baby and had health problems and so he could not pass the exam to be a firefighter, but he became a dispatcher. He lived and breathed the Jersey City Fire Department, and he came in with three firefighters. [They] got out on one side of the truck, [and] Joe got out on the other side and got caught in the collapse. They told me that when they found him at the morgue he was lying between Chief Ganci and Commissioner Feehan, which is a great honor for a dispatcher from Jersey City.'

I go back to work on the pile until after 10:00, and then walk to my car through the glare of the night lights. I have an appointment uptown at midnight; I have been asked to appear on a special news program that Tom Brokaw is doing on NBC. I first met Tom many years ago when I appeared on the *Today* show, and he has recently done me a favor of doing an interview with me in connection with another book of mine.

I can see at the perimeter of the site that the police have been replaced by the men and women of the national guard, who are dressed in camouflage uniforms and are carrying large backpacks. I think about these soldiers as I drive north on 8th Avenue, some of whom seem so young to me, and then remember that I went into the U.S. Air Force on my seventeenth birthday. I was ready to fight for my country, if necessary, and I am certain there are many seventeen-year-olds in today's military who are ready to go to the Middle East.

A crowd has gathered in front of the firehouse on 48th Street and 8th Avenue, the headquarters of Engine 54 and Ladder 4, from which twelve men, including Michael Angelini's brother Joseph, are missing. I stop the car to study the scene before me, for I see something that I have never witnessed before. It is a shrine, a makeshift homage, laid out across the sidewalk, with hundreds of notes attached to it. It is illuminated by dozens of tall votive candles of different colors, and the scene reminds me of a carnival where not the tent, costumes, or prizes match. There is no harmony to anything before me, yet its elements make a definite statement. To me it is almost like a sudden, distracting scream, yet I remind myself that each candle and note was placed here with a quiet reverence. The people mingling outside of the firehouse with a few of the firefighters are not curiosity seekers, but neighbors who are visiting because they know of the great loss. The firefighters are generous with them, taking the time to accept their condolences.

I begin to read the notes. Most are from neighborhood people, many of whom work in Broadway theater, and they are very affecting, from-the-heart messages that convey a simple and fundamental love and respect. There is no rhetoric, no complicated salutations, or compliments, but simply various versions of the sentiment, 'We love you. What would we ever do without you?'

As I proceed to the NBC studios I think of all the things that I might speak about on television. There was great excitement at Ground Zero today when two firefighters were pulled out of a very deep void, and a chorus of cheering stopped all work for a moment. A steelworker next to me yelled out, 'That's another one you haven't killed, bin Laden!' But, in fact, the two firefighters had fallen into the void earlier today, so it was a victory in which the rescuers were rescued. Even so, calls from the families of missing firefighters flooded the department's hot line in response. Also, I could raise the question of why some people were attracted to sites of catastrophes, where they sought to cause even more damage by stealing or creating havoc. Ninety bomb threats were received today, and Grand Central Station, the MetLife Building, LaGuardia Airport, Macy's, CNN's midtown bureau, and the Condé Nast Building all had to be evacuated. I worry, though, that to discuss this phenomenon on TV would only give others ideas to do the same.

What is it like at Ground Zero? would, undoubtedly, be one of the questions put to me, and in thinking over the day I realize that I will not have a good answer. Although no one was found alive today, the bodies of a few firefighters were recovered, and that raises the only question in my mind that is worth consideration: How will the families of the World Trade Center cope with their losses, thousands of families, friends, and loved ones? How?

Day 4, September 14

This morning *The New York Times* published the article I wrote under the headline, A FIREFIGHTER'S STORY. It is a short account of my first impressions at the scene on Day 1. They have cut about a third of the piece, something all writers dread, but it reads well enough. Normally, when I have something published in the *Times,* I call my family and alert them, and friends will call with something supportive to say. But today I simply give it a cursory read, for I want to head down to Ground Zero early and contribute at least seven hours. President Bush is coming to New York today and will undoubtedly visit the site, but since I have put so many things off for the past four days, I cannot be there. President Carter and former President Bush had been able to clearly identify their enemies in the Middle East in a way that the current administration cannot do. Our current president must also confront the dilemma of civil liberties and national security having opposing interests in this kind of war, but he has put his chin up and represented perfectly a strong and determined nation. He has been in tune with the American people since 9/11, and so I truly regret not being able to be near him this afternoon to cheer him on.

The site has expanded even farther today. I walk around the World Financial Center building and see a peaceful harbor. At the service tables and tents various volunteers are handing out everything from toothbrushes to hydraulic jacks. Rescuers are walking back and forth, perusing their offerings, almost as if they were at fair day in a country community. I am sure there must be at least fifty Secret Service men around, securing the site for the president's visit, but I don't see any.

There are still long lines of bucket brigades, but they have been joined by more big rigs around the site now, cranes and grapplers. I have traded in my black rubber gloves for a pair of soft but thick leather ones, and they are much easier to work with for the gripping and lifting I have to do. For others, the teams from the rescue and squad companies and the FDNY-directed search teams, the work is truly dangerous. They are crawling over the very tops of the piles or exploring the five subfloors of the WTC, as far down as the PATH train station deep below Church Street.

An engineer, Hyman Brown, who oversaw the construction of the towers when they went up, has suggested that America should rebuild them as a monument and make even bigger than before, at a cost of $750 million. He has even done the math and announced that it would cost each American $2.67 to build it. I wonder if Mr. Brown gave a thought to holding his suggestions until the spouses, parents, and children of the many victims have had a chance to think about the type of monument they might prefer.

At the Liberty Street side of the site, I begin to carry pieces of long and thin metal from the edge of a mound to a heap of debris by a dump truck. Many of these tractor trailers are fourteen-wheelers, but four of their wheels are lifted a few inches off the ground. When the truck is filled with debris the ten remaining tires compress down, allowing the extra four tires to hit the ground, and the weight is thus shifted onto all fourteen of the wheels.

The pedestrian crossway over West Street is coming down slowly but surely, and soon there will be a direct access from Vesey down to Liberty. I am feeling today for the first time that I am at a construction site, for I see so many more construction workers around Ground Zero. The search and rescue mission is still officially in effect, and the mayor has said we will search this site until the last stone is turned. But it is now Friday morning, seventy-two hours from the attack, and forty-eight hours since we have found anyone alive. The rescuers are still optimistic, however, that at any moment a void will be opened, and voices will be heard, or an elevator found with ten or more survivors. Life can be sustained without water and food long after seventy-two hours; in the great Indian earthquake at Gujarat earlier this year, survivors were found in the rubble still living after ten days. If there is a source of water it is possible to live even longer. I know that if I talk to Lee Ielpi, Ed Schoales, or John Vigiano, all of whom have sons who are missing here, I will learn a lesson in the value of optimism.

At the end of the day I see President Bush's

appearance on TV, standing amidst a crowd of construction workers, cops, and firefighters who are chanting 'USA, USA!'

The president answers through a bullhorn, 'The people who knocked these buildings down,' he says, 'will hear all of us soon.'

A great cheer arises, and I am reassured that the president is not voicing an idle threat. I know that he will take action, and I know we all have to remain patient, as he advises. Yet there is a new, ambiguous feeling I get when I think about the military consequence of Ground Zero, something I have never felt before. Like many of my friends, I am simply waiting for the other shoe to drop. What will be next? Will it be our move or theirs? This is not an ordinary war.

Day 5, September 15

Today the first fire department funerals will be held in Manhattan, in Queens, and out on Long Island. Normally, a line-of-duty funeral brings thousands, and uniformed firefighters are lined at attention for blocks and blocks. But, as a result of the tragedy, the department is segmented, and this is the first indication of the difficulty we will face in trying to lay our members to rest in the traditional way. Father Mychal Judge, 68, is being buried in his Franciscan church in Manhattan, in a mass celebrated by Edward Cardinal Egan and attended by former President Clinton and Senator Clinton. At the same time, First Deputy Fire Commissioner Bill Feehan, 71, winner of two Purple Hearts, is being buried in Flushing, Queens, with the mayor and fire commissioner mourning across the aisle from the Feehan family. The bagpipe band, usually complemented by more than forty players, is divided into three groups as well. As taps are played for Commissioner Feehan, the mayor and the fire commissioner speed off to attend the Farmingdale funeral of Chief of Department Peter Ganci, 54. The department's ceremonial unit has organized each service, and the thousands of firefighters who have been distributed to these three points of the city all have tears in their eyes.

The city's medical examiner's office has confirmed that 39 people have died in the attack. More than 5400 remain missing.

Day 6, September 16

Today the mayor and fire commissioner promoted 168 members of the department to fill some of the ranks that were lost on 9/11. They have appointed as the chief of department Chief Dan Nigro, a man who is as capable as anyone in the world to assume the position of the highest ranking line firefighter in our country. 'Yes,' Mayor Giuliani said at the ceremony, 'they may have taken some of our most precious lives, but they haven't

destroyed our spirit.' These men who have been promoted to chiefs, captains, and lieutenants are resolved to
bolster a department that has been injured but not
crippled under fire.

This afternoon I visit a friend, attorney Leonard
Gordon, who knows of my interest in the New York
Police & Fire Widows' & Children's Benefit Fund. This
fund was established to provide an immediate grant to
the family of a police officer or a firefighter who has
given his or her life in the line of duty, and then to pay
an annual stipend for the rest of their lives. It is a sensible
and necessary supplement to pension and insurance
money that may come to the family, and a cause that has
gained the support of many prominent New Yorkers,
including Michael Bloomberg. Leonard Gordon is on
the board of the Frankel Foundation, which has decided
to donate one hundred thousand dollars to our charity. I
am delighted with this contribution and report that
I will bring the news this evening to some of the fund's
board members.

At 8 o'clock, I go the Mark Hotel on Madison
Avenue to meet with Steve Dannhauser, who is the
president of the fund, and Rusty Staub, the retired Mets
star who was one of its founders. We talk about the
extraordinary generosity that has been forthcoming to
our charity since 9/11 and about our responsibility
to meet with various corporate leaders in the city to
explain what we are doing with our money, and why. We
discuss the needs of the many people who have been
lost, especially the uniformed personnel, and the

feasibility of revising our constitution to include the thirty-seven Port Authority police officers who are missing. Rusty has been working tirelessly on behalf of the charity since 9/11, and I suggest we should establish a specific goal for our efforts. When six firefighters died in a terrible fire in Worcester, Massachusetts, two years ago, each family was given a check for $1 million. I believe this is a reasonable amount of money for a line-of-duty death when we are paying TV morning-show stars $12 million a year, $1 million per half-hour episode for sitcom actors, $25 million per picture for movie actors, and $40 million per year for baseball and basketball players. But we are as of now a long way off the mark if we want to deliver $1 million to each of the 403 families of 9/11 firefighters, police, and Port Authority police.

Steve listens carefully to the discussion, and I know he will consider all sides of every question. He will have lawyers look into charitable law and conduct research into historic charitable grants. He can see that the fund will grow enormously in the next few weeks, and that time and effort will have to be expended now to prepare for this larger responsibility.

Before I sleep I normally look forward to an hour of reading or watching television, but I now try to catch up with the newspapers. My world has become so concentrated on the events of Ground Zero that I have no idea what has happened in the rest of the world since 9/11. Even so, I realize I have no interest in anything other than what relates to the attack on our shore.

Among the many facts reported is one that strikes

me immediately. I read that more than fifteen hundred truckloads of debris and steel, more than nine thousand tons, have already taken the slow route through the Brooklyn Battery Tunnel, over the Verrazano-Narrows Bridge, and to the peak of Muldoon Hill on the top of the Fresh Kills landfill on Staten Island. I wonder where all that steel will end up. It certainly won't be buried, as it is too valuable – I have heard an estimate of about forty thousand dollars for every beam – but it also constitutes evidence, and I wonder if they will try to reconstruct the towers the way the NTSB reconstructs airplanes after suspicious crashes, like the 1996 TWA Flight 800 to Paris that killed 230.

Day 7, September 17

Being Irish, I have always felt a special bond with the Jews, because we are both peoples with a history of diaspora, we both are romantic and literary, and we both share a sincere concern for charity (perhaps because it is so easy to remember our own past). Still, I try to not let my friendships cloud the issue of Israel, for I also feel that we should be supporting democracies all over the world. But American Jews, like American Muslims, are suffering in a particular way since 9/11. They are attached, however unreasonably, to the greatest threat America has faced since Hitler and are suffering the prejudices of people who want to blame anyone handy for the outrage that has been committed upon us.

I have read that a Muslim worker at a midwestern gas station was shot for no other reason than wearing a turban. I want to believe that all the innocents who died on 9/11 were killed in the line of duty protecting our fundamental rights, including the right to practice any religion in whatever way we wish. To kill a person merely for wearing a turban degrades these deaths and all the deaths suffered in the two world wars, which were given in the name of freedom.

Today I e-mailed my longtime friend Peter Quinn, who is an executive with AOL Time Warner and also a novelist of some reputation, and asked if he would arrange for me to meet with Jerry Levin, the company chairman. AOL Time Warner has been generous to The Twin Towers Fund, which was set up by the mayor to assist the victims of 9/11, but I am hopeful that the company will also be open to a direct appeal for the Police & Fire Widows' & Children's Benefit Fund. I learned many years ago that a personal visit with a potential donor is the best way to raise money for charity and will be more effective than a thousand letters.

The New York Stock Exchange opened again today amid much fanfare and backslapping, though the market fell drastically at opening, a discouraging showing after being closed for six straight days. The brokers had plenty of time to figure out what was best for the country, already in an economic downturn, but they decided to cover themselves from behind and sold aggressively. The Dow-Jones declined by 7.1 percent, in the largest one-day drop ever – 685 points – bringing our stock value level back to where it was in December 1998. It angers me much that investors have not been more patriotic, for what a positive message it would have sent to the world if the market had opened to a surge of optimism.

At the site I listened to a group of battalion chiefs at the command center, discussing the potential asbestos problem. During construction of the Trade Center, the Port Authority had used asbestos to fireproof the steel up until 1969, when they banned it as they reached the thirty-ninth floor of the north tower. Above that level the construction engineers began employing a substitute fireproofing made from mineral wool (melted and spun rock) and binders. This could have contributed to the fall of the buildings within such a short burning time. One pound of jet fuel has about ten times the potential destructive energy of an equivalent amount of TNT, and so the crash of the planes into the buildings may have dislodged all the fireproofing over the exposed steel beams, leaving them exposed to the burn. A fire chief expects any burning high-rise building to stand for at least three hours before its structural steel begins to stretch dangerously, at which point they will begin to worry about a potential collapse. But the south tower, which had no asbestos at all, burned for just fifty-six minutes before it fell while the north tower collapsed in one hundred minutes. This is an issue, I suspect, that I will hear more about.

After spending a good part of the day searching and

spotting the grabs of the Cats and grapplers, I met Kevin Gallagher, the president of the firefighters' union. I commented how happy I was to have seen him with his son, Kevin, that evening. The elder Kevin has been spending most of his time with the families of the more than 250 firefighters in his union who are missing, and so it is good that he has come to Ground Zero to give support to the firefighters on the piles.

'Yeah,' he says, 'it was good to have my son there, but I felt bad that Jimmy Boyle was there at the same time.'

I understand completely. Jimmy Boyle is so well regarded by everyone, no one would wish to see him hurt any more than he already has been. Still, to know Jimmy is to be certain that he was overjoyed that Kevin found his own son at the end of Day 1 and that he would never measure one man's happiness against his own misfortune.

'You're doing a great job, Kevin,' I say, standing on West Street, in the midst of the clouds of the still-smoking piles. 'It's not easy, I know, but you'll always be known as a great union president.'

'You know, Dennis,' he replies, 'I would rather be remembered as a good firefighter.'

Day 8, September 18

Soon after I awake, I check my e-mails, and there are almost too many of them to deal with. One of the messages is from Peter Quinn, and it is simply a copy of an e-mail he has sent to the chairman of AOL Time Warner:

Jerry,

I worked in the trade center for nearly three years. I also grew up with a lot of guys who became firemen and cops and who are now listed as missing. I haven't yet totaled up the number of people I know who lost their lives or the families affected because the mathematics is too bewildering and agonizing. Someone even more touched by all this than I is my friend Dennis Smith, the writer, who spent eighteen years in the NYFD. I am sending you the email he sent me because he asked me to and because it's one of the few concrete things I've been able to do since watching from the 29th floor conference room as the towers burned and returning to my office to weep.

One other e-mail that captures my eye, a note from Colleen Roche, who was press secretary to Mayor Giuliani. 'My heart is breaking with the unspeakable loss. . . .' she writes. It is so reassuring to see that people are unafraid to speak from their hearts in response to what has happened. I know that Colleen plays the Irish fiddle, and were she here at this moment I would ask her to play the most soulful Irish tune she knows to match what I am feeling.

While I am waiting for my car at the garage, a

woman notices that I am wearing firefighters' clothing, and she tells me she is a member of the Roadrunners Club, the group that sponsors the New York Marathon, and that Vincent Kane of Engine 22 is one of their members. She starts to cry as she tells me that Vinnie is missing and fears that he won't be here for this year's marathon in November. I tell her that we are still in the search and rescue operation, and so we continue to hope.

At the West Street checkpoint I meet Brian Grogan, a fire marshal, and a police officer stops me to take a closer look at my identification.

Brian reprimands the police officer with a wink of the eye. 'Don't you know this man?' he asks. 'This is the author of *Report from Engine Co. 82.*'

'Never heard of him,' the cop answers, approving my ID.

Brian laughs, saying, 'We should have told him that you write the questions for the sergeant's exam.'

The site seems to have more order this afternoon, and as I walk across West Street I see Assistant Chief of Department Harry Meyers. He is today's chief in charge of the operation, a job that shifts from day to day among the staff chiefs of the department. There used to be twenty-two staff chiefs in the FDNY (assistant chiefs, and deputy assistant chiefs), but since they have begun placing civilians at the head of departments, like the Bureau of Fire Prevention, only eight of these high-ranking officers remain. Harry is one of the highest, and I greet him as an old friend. We worked in the same firehouse in the South Bronx, and such early relationships formed in the fire department almost always last. Sometimes a good friend gets promoted way up the line, and you couldn't ask for a better 'rabbi' if you wanted to get yourself, a son, or a friend transferred. I tell Harry that many men at the site have mentioned to me what a great job he is doing, and I can tell he appreciates the compliment.

Across the street are two other firefighters with whom I once worked, Mike Leddy of Ladder 61 and Nick Liso of Engine 85. Like me, they are volunteering, and I realize how much easier it is for those of us who are retired to work at the site than those still on active duty. Many firefighters stationed in the firehouses would spend every waking moment here if they could, but they are not permitted in the area without a specific assignment. I believe the rest of us are tolerated by the chiefs because retired firefighters have attained a special status in the department, the kind of status that is the result of having paid the dues of the profession.

Leddy is here as a member of the American Red Cross DART team (disaster assist relief team). Because of his years of service in the fire department he has qualified to volunteer at the scenes of floods, hurricanes, earthquakes, fires, train and plane wrecks, and other natural or manmade disasters.

'I know how to do this work,' he tells me, 'so I told them I was available and they flew me out. I've been to three disasters this year so far, not counting this.' I am reminded of all the other volunteer workers I have met at Ground Zero, some of whom are free to walk around with water and soda bottles while others are closely controlled by organizations like the Red Cross and the Salvation Army, and are not allowed anywhere beyond their specific assignment counter.

There are actually two perimeters to Ground Zero. The interior one is usually patrolled by police officers at barricades and is bounded by Church Street to the east, Liberty Street to the south, West Street to the west, and Vesey Street to the north. Only uniformed, hard-hatted, and masked emergency workers and construction workers are permitted within it. The outside perimeter is bounded by Broadway on the east, Rector Street on the south, the Hudson River on the west, and Chambers Street on the north. Many hundreds of onlookers are lined before the police and national guardspeople at its barricades. It is difficult to see anything from these distances, and the police politely but firmly keep people moving.

But the people of New York are in mourning, and I believe they should be allowed to get a closer look at the devastation here, which would enable them to confront their grief in a less abstract way. But with city and federal officials present, there are understandable safety and security concerns.

A line of bucket haulers is working in front of a big crane that has been placed at the West Street edge of the pile. I join them at the front of the line and decide to help the diggers filling the buckets. I begin pulling at some steel sticking up like a thatch of petrified plants and retrieve a yard-long piece of what looks like the riveted side of a refrigerator. Suddenly a man wearing an FAA jacket comes over to me and asks to look at it. 'This is a part of the airplane,' he says, and adds it to a bucket filled with other scraps of steel, and I feel that I have contributed something especially valuable to this effort.

The overpass has been almost completely torn down, and a roadway has been cleared from Vesey Street to Liberty Street. People will no longer have to walk around the World Financial Center buildings to head either north or south.

'Hey,' a police officer says, 'Broadway is dead! Wall Street is dead! So, how are you going to control people who strap gelignite to their privates? Just rip their privates off beforehand.' The banter at the site has been growing more and more scatological, which to me is an indication that the profound shock that everyone here has experienced is in the process of being transformed into a kind of anger blanketed by humor. Moving from shock to anger is a natural progression in the grieving process, and anger will express itself in many forms. While these men have chosen to be ribald, it would take very little for them to lash out physically at something of which they disapproved, or at someone who disappointed them. If it is safe to say that the world is on edge, it is safer to say that the men and women of Ground Zero are on a 'veritable' tightrope, for they must not allow themselves to be distracted from keeping their optimism and determination strong if they hope to find survivors or even remains. It is the very quality that affords these men and women the possibility of attaining genuine heroism – the unshakable belief in their own ability to face and transcend danger – that also makes them sometimes unforgiving.

Yet I remember a comment Joe Dunne once made about the nature of heroes: 'We cry just like you do.' I think the first deputy police commissioner was saying that these men and women are fundamentally down-to-earth people, Americans who could well be your neighbors. But they are trained in and accustomed to dealing with emergencies, an experience that has strengthened and often toughened them.

Interestingly, the area has not changed much in the last several days. A huge piece of the WTC is still piercing the side of the Bankers Trust Building, and another section has impaled the World Financial Center. Both appear to hang as threateningly as the sword of Damocles, though they do not seem to pose an immediate danger. Engineers have already studied the area and determined that isolated structural repairs are required on the Bankers Trust Building, 30 West Broadway, and 3 World Financial Center. Modest exterior repairs are needed on 22 Cortlandt Street. The good news is that it is unlikely that these or any other nearby buildings will have to be demolished.

Much of the area south of Park Place, north of Albany Street, and west of Broadway is also in need of repair for hundreds of windows have been blown out, and roofs have been damaged from falling debris. Airplane parts have been found on the federal office building at 90 Church Street. An eight-page Department of Buildings survey of 195 structures in and around the World Trade Center, including the Millennium Hilton, and 1 Liberty Plaza, has concluded, however, that most of the buildings will soon be able to be occupied.

Later in the afternoon I run into Billy O'Meara. I worked with Billy-O for many years in La Casa Grande and wrote many accounts of his abilities in my first book. Billy-O later transferred to Ladder 31, and then to Rescue 3 as an officer. He's been out of the job about as long as I have, but it doesn't surprise me that he has now taken his place with the firefighters of Rescue 3. 'We were up there searching in the Marriott Hotel,' he says to me, 'and this chief starts yelling "Hey, Rescue, get outta there." These chiefs are so scared something will happen on their watch, doing an Irish jig like that. We need more guys like Harry Meyers, moving things, directing people.'

Billy orates on how the fire department could be better managed. 'Look at Vinny Dunne,' he says. 'He's our collapse expert, right? But he just retired, and so they have him going around notifying wives and families.'

The debate about management continues until suddenly another fire officer, who is engaged in a different conversation, mentions a firehouse in upper Manhattan that has lost a captain and eleven firefighters. The men shake their heads. It is so shocking, still, eight days later, to be reminded of the toll that has been taken.

It makes me angry that so many have given their lives to heroically, and yet it is impossible to know all of them. But we must remember each and every name. Monuments will be erected and plaques inscribed, but I know I can make a contribution toward assuring these heroes have a proper memorial. I determine now that I will write a book about Ground Zero. And on the first pages I will list the names of all those firefighters who are missing, so that this firehouse's eleven men and its captain, and all the other missing men in all the other firehouses, will be seen and remembered for as long as the book is available in a library, for as long as the Library of Congress remains standing. There is no better reason to undertake such a project.

I return home to find a message from Peter Quinn, asking if I am free tomorrow to meet with the chairman of AOL Time Warner. This is good news.

Hank Fellner, a friend in Washington, D.C., has e-mailed me that Peter Brennan of Rescue 4 was scheduled to be in his coming wedding party, just as he had been in Peter's wedding. But Peter Brennan is missing, and once again I see how there seem to be no limits to this tragedy. Hank signs off, 'I keep praying.'

I watch television instead of sleeping, and I see a feature about a man who was severely burned on his face, arms, and legs. He was working on the eighty-first floor of the north tower when the fire enveloped him. He walked down the stairs, hundreds of them, with his friends, each step bringing great pain. When he protested that he couldn't go on, his friends began to lie to him, telling him they were on the fifteenth floor when they had actually reached only the fifty-fifth.

There is more drama in that one interview than I've seen in the movies during the past few years.

Day 9, September 19

I awake this morning to the news that the president has deployed two dozen bombers, tankers, and support aircraft to begin an attack on areas of Afghanistan where it is believed Osama bin Laden is hiding. He has ordered the aircraft carrier *Theodore Roosevelt* and its accompanying battle group from Virginia to an undisclosed location. I know that everyone at Ground Zero will cheer this news.

It is an interesting fact that those who are now being enshrined as America's heroes have views that are generally antithetical to those of the rich and powerful in New York. If you took a survey at Ground Zero you would undoubtedly find that people felt that it was more important to invest in cruise missiles rather than Head Start programs. They would argue that America was the greatest country in the world precisely because it has cared for others throughout its history. While it never loses that generous impulse, our priority now was to protect our families at all costs. They would point out that our wars will not be fought by the children of editorialists, congressmen, or liberal Democrats but by the children of Ground Zero, the children of firefighters, police officers, ironworkers, and crane operators.

These are the matters I am pondering as I leave for Jerry Levin's office. My only role at this meeting would be to convey the specific needs of hundreds of our widows and their children. These families have been torn apart by this cowardly attack, and now face a great challenge to keep together and mutually supportive.

I also remember that Port Authority Officer Dominick Pezzulo is being buried today, and I think of his friend, Will Jimeno, who was with him when he died in the lower level of the south tower. I say a prayer for both of them and hope that Will's leg is mending and that he is feeling well enough to attend the service.

On the way downtown I think about the great amount of charity that has been extended in the last nine days. A law firm gave $5 million to New York University for scholarships for children of cops, firefighters, and EMTs lost in this tragedy. This is generous, but why was it restricted to NYU (my own alma mater)? A lawyer undoubtedly has a significant connection to NYU, but I worry that many children of cops and firefighters won't have access to this gift unless they choose to attend college in the middle of downtown New York.

A fellow from ty.com called me this morning to ask if we on the fund would be interested in endorsing two Beanie Baby puppies they would design especially for us. One would be called 'Rescue,' a Dalmatian, and the other 'Courage,' a German shepherd. I liked the idea, and the fund would receive 100 percent of the profits from these collectibles, so I asked that they send more information. I will talk to Steve and Rusty about it.

Jerry Levin and Dick Parsons welcome us into Dick's office – a beautiful space, as one would expect, for Dick is next in line to be chairman of the company. After we join Peter Quinn, Steve, and Rusty, I begin to tell them what I have seen and experienced in the previous eight days. Until then I had been able to talk about the events of Ground Zero with a controlled state of mind, but now, at the meeting with the royalty of corporate America, I suddenly see the face of my friend Jimmy Boyle, and then that of Michael Boyle, and then I see Jimmy again. Whatever I have been internalizing for these many days now bursts forth, because I can no longer abide the sadness that has come into my friend Jimmy's life. And so I sit among these men in my fire-fighting clothes and let the sobbing overtake me. It would be worse to try to fight it, I think. Dick Parsons puts his hand on my shoulder, a comforting human touch, and after a few awkward moments I am able to continue my story and ask that they support in particular ways the New York Police & Fire Widows' & Children's Benefit Fund. As we prepare to leave, the executives assure us that assistance will be forthcoming, and that is a good promise coming from such an important company.

In the lobby I thank Peter Quinn for his generosity in taking the time to intercede for us, and I thank Rusty and Steve for their convincing presentations of the good work they have undertaken for fifteen years at the fund. We are not just a one-time facilitator of benefits, but will be doing everything we can for these widows and children for the remainder of their lives. We will try to get their children jobs and into the colleges of their choice, and send them checks every year based on what we have available in our accounts. I am proud to be associated with this very worthy effort, and I know that I have to think of other ways to generate money for the fund.

At the site that afternoon, I meet Tom Henderson, who was once the president of the Uniformed Fire Officers Association, the officers' union. I have known Tom since high school days and appreciate his labor union perspective. I mention that I am trying to raise money for the fund, and he goes on a tirade about the money that is being earned in our country by everyone other than firefighters. 'You tell me,' he says, 'why a stockbroker with the same education as a firefighter earns five times what the firefighter makes?'

Tom also tells me about a firefighter from Massapequa. His daughter attends school with a girl whose mother was on the airplane that hit the north tower and whose father was working in his office on the ninety-fifth floor of the same building, and is now missing. That this little girl had had both parents taken from her so cruelly makes me think of the many compounded tragedies of Ground Zero. Earlier today I saw Lee Ielpi, a very well-known retired firefighter from Rescue 2, at the entrance of 10&10, speaking to retired Captain John Vigiano, also a very well-known firefighter from the rescue. Both men have been down at the site since September 11 digging through the piles for their children. Lee has lost his son, Jonathan, of Squad 288. Captain Vigiano has lost two sons, John II from Ladder 132 and Joe, an emergency service police officer from ESS 3. As a father, I cannot imagine the pain that comes with the death of a child, and it is hard to keep a straight thought when I consider what Captain Vigiano must be enduring. And there is Michael Angelini of the fire patrol, who has lost both his brother, Joe, of Ladder 4, and his father, Joe, of Rescue 1. Then there are the Harrell brothers, both lieutenants, one from Rescue 5 and the other from Battalion 7, and the Haskell brothers, Timmy from Squad 18 and Captain Tom Haskell from Division 15. And the Langone brothers, Peter from Squad 252 and Tom, also an emergency service police officer, of ESS 10. How these families must be suffering, for what explanation can they find for such loss in a single day, a single hour?

There is, as one chief said to me, no rhyme or reason for why some were lost while others survived. Ironies and coincidences abound. Lieutenant Mike Warchola was coming down the stairs of the north tower with his arms around a young woman when the building collapsed. He lost the young woman instantly and was himself pinned. He established voice contact briefly with Jay Jonas of Ladder 6, but was not able to hold on. It was his last official tour of duty, and his retirement papers were tacked to the office wall of the Greenwich Village firehouse on 6th Avenue. John Perry of the police department also was literally on the very point of retiring when he went to his last call.

And I think of Captain Timmy Stackpole, who had been burned so badly in a job two years ago and worked so diligently at the rehab center to be able to come back to fire duty. The fire commissioner made an exception to allow Timmy to return, for normally a man so badly injured will go out on a pension. But Timmy was an inspired person and brought inspiration with him wherever he went. He had just returned to duty the week before.

I look past Tom Henderson and see my sister's-in-law nephew, Paul Kelly, with a bright probationary firefighter's front piece on his helmet. It is good to see Paul safe, and his presence here reminds me of a further irony. Paul had worked on September 8 for a firefighter who had to attend a wedding, and the firefighter paid him back by taking his shift on the 11th. With the exception of one man, his entire company, Engine 6, was lost. People switch shifts all the time in the fire department, and these things happen.

I walk from the north to the south side of the site and pay another visit to 10&10. There I meet Lieutenant Sean O'Malley, who on September 11 had bicycled down to the firehouse from his apartment on 86th Street. The firehouse is a mess and is being used as a stopping point for just about everyone at the site. The firefighters who are assigned here are now living like nomads, for just about all the equipment in the firehouse has been usurped for other needs in the Ground Zero operation, including hose, tools, radios, blankets, stationery, and most of their fire gear. Lieutenant

O'Malley tries to keep his spirit up so that he can provide positive leadership for his men, but it is not an easy task in a firehouse that was almost destroyed in the collapse of the south tower and lost three firefighters and a lieutenant.

We go up to the roof of the firehouse where there are several telescopes situated on tripods, and where a few specially assigned firefighters are peering through binoculars across the site. They are spotting the piles for bodies and body parts as the cranes and the Cats pick at the steel and debris. It is the first time I have seen the site from any height, and I am further shocked, if that is possible, for I can see comprehensively how the damage extends for blocks to the east, west, and north.

The site has grown as busy as a mining town on a Saturday night, for everyone seems to be out and about, moving purposefully from one place to another. Many are carrying boxes or bags of water, handing them to every passerby as if they were flyers for a new store. One of the tents has a handwritten sign pinned to its entrance with the words TRAUMA INTERVENTION written on one side and CHAPEL on the other. It is so noisy here, however, I wonder how anyone can have the kind of settled thought necessary to provide assistance with any kind of trauma or to be alone with your God when deafened by the din of a thousand machines.

After lending a hand on the West Street side of the piles for most of the afternoon, I drive up to Engine 33 on Great Jones Street, right off the Bowery, to pay my respects. With one of the largest arches ever built in the city extending from the first to the fourth floor, it is one of the more beautiful firehouses in New York, and since the 11th its front has been covered by a thousand candles and notes. They have lost nine men here; this was Michael Boyle's firehouse, the home he wrote about in his history of Engine 33 and Ladder 9. Be proud, he would tell incoming firefighters, for we have a proud history. In a gesture that I imagine is something like sitting shiva for the Jews, the firefighters are always here, unless they have been called out on a run, as they are now.

On my way home I listen as someone on the radio suggests that senators Schumer and Clinton and Governor Pataki get together to recommend the type of monument that should be erected at Ground Zero. I find myself growing upset at this proposal, for I suppose I have become a little proprietary about this site. I remain thoroughly convinced that the only people who should have the authority to make a decision about a particular monument are the families of those who died here. Surely, a committee should be formed, and the city and state should find a way to pay for it.

I switch the radio off and think about the day's achievements. Steve told me that the NFL is contributing $1.5 million, and Rusty said that the Mets will donate $1 million. The federal program for first responders who die in the line of duty will give each family a lump sum of $151,635. Even with such levels of generosity, however, we are still far away from compensating our families appropriately. A story has recently run about an inequity in charitable contributions, and it seemed to me that the power brokers in the city were concerned that disproportionate amounts of money were going to the separate firefighters' and police officers' funds and to our fund. I notice, though, that a concert is being held in Los Angeles tonight featuring Paul McCartney that will benefit the United Way Fund. I have heard that families of some of the victims, such as the busboys of the Windows on the World restaurant, will receive only $30,000 in compensation, and that is terrible if true. But the last chapter is far from being written on the charity coming to New York, and I only hope that it is evident to our power brokers that the charity comes because of the inspiration brought by the hundreds who went in to help people out.

It has been predicted that life insurers will face claims totaling as much as $5 billion, and total insurance claims are expected to run as high as $40 billion. Some firefighters have a life insurance policy for as much as $100,000, but most have a $50,000 policy that is issued through the union.

When I arrive home, Debbie Spanierman calls and asks if the prestigious Spanierman Gallery could do something for the firefighters and police officers, and suggests sponsoring a special exhibition of nineteenth-century American paintings. This is a wonderful gesture, and I assure her I will talk to the board about it.

Tom Henderson calls to ask if I can help a fire department chaplain from out of town – a man appropriately enough named Chaplin – to obtain the proper credentials to enable him to contribute his services. I give him the number of Malachy Corrigan, who runs the department's Critical Stress Incident Unit and who has been working 24-7 to make certain the firefighters and families of fallen firefighters are receiving the appropriate help.

I also receive a call from Hal Bruno, the chairman of the National Fallen Firefighters Monument Foundation, which funds the trauma intervention programs that are so important to the firefighters. He has sent up a team of workers and psychologists to assist Malachy and the New York unit, for it is a huge undertaking to ensure that the families of 343 fallen firefighters are receiving the necessary counseling.

Malachy and I had spoken the previous day, when he asked if we could help in any way with the financing of his programs, and I called a contact at the Ittleson Foundation to ask for their support. They promised to send ten thousand dollars, and in this case, every penny helps.

Day 10, September 20

Frank McCourt phones this morning regarding an article about firefighters he has written for *Talk* magazine, on which I compliment him, and he urges me, 'You have to go down to Union Square to see the outpouring of human emotion there. It is such a beautiful thing.'

Union Square, named after the preservation of the union, has a history of being a radical gathering place, and I make a note to stop by there.

Sandy Meeker, of AOL Time Warner, calls to tell me the company will highlight our fund in the 3-for-1 matching program they offer their eighty thousand employees, which is very good news, and I am grateful that Mr. Levin and Mr. Parsons have set their promise into motion.

I usually feel that there is no free lunch in this world. But today I was the beneficiary of a salubrious gift. Because my eyeglasses fell into the pile yesterday and got mangled, I took them to be repaired at the General Union Optical Store at 14th Street and 4th Avenue. One of the employees explained that the company was giving free eyeglass replacements for any emergency worker from Ground Zero. I insisted that I would be happy to pay but he refused. It is good to know that everyone, even the eyeglass companies, is taking steps to do what they can to make things a little easier since 9/11.

While on 14th Street, I walk over to Union Square, where a choir of some fifty singers from Nazareth College in Rochester, New York, is performing the national anthem. Behind them is a statue of George Washington with his sword drawn, leading a charge. Over his sword someone has placed a peace sign of the kind we haven't seen much of since Vietnam. I notice that the base of Washington's statue is covered with awkwardly printed exhortations to LOVE and LOVE ONE ANOTHER. Washington's horse seems confused, for despite all the calls for pacifism, the choir's performance of 'America the Beautiful' has elicited tears beneath the umbrellas that have been opened in the light rain that has begun to fall. An American flag hangs beneath the animal's neck, spread between the general's feet. Flowers are scattered all about, and hundreds of votive candles and the smell of incense are pervasive. The atmosphere seems almost Tibetan, though I know these makeshift shrines that cover the plaza in every direction have been created by Americans of every descent. Standing at the center of the entrance to the park is a rough plaster column, 8 feet high, with a bouquet of flowers placed at its top, before which people are kneeling. Before one makeshift altar someone has written on the ground: 'In the flow of life, destruction never has the last word; creation brings a phoenix out of the ashes every time.'

Another sign is more straightforward: WAR IS NO FUN.

I do not see any firefighters or policemen, but sirens are, as they have been for the past week, a constant part of the background city noise. If the first responders came here, I think they would approve, for however polyglot the sentiment, it is all genuine.

It is hard to leave the symbols and shrines, and as I depart I catch sight of a wet T-shirt wrapped around a lamp post base that says GOD BLESS ALL OF THE FAMILIES OF NEW YORK'S BRAVEST. It is the image I will keep in my mind.

In the subway another shrine has been erected, this one among standing wreaths, as if in a funeral home. It contains a photograph of missing police officers Ray

Suarez and Mark Ellis from D-4, Transit PD. The photo of Suarez shows him carrying a woman out of the WTC. It was the last time he was seen. Passengers are stopping here to say a quick prayer.

I talk to an Officer Meehan, who is standing guard over the display.

'We're here,' Meehan says, 'because Suarez and Ellis spent their lives in the subways. And so all will remember them here.'

Tonight President Bush spoke to Congress and announced the creation of a new office, the Department of Homeland Security, headed by the governor of Pennsylvania, Tom Ridge, who got a big hand. I am pleased that the president has consigned responsibility outside of FEMA to handle our preparedness to deal with terrorism, and silently say a prayer for Governor Ridge to protect us well and quickly. Mayor Giuliani and Governor Pataki were sitting with First Lady Laura Bush, and they received the biggest applause of the evening, as if the members of Congress had recognized that these two men had come directly off the battlefield. To me, they are spokesmen for friends like Ed Schoales, Lee Ielpi, Michael Angelini, and Jimmy Boyle, and for John Vigiano and the families of the Harrell brothers, the Haskell brothers, and the Langone brothers.

As I try to sleep tonight, I consider the questions that the firefighters have been discussing at the site. Will we go to Iraq as well as Afghanistan? Where will a war on terrorism finally take us?

Tom Henderson calls early this morning to ask for a lift down to the site. A fire officer's information truck has been stationed on West Street to make services like phone and computer communications available to the fire officers, and Tom has promised to man it for a few days.

I tell Tom I want to stop at Ladder 13 and Engine 22 on East 85th Street first, to pay my respects to the men who have lost nine firefighters, and he agrees to meet me there. At the firehouse the shoes of firefighter Tom Hetzel are placed at the side of the apparatus floor. They are unmistakably Tom's for he characteristically cut out the heel on one and wears it like a flip-flop. His shoes stand waiting for Tom to return, and everyone in the firehouse knows it. But ten days have passed since the attack, and talk has begun among the firefighters of changing the operation from a search and rescue to a search and recover mission. It becomes more and more difficult to keep optimistic that survivors will be found.

I speak to Greg Mattingly, tall, with a red mustache, and together we study the photos of the nine men posted on the outside wall of the firehouse. I notice among them Captain Walter Hynes, whom I have met several times and remember well. He had worked himself through law school and also studied to become a lieutenant and a captain – a not insignificant intellectual challenge. 'He was the biggest pain in the world when we had a successful job,' Greg says to me, 'but if you were ever down and out, his arm was the first around you.'

'It's so tough, Greg,' I answer. 'And how are you keeping straight?'

'To tell you the truth, I am like a rubber band most of the time, blank for hours, and then something happens, we get a run, or someone comes in the firehouse like that kid there. [I use] anything [I can] to focus my mind again.'

The child he refers to is Kelly Carpentier. She is about seven and is wearing the little checked pinafore of the Convent of the Sacred Heart School. She has her allowance money in her hand, a twenty, and begins a significant discussion with her mother. She points to a big jar by the housewatch desk, and Kelly pushes her bill down among the many donations that have already been given. The housewatchman says, 'Thank you,' and after Kelly has another whispered conversation with her mother, the two leave. My friends Nick Liso and Mike Leddy have joined Greg, and Mike is talking about the Red Cross.

'I know how to do this work,' Mike says, 'and I told them I was available. So far they've flown me to twenty-six disasters around the country.' He is typical of most of the men I have met at the site – a trained emergency worker eager to use his skills.

A big celebration will be held at Shea Stadium tonight, before the first Mets game since September 11, where Liza Minnelli will sing and Julia Roberts and other stars will be in attendance. The salary of all the Mets playing tonight is being donated to the New York Police & Fire Widows' & Children's Benefit Fund. Even

Rusty and Steve will kick back at Shea Stadium for the event, following the mayor's urging of New Yorkers to live a normal life of work and entertainment.

At the site, I see Paulie and Big Pussy from *The Sopranos,* and I have heard that other stars like Robert De Niro, Bette Midler, Muhammed Ali, and Candice Bergen also came down to show their support. One fireman remarked, 'I heard that Kathleen Turner showed at St. Vincent's Hospital dressed all in white and ready to go to work. What a stand-up broad!' At the center of West Street a large group of firefighters has gathered at the edge of the site, waiting for the ironworkers to lift some big steel away. It seems so appropriate that we are eating away at these awful mounds to reach a street called Liberty.

Suddenly I notice a worker run up a sixty-foot girder with almost animal grace and surefootedness. It is an amazing sight, almost like watching a circus feat. The steel is hanging at a 30-degree angle, attached at its center to a cable coming down from a crane. The ironworker, Tommy Beattie, is from Local 40 and trailing behind him is a length of hose connected to the cutting torch he carries. The hose is at least fifty feet long, and it runs down and over the piles. He straddles the steel as if it were a bucking horse, legs firmly around it.

The high end of the steel beam is hanging from the corner of what is left of the Marriott Hotel, about five floors up. Tommy has climbed to about thirty or forty feet above the piles, and it is impossible not to remember that below him are broken floors of razor-sharp pikes, knives, and stakes. To fall into these piles from that height would almost certainly be fatal. The crane operator is trying to shake the steel as Tommy is cutting it, and what seems unfathomably dangerous to those observing him seems almost elemental to him.

Suddenly, the steel is detached from the Marriott, and the beam falls to distribute its weight equally. I hold my breath as I watch Beattie holding tightly as the beam seems to bounce, truly like a bucking horse. It finally settles, and Beattie nonchalantly climbs back down with as much élan as he ascended.

The ironworkers have gained the respect of every fire-fighter at the site. Beattie, who is 31, comes originally from the west side of New York, from the area called Hell's Kitchen. His family is in the theatrical union, and Tom's father moved them out of New York to Long Island.

■ One thing about ironworking, you only fall once. And it's not the fall that hurts so much as the sudden stop. But I was born with magnets on my feet.

I got into ironworking through guys I know, friends. I was a teamster for a while, in the theatrical union, but it just wasn't for me. I'm not good indoors, and it's just way too much time away from the family. To get a union book in the ironworkers, you have to go through the apprenticeship program, a four-year apprenticeship. You start off as a coffee boy, round tools up for everybody, and you slowly work your way up to the iron. When you go down to do the application, you take an aptitude test. Doing ironwork, you've got to have half a brain in your head because it's not just your life [at stake] – you can wind up killing somebody else.

I'll tell you who I tip my hats to there at Ground Zero – to the firemen. And I can speak for every ironworker here. What a phenomenal bunch of guys. I can't say enough good things about them.

I was up on the Outer Bridge when it happened, on Day 1. I couldn't get into the city [until] about 11:45 that first night. There was more devastation than I ever expected, and I hope I never see anything like that ever again. I couldn't believe it till after we put that big 500 up, the real tall boom. When I went to the top of that in a basket, 420 feet up in the air, and [looked] down, I was mortified.

I've never brought a camera; I don't need pictures. It's all in my head.

I've always been proud to be an ironworker. But the worst part is that it took something like this for people to even acknowledge us. You'd never put an ironworker sticker on your car, because even the cops would pull you over just on general principle.

We're not in the newspapers – we're not into that. I'm not looking for a pat on the back. I got a 'thank you' from a fireman, and that is all the thanks I'll ever need in my life. We're just men.

The toughest part isn't burning people out, cutting people out, finding people. The hardest part is seeing these poor firemen's wives and kids. They'd come up to the site. I don't even know what I could possibly say, besides nodding my head. I had to look the other way, not turn my back on them. It's just that there's nothing in the world I could possibly say to make this any better.

A fireman's wife came here, crying, asking me to please just find something. I said, 'I'm sorry.' I nodded my head and went back to work. That was the hardest thing. ■

We are called in to continue a search, the same search that was started on Tuesday night, 9/11, but now we are distressingly far from that date. I am still wearing the same cloth dust mask, and I don't want to walk the distance to the command center where there are actual mine filter masks. I remind myself to get one as soon as I see one up for grabs.

I am on the mound with John McCann of Ladder 26 in Harlem, which has lost Lieutenant Bob Nagel. John is what we used to call a bruiser – stocky, in great shape, the kind of guy you will respect in a bar or an emergency. He is on his knees, digging with his hands, and I am immediately behind him, shoveling out.

Earlier in the search effort we needed the dogs to determine if there was evidence of a human being, but now, after ten days, the bodies have begun to deteriorate and the odor is unmistakable. It is a harder smell than sulfuric acid and ammonia mixed together, a human smell that few humans know. McCann digs furiously, and I wonder how he keeps his strength up. He finds a body, and a Stokes basket is brought to the site. It takes three of us to lift the remains with our tools and place it in a body bag. It is a civilian for there is no evidence of fire clothing or equipment in the area. Someone zips up the body bag, and the Stokes is carried away.

Through the smoke still rising from the site, and the remaining sides of the two towers, which are just a few stories tall and leaning in opposite directions, I can see the Woolworth Building, which for many years was the tallest building in the world. The configuration of the steel of the World Trade Center creates a trompe l'oeil, and it appears that the Woolworth Building itself is leaning. I think people will come to view this structure differently now, as we will all high-rises. Some of the firefighters have argued that we must immediately replace the twin towers with a project higher than the towers in Kuala Lumpur, that we should create the highest structure in the world. This will let the Muslim extremists know that OUR FLAG DOES NOT RUN – a sentiment I saw expressed on the front of one of the huge Caterpillars down here. Others argue that this is now holy ground, and it should be converted into a field so that people can walk over it for contemplation, as at Gettysburg. In the end, a committee appointed by politicians will determine what use will be made of the sixteen acres, and it is the hope of everyone I have talked with that this committee will acknowledge the right of the grieving families to have the right of approval. In the meantime just the smallest dent has been made in removing 1.2 million tons of debris from the site, and mountains have not yet been filtered.

The bucket brigades, for the most part, have been discontinued. Now the grapplers and Cats pick at the piles with firefighters stationed around them spotting. It is a faster process than picking through the debris by hand, but it is not as exacting. When you are on your hands and knees here you won't miss a finger, or a section of a scalp, parts of a body that can be easily overlooked when a Cat picks up a large load of debris.

I return to the command center through the thick muddy road that West Street has become. I find the Uniformed Fire Officers Association trailer empty for the moment, as Tom Henderson has gone to get his daily mayonnaise sandwich. On the makeshift desk I see something that momentarily stuns me. It is a book headed MOS MISSING, and it is filled with the photographs of all 343 firefighters who have been lost. I open it and the first photo I see is that of Paddy Brown, one of the greatest. I look through it, nine photos to a page, and I recognize many faces, though there are many I did not know personally. I see the name Christopher Santora, and I wonder if this is Al Santora's son, Al Santora who headed the department's research and development office and who knew the engineering of firefighting as well as anyone. Here is Captain William Burke, and I wonder if this is the Billy Burke who is the son of Chief Bill Burke, who proudly introduced me to his son so many years ago.

Each of these men has a family that must be cared for, and each will have to be buried or memorialized. *Memorialized*. This is a term we are not accustomed to using in the fire service. Throughout our history, we have always brought the body out. We don't leave our men behind under any conditions. But here at Ground Zero, the buildings came down with such force that most of the objects and people in the buildings have been atomized. As of today there are 241 confirmed dead and 6333 missing in the two buildings.

An army of about two hundred police officers is crossing West Street and heading up Liberty. A battalion chief standing near me says of them, 'They want to be here, they want to work. So we have to divide up the work for them.'

One thing that has continued to amaze me is how completely the world changes just one block to the west on the other side of the World Financial Center. A marble and stone walkway runs alongside the Hudson River, but today no one is walking there. Standing in the softly shifting waters of the Hudson is a large ship, a sleek and modern yacht. The boat is called *The Spirit of New York,* of the Spirit Cruises line, and it strikes me how little this ship represents the spirit of the city at Ground Zero. The spirit here is now action personified.

Debbie Spanierman has sent a fax at home explaining that she would like to have the exhibition in November and donate the entire admission and 10 percent of the price of paintings sold to the New York Widows' & Children's Benefit Fund. I asked Steve about it, and he thought it was a good and generous offer.

Day 12, September 22

At the quarters of Ladder 16 I meet Rich Ratazzi. Like many firefighters, Rich usually carries a small throwaway camera in his pocket, and he used one to take a photo of Lieutenant Ray Murphy at Ground Zero, just minutes before the lieutenant was lost in the collapse of the north tower. It is a fascinating document, even though all it pictures is Lieutenant Murphy's back as he walks away. It is a record of destiny, because I know these firefighters felt a danger there the likes of which they had never known before.

The night before had been a typical tour for Ladder 16. They were busy until midnight, and then there was just one run afterward. I asked Rich what the morning was like.

■ We were sitting in the kitchen. The TV wasn't on, but we heard over the loudspeakers that there was a third alarm on arrival at WTC. We turned on the TV and saw the tower burning. It was about ten to nine. At the housewatch desk we listened to the fifth alarm come in. They called Engine 39 at eight minutes to nine. I went in on the recall. Lieutenant Williams wouldn't let anyone not working get on the fire truck. 'If you are not on my riding list,' he said, 'you don't go.'

Robbie Curatolo was just back from his honeymoon, his second tour. So Robbie, Lieutenant Ray Murphy, and I asked a cop to give us a ride down to the site. He said he would, and we grabbed two spare masks at the firehouse, and at West Street we picked up some spare cylinders. We were walking down West Street, looking for Ladder 16. The command post was at the loading dock, under a crosswalk. There were lots of rigs all around. All of a sudden it sounded like a plane coming down. We couldn't see past the Marriott, but we looked up and the Marriott blew out at us. Lieutenant Murphy and Robbie and I turned around and ran. They jumped on a rig, and I went on the back step of the Rescue 1 rig that was facing south and covered myself. Everything just got black, and it was quiet for fifteen seconds. Then someone yelled for help. 'Keep yelling,' I said. I led him out to Vesey. We just about walked into a rig. When the cloud started to get lighter, I went back for Lieutenant Murphy and Robbie. I had a disposable camera in my pocket. There were big pieces of tower 1 now in the middle of West Street and I wanted to take a picture of it as I walked to catch up with Murphy and Rob. I shot a picture then. They came across an engine company chauffeur who had a busted shoulder. Lieutenant Murphy told me to leave my tools and to take this guy to an ambulance on Vesey Street. I did that, and as I turned back to catch up with them, my tools were gone. They were never found. I met a chief, told him I was missing two guys from Ladder 16. He said 16 Truck went to the load-ing dock.

'Sure?' I asked.

'Yeah, Ladder 16 was at the command post waiting for work.'

I went to the loading dock and out back by the marina. There were sixty guys there, but no one from Ladder 16. Lieutenant Williams was two floors below with an EMT who was hurt. I took a mask from some firefighter and went to Vesey and West. When I got to the parking lot on West, the north tower began to come down. I looked up and saw it crashing down. There was a lot of smoke. I saw one story fall and I began to run up Vesey. I got blown off my feet by the entrance to the Merrill Lynch Building. I got up and ran into the building and shut the doors behind me. There was a black cloud that came up the street. It just seemed to envelop the street. It didn't seem to move then. I went back out Vesey to West and told some firefighter I was missing two guys. I saw Ladder 5 teetering off its wheels. A guy up on the sixth floor of #6 threw a chaise through a window. He must have had his door closed because there were no clouds of dust behind him. He was in a white shirt and black tie. Clean. We yelled, 'Don't jump!' and headed for the crosswalk. Someone was yelling 'Help!' We came to a photographer with a broken leg leaning on a battalion car. We found a backboard, and put him on it. I gave him some air out of my tank, and he threw up, starting to go unconscious. We took him over the piles to Vesey Street. He was about three hundred pounds. Four of us carried him to a structure and EMS took him away. I began to have trouble then. My eyes were hurting, and it was hard to see because of the concrete dust. There was a ton of fiberglass in the air, and I took on a bad headache. I walked up to the Stuyvesant High School and came across Assistant Chief Callan. I told him I was missing Ray and Robbie. A woman came over and gave me oxygen for fifteen minutes, but then I threw up. I was very light-headed. I went into a bathroom in Stuyvesant and threw up some more.

I met two guys from E39. They brought me to EMS, the Hezbollah ambulance. We went to St. Vincent's Hospital. I called my wife. The hospital told her I had eye and respiratory problems. I took the phone, and I told her that people wanted to use the phone. I met Chief Sal Cassano in the emergency room, and he drove me all the way up to the firehouse on 67th. I stayed there for an hour and went back to Ladder 35 for two hours, for they lost a lot of men there. I was still on duty. The recall ended at 10 P.M., and I stayed up in the kitchen until 5:30 A.M. I went to the TV room. Guys were sleeping all over the place. I worked twenty-four hours on Wednesday. That night at 8:00 we went to the site to dig until 4 A.M. 'There's gotta be people alive. There were big voids. Some of these voids were fifty feet down.'

No one gave up hope that Lieutenant Murphy and Robbie were okay. They found Robbie one week later, and they stopped us from digging. A week later they found Lieutenant Murphy by the Winter Garden, underneath a pile. ■

387

Rich and Lieutenant Murphy and Robbie Curatolo just missed the fire truck that had gone down to the site from 67th Street. Their next choice was to go to a staging area over at Ladder 35 and to wait there for a bus down to the site. If they did that, they probably would have been on the same bus I took down to Ground Zero on Day 1. But they met a cop on the street and asked him to drive them down to the site, and the cop, like every cop that day, did what he could to accommodate them.

Lieutenant Wick from Ladder 16 accompanied the bodies of Ray Murphy and Robbie Curatolo in an ambulance to Bellevue. He wanted to open the body bag to view Robbie, who had been assigned to him as a probationary firefighter, but the other firefighters wouldn't let him. I am reminded of the speech Lieutenant Wick gave on the bus as we headed south that fateful day: 'We will see things today,' he had said, 'that we shouldn't have to see.'

That night, after they realized that their brothers Lieutenant Murphy and Robbie Curatolo were missing in Ladder 16, the firefighters of Ladder 16 and Engine 39, Lieutenant Jim McGlynn among them, were huddled in consternation on the apparatus floor. There was an awkward silence until the probationary firefighter Pat Connolly, the largest man among them, just put his hand out in the middle of them. He said, 'Brothers, we need to say something together.'

The other firefighters put their hands on top of his until all their hands were joined in this way. Then they recited 'Our Father, Who art in heaven, hallowed . . .' and their voices echoed and reverberated from the tin of the fire trucks beside them, trucks laden with the grime and the gray dust of Ground Zero.

I drive down to the site and notice that the police now have only two security checks on the East River Drive. The city has put into effect an ordinance that no car with a single occupant will be allowed to enter Manhattan in the mornings, which has considerably decreased the traffic.

Someone has put a huge sign across the World Financial Center reading WE WILL NEVER FORGET. It ripples gently in the wind above a thirty-foot-wide American flag. I appreciate this sentiment, and the fact that it has been stated in twenty-foot-tall lettering. But I can't help thinking, we do forget. Ultimately, we find that it is almost always in our interest to do so, for our own benefit as a society.

I love the way firefighters sit around the site in little groups from their own firehouse, being most comfortable with those they know best. Walking across Church Street today, I see a contingent from La Casa Grande, and I join them for a while. Some of them are young, in their early twenties, and they so remind me of myself and men like Harry Meyers, Ed Schoales, and Vinny Bollon (the current secretary general of the International Association of Fire Fighters), when we worked together in that firehouse on Intervale Avenue. We gathered in small groups then as well, working the fires together, picnicking together in the summertime, and partying together in the wintertime. When there was a crisis in your family or a wonderful event like the birth of a child the first place you called was the firehouse. When there was a funeral, you met first in the firehouse and attended as a group.

I busy myself for several hours at the digging point of a Cat on the West Street side of the site, and then I wash up as best I can and change into a shirt and tie to pay my respects at the wake of Captain Timmy Stackpole.

It's night when I arrive in Brooklyn, and the traffic on Ocean Avenue is backed up. I am completely lost, turning at every other block trying to find a familiar landmark. Finally, I find the funeral home in Marine Park. Tara Stackpole, Timmy's wife, is sitting before the closed coffin at what seems the exact center of the room and greets the hundreds of firefighters who have lined up. Tara had already been through a great deal of pain and agony in the past as she lived through Timmy's rehabilitation from his severe burns, yet none of this is apparent in her face. She has the character of a strong and determined Brooklyn woman, but is warm and accessible.

Awards and photographs are displayed around the room; Timmy might be the most photographed man in the history of the department. There are forty or fifty big flower arrangements, some of them in the shape of the Maltese cross, the symbol of firefighters. There are several rows of mass cards. I see Joe Murphy, who is engaged to Timmy's sister and who is the band president of the Emerald Society Bagpipers. I tell him I have heard that Charles Osgood is going to do a story on the bagpipers, and he is pleased about that. But, he tells me, the band is being pulled in all directions. Its sixty members are trying to make an appearance at every funeral and memorial service, but there have been three or four every day, with twenty-seven scheduled for next Saturday alone.

The bagpipers are a special breed of firefighter. It might be thought that these men have comfortable jobs, like members of a military band, but in fact most of them come from the busiest firehouses. They have dedicated a large part of their lives to providing the appropriate music for the occasion, and at least one bagpiper will play 'Amazing Grace' at every service. They do it for the families, because they know it could be their family in mourning, and for one firefighter, Bronco Pearsall, a drummer with the band, that is indeed the case.

Joe tells me that I just missed Mayor Giuliani, and I think there must be a dozen Rudys racing around the city. He tries to appear at as many services as possible, because he knows how important it is for the families that he is there. I think at heart Rudy is a cop or a firefighter.

The Brooklyn Expressway is snarled with traffic, and I am stuck for almost two hours. All I want to do is sleep, but I listen to the radio and hear the mayor talking about the search and rescue effort. It is obvious that he will not let this mission be turned into a search and recover effort, for he believes that the families should continue to maintain hope.

Day 13, September 23

Seated in St. Jean the Baptist Church, I almost feel as if I am in costume. I have hidden my FDNY sweatshirt beneath a sports shirt, but my jeans are filthy, as are my shoes. I know that there is no dress code in heaven, but I cannot help thinking of how my mother would have disapproved of how I am dressed for church today. I would have explained to her, 'But, Mother, I'm going down to Ground Zero.'

And she would have replied, 'Would you dress like that if He asked you over for dinner?'

I think of my friend Jimmy Boyle during the service and say a few prayers that he and Lee Ielpi and John Vigiano and Ed Schoales find their sons soon. Jimmy called this morning and mentioned that his dental hygienist happens to be Donna Hickey, wife of Brian Hickey of Rescue 4, and I said, 'That is one for the small world department.' But this observation was just a preface to his telling me that he had received a long and very moving letter from another firefighter, one who worked with his son Michael. He described the letter, which spoke almost poetically of how so many firefighters had come to rely on Michael in so many ways.

As the psalm is read – 'Praise the Lord, Who lifts up the poor' – I suddenly realize how little I have thought of the plight of the poor since the first day of the disaster. Surely, in a situation like this war against terrorism the poor will certainly grow poorer, whether they live in the South Bronx or in Afghanistan. We have to be diligent in our efforts to help these people in need. I know that in these very complicated days they can all too easily be forgotten. How can we help the world's poor and also help our own? To my thinking, we need to develop a community of caring, a community that extends nation-wide, and even globally. The local firehouse, I think, is a hot spot of caring in a sometimes cold world. I heard a story about Michael Boyle's finding a hungry person on the street and bringing him into the firehouse for something to eat. Someone told me recently that he had once given a cashmere sweater to Father Mychal Judge, which the following day the priest gave to a homeless person on 34th Street.

At the site I run into Battalion Chief Ben Cassidy and Deputy Chief Tom Kennedy, both from La Casa Grande and both retired in the past few months. I think of the many alarms I have responded to with these two men when we worked together in Engine 82 and Ladder 31. Not tens or scores or thousands, but more. And I recall a visit I made to the head officers of the Soviet Fire Service in Moscow back in the seventies, where I was introduced to a man who with extra-ordinary pride told me that in his life he had been in more than one hundred fires. I congratulated him and smiled inwardly, for in my time, a hundred fires was a good weekend in the South Bronx. Frankie Griffen, also a chief, joins us. He and Cassidy and Kennedy have done well in advancing up the ranks in the department, and I wonder, if just for a fleeting second, if I would have passed those exams along with them had I joined a study group the way my friend Dan Potter has. Then I think of Pete Bondy of Rescue 2 in Brooklyn, who once said to me, 'I just want to be a can man.'

Naturally, I thought he meant because the can man, the lowest ranking position in a ladder company, is the one who gets first and closest to the fire with his two-and-a-half-gallon can of pressurized water. 'To be at the door of the red devil?' I ask.

'No,' he answered, 'so I won't be an officer and have to order my guys to put their lives on the line.'

American journalists have been at the top of their game during this man-made disaster, and no one has been more comprehensive in its reportage than *The New York Times*. The paper ran an image of a laser-based analysis of the site in a computer-generated side view of the collapsed buildings, and for the first time I have a clear understanding of just how much work there is to be done and how these square buildings were turned into round mounds of debris. It will take a year to dig through the fifty to sixty feet of these piles that are now visible, and then through the additional sixty to seventy feet below ground level.

I take a walk around the twelve arches of what I call the south cathedral wall, which is still standing. I saw from *The New York Times* graphic this morning that the bottom of these arches extends another thirty feet below the surface of the rubble. The firefighters who are present seem to be helping the construction workers move equipment, or have joined one of the many digging crews around the site. On my way to the

Church Street side to see if I can assist anywhere, I meet three firefighters from the Salt Lake City County Fire Department Urban Search and Rescue team. They are covered with dirt and tell me they have been deep below the piles, in the subterranean tunnels and subbasements. They found automobiles completely untouched, and abandoned subway cars, presumably from the PATH trains. In all the destruction, there were no bodies.

I walk up between buildings 4 and 5, and at a concrete platform overlooking the golden globe and the absolute middle of the site I witness the most incredible thing I've seen since I've been coming here.

Before me is a pyramid, an amazing pyramid of smoking rubble, reaching about 100 feet high. Hovering above it is the arm of a crane, maybe 250 feet high, with a basket being lowered from it. In the basket are three workers guiding a group of ten or eleven firefighters and construction workers who are up near the top of this mountain, digging out, piece by piece, some crevice or void. It is an incredibly dangerous place to be, for not only is the incline on the side of this mountain nearly equivalent to that of the great Egyptian pyramids, but huge pieces of steel jut out all around them. They are all very attentive and aggressive – two qualities on which they pride themselves.

Just beyond the pyramid is the very tip of the Verizon Building, which has been topped with an American flag, as has the north tower of the World Financial Center. All around us the smoke of the piles blows across the steel fields, and as it rises it momentarily obscures the flags and renders them in the background as if in an artistic photograph. On the south side of the site, there also are two forty-foot flags hanging on buildings of the World Financial Center, surrounding us with the American colors.

I walk around to Liberty Street to the base of the crane that has lifted the workers and find myself among the men of Rescue 3. I greet Tom Conroy, Mickey Conboy, and Paddy Hickey. Conroy, retired from Rescue 3, is a man I worked many, many fires with, and I have known Mickey Conboy since he was a probationary firefighter. With Paddy Hickey he is preparing to go up in a basket to the top of the pyramid, which is now just north of us. Along with two other firefighters from rescue Mickey steps into the basket and is lifted to the top of the pyramid, where digging has been going on in teams for several hours. I wait for the basket to come down again to see if I can get on the next trip up, but it remains elevated for more than half an hour.

When the basket is finally lowered, Mickey Conboy and the firefighters have two bodies with them. They call for Stokes baskets and for flags. A chief comes over with body bags, but the bodies are already zipped in their plastic cases. The first one is in a folded-over bag that measures about thirty-six by thirty-six inches. It is a yellow plastic case, and it is square, yet we know it holds a body, whatever the form, whatever the condition, and we know it is a firefighter. The bag is placed in an orange Stokes basket with several handles along the rails on each of its sides. An American flag is draped over the Stokes, lengthwise, and a chaplain approaches. The men of Rescue 3 do not know the identity of this victim, but they are very certain of the other and stand back. The chaplain begins to pray over the remains. Everyone in sight removes his helmet and places it over his heart. Then Tom Conroy and I are asked to pick the Stokes up, carry it over to an emergency medical service golf cart, strap it in, and accompany it to the TM, the temporary morgue.

We do so, taking seats on either side of the Stokes, holding down the American flag as we make our way around almost the entire site, from near the quarters of Engine 10 and Ladder 10, down Liberty, across West Street, and up Vesey Street. The medical technician drives slowly, with respect. A wave of police officers, construction workers, EMTs, and firefighters stand at attention and take off their helmets as we pass. On Vesey Street, a double line of firefighters has mustered, and as we drive through them they salute. The cart stops at the entrance to the TM, and Conroy and I pick the Stokes up and carry it inside. Two police lieutenants who are sitting at a desk quickly rise to attention when they see us and salute as we pass them. All the while I am conscious of the fact that I am carrying the remains of someone who was dearly loved and who will be dearly missed for so many years ahead. No matter that he is bound up in this odd plastic coffin, square, unsuited to any human body. Yet whatever the state of these remains, I know that the blood of its mind and soul still dwells in the physical form. Whatever the circumstances of its arrival, this is God's child leaving Ground Zero.

A doctor greets us at the entry to the interior section of the tent. Three perfectly clean stainless-steel tables stand at its center, and we are directed to lay the Stokes on the middle table. We do so and step back as four men enter from a farther section. One, a Greek bishop, is dressed in a long robe of black with a large golden cross tucked into his sash, two are Catholic priests in black suits, and the fourth is a conservative rabbi with forelocks at his temples. They all join in a generalized prayer asking that God receive His child.

In the street, I see the members of Rescue 3 bringing in their fallen comrade, three men carrying, with an officer leading. Because they are bearing a member of their company, they have decided to walk the distance through the site, and the double line of honor has grown as long as two blocks, each member of the service at attention, each in a rigid salute. Conroy goes to help them. I take my place in this line of my fellows, lifting my hand to be a part of them, waiting in my thoughts until they pass.

Day 14, September 24

The politics of New York seem to have shifted a little because of the attack. Mayor Giuliani, who has become such a positive influence throughout the country, has made what I see as the first mistake today by letting it be known that the residents of New York should decide if he should continue as mayor into another term. The people of the city, rightly or wrongly, decided some years ago that a mayor should serve a limit of two four-year terms. *Many New Yorkers don't like this law,* I think, as I drive to the site. But, it is our responsibility as citizens to change it, and not to find a way to circumnavigate it. Because the mayor is first and foremost a citizen with the same responsibility, I am hopeful he will quickly abandon this proposal, even if there is a significant write-in campaign in support of his remaining in office. If Rudy Giuliani leaves at the high level of esteem in which he is held at this moment, he will be remembered as the most important mayor in American history.

The FDNY has now divided the disaster site into four construction sectors. From white tents in each, senior fire officials coordinate the efforts of hundreds of construction and emergency workers, with overall responsibility assigned to Assistant Fire Chief Frank P. Cruthers. One construction firm from among the largest in the world will be given the contract for site excavation in each sector.

The cathedral wall on Liberty Street now has just eleven Gothic bays, for the twelfth is gone, cut away by the ironworkers. The piles seem to be getting lower each day, but this wall still stands in American defiance or, perhaps, American pride. The wall has become a key gauge for assessing the change in this site, the progress of the cleanup, and the continuity of the rescue effort. It is now bent toward Liberty Street.

I see civilians in the area for the first time today.

Construction crews seem to generate even more of themselves, as site and neighborhood housekeeping is going on everywhere. Scaffolding is rising around the Brooks Brothers store on Church Street. Trucks are washed down as they leave, so they don't track this dust into the rest of the city. Streets are shoveled and swept regularly. New streets are being manufactured over old ones.

As I walk up to the site, the fire commissioner's car passes me, and I realize that he might have just come from Timmy Stackpole's funeral. The department and city officials have made an effort to attend all the ceremonies, but with 403 firefighters, police, and Port Authority police to bury, even if they appeared at five funerals or memorial services a day, it will take months to get through them all. It is almost as if a Homestead Act has been enacted at the site, for a tent has gone up on every bit of available space. Everywhere. Each is personnel by three or four people, and even the Miami-Dade Urban Search and Rescue has a tent for the 72 firefighters they sent up here. South Florida USAR has also contributed 72 firefighters, and 76,000 pounds of equipment sent up on three 18-wheel trailers, including hydraulic saws and cutting and wrenching tools. A crowd of Puget Sound firefighters surrounds Susan Sarandon, who has come appropriately dressed in a blue-striped shirt, rough jeans, and a hard hat. They do seem happy to see her, at least as happy as they were to see the cast of *The Sopranos*.

Buildings 4 and 5, the seven-story buildings on

Church, are burned-out shells, and an amazing waterfall of steel hangs precipitously from the roof of the latter. That these structures remain standing gives testimony to the pancake nature of the collapse of the towers, for if any part of them had fallen east, they would have taken their neighbors down as well.

These two buildings bordered the entryway to the WTC Plaza, the courtyard in the middle of the area between the north and south towers. Farther toward the periphery of the area were the Marriott Hotel and building 6, another matching seven-story structure. In the middle of this courtyard was a large fountain, at whose center was a pedestal holding a huge golden globe, twenty feet in diameter. Miraculously this globe was not destroyed, though it was severely dented and has fallen fifteen or twenty feet into the concourse area that ran beneath the plaza. This whole area is now a deep pit, and I can see the section from where John McLoughlin and Will Jimeno were rescued.

Not far from here the firefighters have found a fire helmet, and the chief of the 27th Battalion, Mike 'Puzz' Puzziferri, climbs down into the pit and searches along with them. This is how it works here: If something is discovered that belongs to a firefighter, a small army of his brothers comes in to dig out the area. If a gun or a holster is unearthed, a small army of cops is brought in to dig. Soon a bunch of IV (intravenous) lines are found, and Chief Puzz guesses they belonged to an EMT.

A woman appears at the site with a friend or a relative. They must have political connections to either the police or fire department, because they have stopped in the dangerous area at the front of the pit. The woman is very attractive, but has clearly suffered greatly, for her husband worked at Cantor Fitzgerald, a bond brokerage firm that lost seven hundred employees in the north tower. In her hands is a little bottle of holy water that she and her husband had brought over to pour on his mother's grave when she died. Now she is putting this holy water to a use she never expected, and wants to sprinkle it over Ground Zero in some memorable way. I want to assure her that it doesn't matter where she performs her private ceremony, for people have died in almost every square inch around us. But she seems to want to make her way to the middle of the site itself, the very area that the cops and firefighters have been climbing up and down ladders to reach. My heart is moved as I watch this young woman, and I wonder if she has children. How proud they would be of their mother if they could see her as she now leans over the edge of the concrete platform on which we are standing, and releases the holy water down into the pit.

About twenty NYC police climb down into the pit, among them a female officer with her hair pulled back by a red ribbon. There are so few women in the rescue effort that it is heartening to see her. I would like to think my daughters could do this work.

Today I noticed a tent labeled ANIMAL-ASSISTED CRISIS INTERVENTION. In a happier time, this might well refer to something as innocuous as a petting zoo. But these animal trainers take their work very seriously, and

I know how effectively an affectionate dog can help calm a distraught person. Fourteen certified teams of trauma response dogs have come from Oregon, and I meet Cindy Ehlers and her dog Tikba, a beautiful, keeshond who is wearing mukluks to keep her feet dry. Both animal and trainer belong to the Hope Crisis Response of Eugene. Someone donated plane tickets so that they could come here, and the Marriott Hotel gave the Red Cross the rooms where they are staying at 44th and Broadway.

A very large prayer tent has been erected on the corner of Dey and Church, and I stop in to say hello. A woman hands me a medallion on a patriotic ribbon and asks, 'Do you know Jesus?'

'I went to St. John the Evangelist School,' I tell her. 'If I don't know Jesus they should take my diploma away.'

A large group of priests, ministers, and rabbis has gathered at Ground Zero. I see one man on his knees, making a confession at the back of a chapel tent away from the traffic of passersby.

Liberty Street has now been cleared sufficiently to function as a wide boulevard. The three buildings on Liberty Street between West and Church streets are completely burned-out hulks, but they are still standing. As time has passed and assessments have been carried out, it has become clear how widespread the damage is. Totally collapsed are buildings 1, 2, 3, 7, and the Greek Orthodox Church, St. Nicholas. Partially collapsed are buildings 4, 5, and 6, 130 Liberty, and another across from 130 Liberty. Structurally damaged buildings are the

Winter Garden, buildings 2 and 3 of the World Financial Center, the Verizon Building, 30 West Broadway, 90 West Street, 130 Cedar, 184 Broadway, and the south extension of 130 Liberty. Buildings that need repair are the Embassy Suites on Vesey Street and fourteen other structures. More than sixty other buildings throughout the downtown area require cleanup.

Today Ground Zero resembles a construction site where everyone is working overtime, and the foremen will not permit a second of rest. The construction trucks arrive empty and depart fully loaded as efficiently as if they were on a conveyor belt. Much of the collapsed Marriott Hotel has been taken down, but three stories still remain, and Cats are picking at it. Just in front of the Marriott has been constructed a colossal crane situated on a platform of hundreds of logs. This is called a four-day crane, for it has taken that long to assemble. It is about thirty feet wide at the base, with tracks that are eight feet high and five feet wide, and has a base filled with twenty huge iron counterweights that prevent the machine from tipping. The small cab where the operator sits is the only human-scale feature of this enormous machine, which almost makes it seem a mistake in the design. Across the field of piles at West Street, several other cranes and a dozen or so Cats and excavators are at work.

I think of my grandson Henry Patrick Smith, and how this 3-year-old loves trucks more than doughnuts. If only he could see this scene, with its panoply of heavy equipment, he would remember it all of his life.

I join a group of eight or so men taking a digging break. A firefighter from Rescue 2 is in conversation with Lee Ielpi, and is telling him, 'I have been here every day since 9/11, and I promised my wife a trip anywhere. The only thing is, we won't go by plane.' Another fire-fighter says, 'I saw the police and fire commissioners on TV in T-shirts and jeans, and I think about all the chiefs who yell about us being out of uniform when we don't have the right shirt on. What's their excuse? "It's an emergency." Well, we're in emergencies, too, 24–7.'

'But they don't care about us,' another firefighter adds, 'because the fire department is not an income-producing operation, but just agents of the insurance industry.'

Still another firefighter, this one from Ladder 14, comes over and greets me, 'Hey, Dennis, I didn't read the book, but it's good to see you, brother.' This is his way of stating, 'I'm aware that you are well known for writing that book, but here it doesn't matter.' And I like this attitude, for it reminds me of a quality I have always admired in firefighters: They believe in the egalitarianism of their work. If you are doing your job, you will always be one of them. But, if you begin shirking, you are immediately out.

A firefighter from Rescue 1 remarks, 'Terry Hatton's father called my firehouse, and he said, "Don't let them throw you outta there. Take our boys home." ' This is an indication to me that people are starting to accept, as we enter the third week since the attack, that the chances of rescuing anyone have significantly diminished, and that the authorities will soon begin to focus only on the demolition and excavation aspect of Ground Zero. I know this will pose a great problem for the firefighters, for they believe that these mounds have to be sifted carefully to find any part of a body that might bring a family the peace of having a funeral, and to spare it the open-endedness of a memorial service.

I was told recently about a gesture that had been made for one of the firefighters who died that could not have been possible at a memorial service. Lieutenant Geoff Guja had been promoted out of my old company, Engine 82. They loved Geoff in La Casa Grande, for he was a true free spirit. He lived on a houseboat, because there was nothing that pleased him more than being near the water and the beach. At his wake in Long Island, his wife, Debbie, and some of his friends decided that they wanted to have one more beer with him on the beach, and so they carried the casket out and loaded it on a truck, bought a case of beer, and drove to the beach. They sat on the shore, told some stories, drank the beer, and, after a solemn toast, returned the casket to the funeral home.

Looking across West Street, I saw a new sign across the wall of the World Financial Center that reads: THE STRENGTH OF THE WOLF ISN'T IN THE PACK, BUT THE STRENGTH OF THE PACK IS IN THE WOLF. This is a typical expression of the strong patriotic fervor that exists at every corner of Ground Zero, and without this kind of sentiment throughout the country we would be a nation in need of inspiration.

Just in front of this sign is the last remnant of the pedestrian walkway that connected the Winter Garden to the north tower over West Street. It is a single concrete bastion, and a set of night-lights sits incongruously at its top, like a sculpture.

On Vesey Street Verizon has installed a trailer with five phones attached to its side, with a sign announcing free telephone service. It is a very generous company, and I remember that it has given $10 million to The Twin Towers Fund. I try to make a call, and find that it lasts only three minutes before cutting off automatically. I then call Tom Dunne, my friend who is a vice president of the company, with a completely fabricated story about a firefighter I know who tried to call his psychiatrist for a ten-minute therapy session. The firefighter felt it was helping, but then they were disconnected, which forced him to go over and pet the crisis intervention dog for solace.

It took Tom a few minutes of defensive explanations before he realized I was joking.

The very idea of being in a joking frame of mind is a healthy change for me, I realize. This time down at Ground Zero has been unrelentingly single-minded. The rescue workers went through a hopeful stage, and then a disappointed one, and have all come to one of acceptance. Yet while everyone is making progress in that respect, I fear that most firefighters have not yet integrated the full horror of the situation into their day-to-day life in a way that will enable them to genuinely move forward.

In the early evening there is a whirl of activity on

407

West Street and Liberty, because they have found a fire-fighter's body. It is that of Tom Holohan of Engine 6, my sister's-in-law nephew's company. The firefighters of Engine 6 arrive, and we line up on either side of Vesey Street as they carry Tom's remains through the line of honor, firefighters and police officers at full attention and salute.

After completing the day's searching I begin to make my way to my car, but I am greeted by a man I haven't seen since he was a 10-year-old boy in Garrison, New York – Bill Schauffler of Engine 73. It is surprising not only to see Bill on the job, but also to discover that he is assigned to the South Bronx. He was in my son's grammar school class when I was writing *Report from Engine Co. 82,* and now he is a member of a fire company I wrote about, and I am struck once again by how circular life can be.

As I walk past the second checkpoint on West Street, I see a young Asian woman weeping at a fence where the police have blocked off the entrance to Battery Park City. She has two very large suitcases at her side, and I ask if I can help. She tells me that a cab just dropped her off here, a half block from where she lives, but that the police will not let her in the gate, and have instructed her to go three blocks down, where there is another police barricade. I speak on her behalf to the police officers, who explain that they cannot make an exception, and in these days of heightened security, orders should not be questioned. I tell the young woman, Letitia Wong of the Accenture Company, to wait for me, and I pick up my car and drive her to the tip of Manhattan Island, where there is a barricade manned by national guardsmen. I show a sergeant from Auburn, New York, my identification, and as he reads it, he exclaims, 'Dennis Smith! I wish I had your books here for you to sign – I read every one of them.' Letitia is as surprised as I am by this connection, and after the guardsman bids us by I drive her another three long blocks to her apartment building. She could not have made it herself carrying those big bags.

Battery Park City has just recently been reopened, and its residents are busy trying to clean the place out. Jean and Dan Potter live in these buildings, and I know they have had to hire a professional cleaning company to come in to remove the inches of dust from every single item in their apartment.

Day 15, September 25

The director of one of our museums has suggested that the three enormous steel skeletal structures that have become famous in photographs – a three-piece section of tower 2 that nearly ended up in the middle of West Street – should be saved for a monument. This could be a good idea if the fragment were used as a backdrop for sculptures of the victims and the heroes – firefighters, police officers, EMTs, and ironworkers. But I fear that if it stood by itself it would be a souless monument, like much of modern art, not celebrating the human.

Like many New Yorkers I have little interest in the Democratic primary, which has been rescheduled for today from September 11, but I do go to vote to honor the franchise. I am hopeful that Michael Bloomberg will win on the Republican side, because I am convinced the city needs a leader who knows how to run a large company, to manage intelligently, and to refine rather than expand government.

This afternoon I pay a visit to the quarters of Engine 40 and Ladder 35 on 67th Street and Amsterdam, the staging firehouse where we changed buses on 9/11. They lost eleven firefighters here, and only one of the men who responded that morning returned.

A wide array of letters, candles, notes, flowers, pictures, and photographs has been placed outside of the firehouse, which is part of the complex that was built when Lincoln Center was created. Among the tributes is a monument with a bronze plaque from Waller County, Texas, and a very large sign from the Sylvan Hills Middle School, which shares many touching thoughts, but neglects to mention where the school is located. I notice Michael Boyle's photo hanging on the wall, and I remember, because he was a probie here, that he played softball with this firehouse in the Broadway Theater League. 'We feel he is a part of us, too,' a firefighter tells me. In the kitchen I have a cup of coffee and select one of a thousand delights, for there must be more calories sitting on the kitchen table of this firehouse than on any kitchen table in the world. Most of my conversation with the firefighters has to do with protecting ourselves. 'They are now searching every truck that is coming across the bridges,' one of the men says, 'and that's good. We have to secure our own country first, and then we can take care of these other countries who harbor terrorists.' There is a lot of discussion about anthrax, and about whether we should issue masks to all our school-children, and to our families. The anxiety reminds me of the fears about atomic bombs and radiation when I was a child, and how we all had to crawl under our desks during air-raid drills in the early fifties. The fears of different times are responsible for creating different memories, and the terrible events of these days are what our children will carry into the future.

As I am leaving this firehouse, a note on the wall catches my eye, because it is signed by 'The Seinfelds, Jerry, Jessica, and Sascha (6 months).' It thanks the firefighters for their priceless friendship and courage, but it is addressed to the firefighters of Engine 35. This brings a smile to my lips as I think of all the famous *Seinfeld* episodes that are based on a misunderstanding. The Seinfelds do not realize that Engine 35 is way up in Harlem, and that it is Ladder 35 who is their neighbor.

It is a busy night at Ground Zero, and I am struck by the eerie timelessness that the artificial lighting with its constant 4 P.M. glow lends the site. I head directly to the corner of Liberty and West streets, but find that Liberty Street is closed, and patrolled by national guard members sitting indecorously in swivel chairs. A crane has been set up, and is beginning to pull at the south

411

cathedral wall, attempting to snap it into two parts. It bends a little at the top, but the bottom remains firmly planted, not surprisingly since these walls go straight down for another seventy feet.

I see Tom Conroy of Rescue 3 who mentions in passing, 'I'm at an age where I should be dreaming of beaches and palm trees, and all I see in my sleep are I beams.' On the West Street side of the pile, there is a large and very flat area leading up to the Marriott Hotel footprint. I see Chief Bob Calley there, and go over to say hello. Two firefighters on a two-and-a-half-inch hose line are cooling the smoking pile just before them. Although fifteen days have passed since these fires first began, they still flare up regularly, for they are deep-seated, very hot, and self-propagating. The firefighters introduce themselves, and I discover that I worked with each of their fathers in the 6th Division. One is Joe, the son of Joe Martin, and the other is Ted, the son of Paul Knapp. In the fire department, the tradition is that one never gives up the nozzle unless ordered by an officer to take a relief, or to give it to another firefighter out of pity, and it is this latter, I suspect, that inspires Knapp to ask if I would like to take over from him for a while. I haven't had a charged two-and-a-half-inch hose in my hands for many years, but to me it was always the best responsibility to have in a fire – to be on my stomach and to have the officer and the men shouting, 'That's it, you got it, move in, a little more, get the ceiling, get the ceiling, watch the windows, you got it now. . . .'

And so, I take the nozzle from Firefighter Knapp and

continue to direct it at the fire, doing something that is as natural to me as throwing a baseball, something that I never thought I would be doing again. Yet here I am, perhaps keeping the flames from burning another corpse beyond recognition, suppressing them enough to allow the firefighters to go into the pile and pull and tear until they find something valuable. This is neither difficult nor, at least in this context, even skillful work, but merely holding a nozzle in your hands does convey a certain power, for it articulates and defines the firefighter. It is a good feeling, and after a while, with gratitude, I return the nob to Ted Knapp.

It has grown suddenly cold, and the surprise chill is made worse by a wind coming off the river. I take the sweatshirt that is tied around my waist and put it on. A nearby group of firefighters from Ladder 31 is all in T-shirts, so I direct them to a supply depot that I know has yellow rain jackets that will at least protect them from the wind. I tell them that I regret that our dinner did not work out a few nights ago. Chiefs Ben Cassidy and Tom Kennedy, Lieutenant Charlie McCarthy, and I had been planning to sit down to a quiet dinner with the men of La Casa Grande, but when word got around that we were coming, these old firefighters of note, more than forty men called to say they were going to stop by. That threatened to become too much like a celebration, so the officer on duty rightly suggested we do it on some other evening. In these times, certainly, none of us wants to do anything that appears to be an entertainment.

Many things at the site are changing. A road is being

constructed down into the pit from Church Street, which will enable trucks to go into the very center of Ground Zero to remove debris. On Liberty between West and Church the 'Green Tarp Café,' named after the green plastic tarpaulin that is keeping the dust out of its abandoned storefront, is doing a thriving business. David Bouley, a famous New York restaurateur, has been volunteering down here since the beginning, offering his award-winning dishes to the rescue workers each day and night. He has recently contracted with the Red Cross to continue the service, and the café has become the corner of congregation for the rescue workers, a place to meet friends and co-workers. The staff is all volunteer and tonight I meet a woman named Deborah Trueman, who by day is a producer at *Dateline NBC* but who also has been putting in nights here. It is now about 2 A.M., and she is going strong. I also meet a woman named Victoria, who is a marketing specialist for Moët & Chandon and who works here every night until 6 A.M., then goes into work at 10:00. I so admire these men and women who are giving so much of themselves to help out.

The backhoes, grapplers, and Cats are all at work on the piles. I count seven of them, all balanced at odd relationships to what would be level, picking at the piles like buzzards. Their jaws are open wide, and I can only think of the souls each is removing with the rubble. The trucks take their loads into the Hamilton sanitation dump in Brooklyn, and if it is busy there, they go to the Staten Island landfill.

I decide to take a look at Trinity Church, and walk

down Greenwich Street, next to 10&10, where fire escapes are piled high with six feet of debris. At the back end of Trinity Church, I see that the trees of its grave-yard have survived, and think of Wall Street, just two blocks from here. Trinity Church, built in 1698, was the tallest structure in New York for 150 years, until the birth of the skyscraper.

I decide to call it a night and head down Broadway. The street is lined with trailers for Con Ed, Verizon, city services, and construction company offices. Each is humming with its interior motors and machines. There is no difference between midnight and morning here, and it is now near 4 A.M.

The last thing I notice is the scaffolding that has gone up on many of the damaged buildings surrounding Ground Zero.

The city is beginning to repair itself.

Day 16, September 26

The excavation cost at Ground Zero is running at $100 million per week. Besides the emergency workers, more than 1000 construction workers have been assigned to projects from four different companies: the Bovis Company, which is based in Australia but has done much work during the Giuliani administration; the Turner Construction Company; the Tully Construction Company; and Amec Construction Management. There are now more than 150 pieces of heavy equipment at the site, including at least twenty cranes and the Caterpillar 345 Ultra High Excavator, which was brought in from Illinois — a machine with eighty-foot arms ending in shears that can cut through steel beams. The management of the site excavation is being directed from downtown Public School 89.

As of today, 115,000 tons of material have been removed from the site in varying daily tonnage. There have been many interruptions in the work, whether to recover the victims, because of problems with the slurry wall, or concerns about the large freon tanks that fed the vast air-conditioning systems of the WTC and which are housed on the third subbasement level.

Charity has continued to come into New York at a remarkable pace, and from all over the world. The Red Cross has already raised $200 million, the United Way's September 11th Fund has raised $120 million, and $70 million has gone to the mayor's Twin Towers Fund for families of uniformed personnel. The Uniformed Firefighters Association Fund has raised about $25 million, and the New York Police & Fire Widows' & Children's Benefit Fund has raised about $30 million, and I have great hope that the Beanie Baby puppies 'Rescue' and 'Courage' will add upward of a million to that.

It seems that everyone in America is trying to do 'the right thing' during this crisis, and the response varies from person to person. An attorney called me about a sculptor who is making an important piece to commemorate the courage of firefighters, another person is making a quilt, and still another is sending her tax refund check to the Widows' & Children's Fund.

A conversation I had last night with a captain who lost two of his men in the collapse nags at me today. He was so angry that he talked nonstop for twenty minutes about 'the job' and how 'the job' is not what it used to be under the best of circumstances, and, now, with all this, it can never be the same. This was a time for the leaders of the fire department, he argued, to reconcile any of their differences in policy and to attain order. 'After all,' he said, 'it's all about lives being at stake.'

There is a definite morale problem in the nation's firehouses and because fire departments, except where they are volunteer, are almost always unionized, there are usually labor issues involved as well. In New York there is the additional factor that 343 of 11,495 firefighters and officers have been killed in the line of duty. This is not an issue like any other, yet I think it will magnify every other problem two- or threefold for years to come.

I have found that during a crisis the public view of firefighters is almost always a positive one. They are praised for their heroism, their selflessness, their sense of duty. When the crisis has passed, however, they inevitably tend to be regarded as unskilled workers locked in an ongoing labor dispute with a tough chief or city manager, a view that tends to filter down through the press to the taxpayer. Firefighters, meanwhile, have to continually defend themselves from accusations of indolence because they are not fighting fires 100 percent of the time – this despite the fact that they have to be prepared 100 percent of the time to answer an alarm, and they never know what they will find when they get there.

Vinnie Bollon, the general secretary-treasurer of the International Association of Fire Fighters (IAFF), called and we talked about the loss of the new men on the job, the seventeen probies. As Vinnie correctly pointed out, 'They may have gone down there as probies, but they came out full-fledged firefighters.'

At the site, they have again created a frozen zone, while they make another effort to pull down the south cathedral wall. This time they have several of the largest machines, with their cables attached, pushing and tugging at it.

I meet Harry Meyers, who is the chief in charge tonight, and who is trying to figure out how to get the north cathedral wall down. An ironworker is cutting it, and a crane pulling at it, but it just shakes.

'Maybe we should use cutting charges,' Harry says. 'I saw them down those huge gas tanks in Maspeth a few years ago with cutting charges. But everyone around here thinks that cutting charges are expensive. Well, they're not.' Harry is in shirtsleeves, and I suggest that he put a sweatshirt on, for it is cold tonight. 'After I sweat a little,' he says, 'from doing my walk around. It will take me forty minutes, and I'll see you later.'

Inside the white tent of the command post is a

blown-up photograph of the site, taken from above from a plane or chopper. It measures four by five feet and is titled 'Ground Zero with 2 Block Radius.' The golden globe stands at the exact midpoint of the photo, with equipment surrounding it on all sides. It is an impressive reconnaissance image, and one that adds another dimension to the terrible act that has been committed. The top of the north cathedral wall looks like a series of battlements through the smoke that seems to continually rise, visible even in a photograph taken from this height.

At 8:30 P.M. a group of about forty relations of missing firefighters approaches, walking on the west side of West Street. The noise of the generator, cranes, trucks, and Cats is heavy in the air, and the fire department chaplain has to shout above it as he explains the configuration of Ground Zero to the men, women, and children. Most are crying heavily, remembering. It is good for them to be here, and it is equally good for those working here to see them, to put a human face to their efforts. A firefighter goes up to a young boy, a robust boy of about twelve, and after they talk awhile, they hug and cry together.

A new shack has gone up today, and like everything that has been erected in the area, it conveys an air of impermanence. It is simply a bare wooden lean-to, and its sign has been spray painted, like graffiti: NYC DDC [Department of Design and Construction]. It is ironic that the Department of Design has selected for its temporary housing a structure so fundamental that it could almost be said to be without design.

Over by the south cathedral wall a basket is hanging from a crane, and from it two men are making a vertical cut in the steel. The sparks fly in every direction as if they were choreographed. This is the beginning of the wall's end, and its demolition will be a milestone in the work of Ground Zero. Like all walls, once it is down it no longer contains, and that symbolism will be good for the world. Finally, Ground Zero will be a sight to be seen and understood, it will move from being an image to a reality.

I work for a few hours on the Vesey Street side of the West Street piles, spotting the Cats, watching in the glow of the night-lights for any odd shadows. The sound is unrelenting, the cranes, the Cats, the night-light generators. I don't know why everyone here doesn't have a headache.

I go over to the Green Tarp Café, and it is as busy as I have ever seen it. Not only have the firefighters and police officers gathered here, but many of the tent volunteers as well. And as I enter, being careful not to trip over the broken stone of the two-step entryway, I can see why: The actors Joel Grey, Harrison Ford, and Bernadette Peters are behind the counter, serving spoons in their hands, and Chef David Bouley himself is in the kitchen, preparing the fare. I look around for the television cameras I expect to see, but reassuringly, there are none.

Day 17, September 27

The Pentagon has released the photographs of the nineteen men who hijacked the four airplanes, and I study
every face carefully. *What am I looking for?* I ask myself.
One, Ahmed Alnami, could be a movie actor, he is so
young and stylish. I am searching for the personification
of evil and sinfulness, but I cannot even discern the
face of delusion in this young man, and I remind myself
that evil will never be found in outward appearance. The
evil lies in the belief that, through an extreme inter-
pretation of a religious book, the murdering of
innocents is a justifiable act.

Before going to Ground Zero today I decide to visit
some of the firehouses that were the very first to respond
to the catastrophe. The Duane Street headquarters
of Engine 7 and Ladder 1 is also the headquarters of
Battalion 1, and its architecture is among the most interesting in the department, for it is large enough to feature
a tennis court on the apparatus floor. This is the firehouse
from which Joe Pfiefer responded when he was called to
investigate a gas leak on the morning of 9/11, along with
Captain Dennis Tardio of Engine 7. A large list of the
twenty-six funerals and the memorial services that are
scheduled is posted, and there are more of the latter.

I have a cup of coffee with the firefighters and talk
for a while. Someone mentions Chief Ganci and how he
shot a seventy-four at Swan Lake Golf Club just a few
weeks before at an outing for the headquarters chiefs
and their drivers, making him the toast of the trip.

Another man points out that there was only one heli-copter at his funeral, despite the fact that at department funerals, four or five helicopters are typically flown over the ceremony as the casket leaves the church. The police aviation squad, it's clear, is as pressed as the bagpipers. One of the men speaks of Captain Jim Corrigan, who retired from 10&10 in 1994 to head up fire security in 7 WTC, and just a few weeks prior to 9/11 was asked to take the fire safety director's job at the WTC. He was lost along with a recently retired FBI agent, John O'Neill, who also just started on the job at the towers as head of security. Some people go left to live, and others retire to a new career. There is no rhyme or reason, as people have often remarked.

My next visit is to Ladder 15 and Engine 4, which is on the East River just where the FDR Drive ends before the entrance to the Staten Island Ferry. Engine 4 is the company with which Ed Schoales's son Tom was working when he was lost, along with eleven others. The mood here is somber and sad, small jokes are made here and there. Five firefighters are sitting around the kitchen with a young, very attractive woman. She is the ex-girlfriend of one of their lost colleagues, and has stopped in to say hello to old friends. Again, the firefighters are discussing the con-ditions of 'the job.' One is arguing that the firefighters should no longer be expected to do all the work they once did, like going to the training center, or conducting certain kinds of building inspections. 'Sure,' this man says, 'we are a 'can-do' firehouse, but it shouldn't be business as usual. We have twelve families here to take care of.'

I head for Ladder 6 and Engine 9, the firehouse Jean Potter went to when she was looking for her Dan. It is in the heart of Chinatown, on Canal Street, and when I arrive I see a television crew from NBC shooting a documentary. This is also the headquarters for Jay Jonas and his men, who all feel themselves most fortunate for having survived the collapse. A small shrine stands outside of the firehouse, and I comment on it to the men, for many of the letters are in Chinese.

'For the first few days,' one firefighter explains, 'there was nothing, no flowers, no notes. But, then [the story] must have been published in the Chinese newspapers, and a horde of people began coming to the firehouse with dumplings, fish, Chinese cookies. It's beautiful, especially when you know how the Chinese people keep to themselves.'

It is about 5 P.M. when I arrive at Ground Zero, and I meet a battalion chief riding a three-wheel, all-terrain vehicle. He tells me angrily that the fire department is going to reduce its presence at the site starting tomorrow, and that only 150 men will be allowed here. The police will now take over responsibility, for it is still a crime scene. 'I hate to give it up to the PD, and leave all our brothers to them, but what can you do?' I am surprised by this news, because as I look over the site I can see that several fires continue to burn. And if the fire department is not in charge of dealing with them, who will be? I walk around the Liberty Street side to Church Street, and then to the overhang concrete platform by the pit where a police captain with a nameplate that says

Joe Cordes is standing. I ask him how many Joe Cordeses could there be in the world, and if his father is Battalion Chief Joe Cordes. He indeed is, and I give his hand an enthusiastic shake. His father and I worked together for many years, and I am so pleased that the family also got a police captain. To me it is equivalent to having a doctor or a priest or a battalion chief in the family, which, in the neighborhood I grew up in, are all ranks of high success.

After working on a bucket brigade for a while, I run into First Deputy Police Commissioner Joe Dunne on Church Street, and tell him how glad I am to see him, particularly after the rumors of his being missing on 9/11. His only response is to kiss me on the cheek.

Joe is escorting the Irish ambassador, Sean O'Huiginn, who has come from Washington. Sean has been a close friend of mine for twenty years, and often we greet each other with a hug. He introduces me to the Irish minister for foreign affairs, Brian Cowen. I walk with them into the pit area, and the Irish officials are visibly moved by what they see. So many people are asking what caused this field of ruination, and so I turn to the foreign minister, saying, 'The implications of radical Muslim teachings must be discussed publicly, Minister. Someone should invite important Islamic thinkers, imams, sheikhs, and philosophers together, in some safe forum, to discuss the origins of the kind of hatred that results in this. We are just kidding ourselves until they use their intellectual and spiritual influence to stop the regular harvesting of young men who want to commit suicide to go to heaven.'

'Yes,' the foreign minister agrees, as we carefully walk through the rubble, 'it should be considered by all.'

Standing next to 10&10 is Sergeant Heaton of the Buffalo unit of the national guard. He was activated from his job in a beer distributor, and his employer gave him one day's pay. He is staying on a ship tied up on the Jersey side of the river, the USS *Comfort*. 'It is inconvenient,' he admits. 'I have a wife and children, but I think of what everyone went through down here, and I will never complain.'

The sergeant tells me that he personally stopped an Arab man on the site, taking photographs, and found that he was carrying three pieces of false identification and ten thousand dollars in cash, all of which he confiscated. The man bolted, but was found fifteen minutes later with yet another camera, taking more photos. I had been told earlier that the south cathedral wall would be pulled down at 8 o'clock, but it is near midnight, and it is still very much in place, having resisted every pull and tug. Maybe tomorrow it will come down. Maybe tomorrow many things will change with the reduction in the fire department presence, if the firefighters feel the search for the recovery of the department's fallen heroes has been reduced to a rough sift through the debris. I am certain that all firefighters will keep in mind Chief Hatton's plea, 'Bring our boys home.'

Day 18, September 28

Cars are parked in parallel around the firehouse on Sixth Avenue, off Houston — an unusual sight in New York, but another small indication of the extraordinary times we are living in. Among them is a firefighter's personal vehicle, a blue pickup truck. It is new, but battered and bent, and its front window is blown out. The firefighter had used it to drive a few others down to the site before the collapse of the south tower. On its side is a sign that reads LADDER 5.

I have been wanting to stop in this firehouse to pay my respects, for it was here that John Drennan served as captain when he was trapped in a flashover fire in 1993. He lay in a burn center bed for forty days and forty nights, attended every minute by either his wife, Vina, or his friend Paddy Brown. (Paddy was in Rescue 1 then.) I knew John Drennan because he loved Irish music, and every once in a while he would show up at Paddy Reilly's, a gin mill on 28th Street and 2nd Avenue where I played Irish music on Tuesday nights. But today, I am here because this house has been deeply rocked by the disaster at Ground Zero, having lost eleven firefighters. Lieutenant O'Neill offers me a cup of coffee in the kitchen, and starts to talk.

■ **Many people are having problems. I wrestled with this a thousand times. Why did this happen? How did this happen? Why am I here? The night before, I got a phone call telling me that my mother was going** through emergency brain surgery. My wife made a reservation for me the next morning for Melbourne, Florida. I didn't want to go; I had a feeling that I shouldn't be going. The firehouse kept coming to my mind. Why would a firehouse be coming into my mind? On Tuesday morning I went to the plane, looked for my seat, and found that someone was sitting in it. A flight attendant came up to me and said, 'Oh, you know, since you're a nice guy, I'll give you first class.'

I still have this feeling that something's not right. I just don't know what it is, and can't get it off my mind.

When I got off the plane in Daytona, there was security everywhere. One of the men told me that terrorists flew planes into the World Trade Center. I say, 'You gotta be kidding me.' There is a TV guy next to me, who says, 'You're a fireman from New York City? Can I film you watching this on TV?'

I said, 'No, you're not going to do that.' I went to rent a car, because it's about an hour-and-a-half trip from Daytona to Melbourne. They won't let me do so, because my license expired two weeks ago. Meantime, all planes are grounded. So I pull out all my badges, my ID. 'I gotta go there,' I keep saying. 'I gotta go back.'

The clerk plugs some numbers in the computer and gets me a car. I drive to Melbourne and see my mother and talk to her. The doctor says, 'There are two aneurysms, one's bleeding.' I continue to watch

427

the news, and try to call the firehouses on the cell phone, but couldn't get through. I had to call my old house, Ladder 80 in Staten Island, and the news wasn't good. A bunch of guys were missing. The surgery was completed, the doctors told me my mother was doing well, and I finally got through to my firehouse, and learned which guys were missing. I waited till morning, to make sure my mother continued to be okay. I then got back into the car and drove fifteen straight hours until I got home.

Now, just before the collapse, Ladder 5 and Engine 24 were both on the thirty-seventh floor of the north tower. The engine came up the B stair, the truck had come up the C stair. I think a Mayday was given, the way the engine officer explained it, with orders to evacuate. As they were getting out, the engine started going back towards the B stairs, and the truck went to C. Engine 24 all made it. A couple of guys actually jumped from the second floor into the courtyard plaza of the Trade Center.

Somewhere along the line the guys in the truck said they were working with somebody with a wheelchair, so that's what slowed them down. We found them on Friday, three days later. We knew they were in the C stair, so every day we went to that pile and we tried to find it. We were getting closer each day, following elevator cables. We knew there was a course there, and kept saying, 'We're in the right spot.' We stayed at this spot until we found a man from Squad 18, and a number of bodies around the corner from there. One coat said MCGOVERN, from Battalion 2, and in its pocket was the chief's riding list. Then someone calls over, and says, 'Five,' because he saw my helmet. 'We got a helmet from one of your guys.'

So now we start digging at that spot, and we find Lieutenant Warchola, from our company. I get on my cell phone and call the captain. I was off duty — I think the majority of the guys that were working there were all off duty — so the captain and all the guys who were on duty came down with the rig to get our men. This is how we do it, your company takes you out. And one by one, we take Lieutenant Warchola out, and a probie, Andy Brunn, who was one of fourteen new trainees. He had been a sergeant on the police department when he was appointed, and it was his first day in the ladder company after doing his seven weeks in an engine company.

We found him right next to the officer, and we thought, *What a good man — here he is, the can man* [the most junior man in the company who carries the two-and-a-half-gallon water extinguisher], *right next to the officer, just where he's supposed to be.* Then we pull out John Santore, Tommy Hannafin, and Louie Arena. With Chief McGovern, that made six guys from this firehouse that day. I think the greatest tribute paid to these guys is when they were carried from that eight-story pile. We had to bring them down this eighty-foot valley of bent steel and then up another, maybe, four-story mountain. Then we went down again, four stories, then across the wide expanse of West Street, still deep with steel and debris, through the big entrance of the World Financial Center, and then through to the back of the Financial Center, and out to Vesey Street to where the temporary morgue is. And through that whole distance, our guys carried the chief and the men of Ladder 5 on Stokes baskets!

There was an American flag over each one, and each had his helmet laying on his chest, and they were carried through a double line of firefighters, more than three hundred firefighters, end to end. And as we passed, each firefighter took his helmet off and covered his heart in tribute. There was steel and concrete everywhere, and it was a tough, tough trip, a long ordeal to bring them down. But what a tribute, what a great honor to be a part of it.

It took five hours, and in that period that line stayed. Nobody left. These men died in that fire and that collapse, and they came down with a 5-Truck helmet on their chests. And, everybody in the line was saying, 'Go 5.'

It was the saddest day of my life, and the proudest.

It is also a blessing that we could bring them home to their families, bring closure to them. I didn't realize how important this was until we had that first memorial service, and it turned out to be the first memorial service in fire department history. That service was for Gregg Saucedo, from this company, the one member of the group who they didn't find. It was a strange feeling – no casket, and just a photograph on the altar. None of us has ever been involved in anything like this before, without any closure to it. I kept asking myself, 'Well, are we going to have a funeral in two weeks when we find him? Are we never going to have a funeral?' There's no grave to visit. If you believe in cremation, there's no urn to look at. I'm a big pet lover, two of my pets have been cremated, and to me there's a comfort, when they're there with you. My father just passed in July, and we've gone a number of times to visit the grave. And it gives you a feeling of comfort.

But, you know what? We brought these guys back.

Chief Richard Prunty was from this firehouse, too. He communicated in that collapse all the while before he died. He was found just below where Ladder 6 and Engine 39 were. He said he couldn't feel his legs. On Wednesday night, thirty-six hours after, his cell phone was strangely activated. It wasn't used, just activated. The button was pushed or something, maybe just rubble was being moved and the body shifted. The telephone company called and said the phone had been activated. It was confusing for the family, and strange. I don't know how all this is going to affect this job, once this great attention passes. People in the neighborhood have been phenomenal, people throughout the city and the country are phenomenal – the cards, the letters, the flowers, the donations. The attention takes you away from the pain sometimes. But what happens when that's gone?

When you're sitting in the kitchen by yourself, and you start remembering? Because of the things you've seen, you can never erase those pictures in your head. Over the course of twenty years I've seen death, and you kind of displace it – it's not real. Men in this truck have died before, and that's hard enough to deal with. But, when you've carried five of your friends down, it's – incredible.

I woke up the following morning, and my foot was blown up. I had an infection from climbing on the piles, and a blister, and cellulitis. After the funerals, I went to see my wife, an emergency room nurse, at the hospital. They admitted me, and had to calm me down. There was a doctor there, a resident who was talking to me, who wanted to hear the story of what I'd been through. He was a doctor, and could handle this, so I told him everything I'd seen, even the horrible stuff. I was trying to release some of the pressure I had, and talking to this man could be helpful.

So he left after hearing everything, and I went for an X ray. When I came back, this doctor was now crying, talking to the psychologist in the hospital. A nurse said to him, 'Hey, don't be crying around here. This guy has it a lot worse than you. Get out of here.'

The next day I got home, and that afternoon my wife called. She says, 'Remember the doctor who saw you yesterday? He called up looking for you. Not knowing that I'm your wife, he says, "How's the patient doing, the firefighter?"'

'Oh, he's fine,' my wife said to him.

'Oh, I was thinking of him all night,' he said.

My wife said he kind of lost it. He had become freaked out from the story I told him, because he lost a friend from the tragedy. So because he lost his friend, and because I told him of the whole ordeal, it hit him in a way that hurt him. He didn't show up for work for a couple of days.

I don't know why it happened like this. I told my mother, when she started recovering, 'You saved my life. God works in mysterious ways.' Her speech was bad, but she said, 'I know.'

Through all that surgery she prayed.

But the battles are changing. Dealing with terrorism means it's not a conventional war. We're the first ones on the scene, doing the bulk of any of the work. I have had CPC [chemical protective clothing] training, when I was in 80 Truck. I was put in gear to learn what these chemicals are all about. The more you know, the more scary it is. And your job in a CPC company is triage: You go into an environment in a triple-layer suit and find the bodies that are lying there. You kick them. If there's no noise, and no movement, you go to the next body. They give you a package of antidote to take in with you, and if you feel anything strange, you take the antidote and you just drag yourself away. When you go through that, all you're expecting is terrorism. It is frightening. The city doesn't have the money to equip us with this stuff. They're not ready. Our masks are not good enough. You need the whole suit. Any opening and the killer chemical will go through. All this stuff started after that first terrorist attack in '93 at the World Trade Center. That's when OEM started, that's when CPC started, and you know what? They went to a certain level, and they didn't go far enough. I think they probably will increase the number of CPC companies now. ■

As I am leaving the firehouse I ask Lieutenant O'Neill about the battered pickup. 'That's Craig Monahan's,' he says.

■ Craig drove down to the scene after the second plane hit. When the second building came down, the truck's windshield was blown out, the paint started on fire, and the headlights melted. At the same time, our ladder truck got completely destroyed by the fire. So Craig took the 5 sign off the rig, the one that goes alongside the FDNY, and puts it on the pickup truck.

That's now our truck, and we drive back and forth every day to the scene in it. *Good Morning America* did a little piece on the truck company and showed us responding in the pickup truck. The town of Jackson, New Jersey, saw this, and they want to donate a new rig to us.

We parked it down there the day that President Bush came, and they began to tow it away, until one of the guys far from the piles, a firefighter, sees them. 'Hey, that belongs to Ladder 5! That's their fire truck.' They towed it back to the firehouse. Yesterday, we went down to the New York Stock Exchange in the pickup. Some of the guys from the company rang the opening bell, including Craig Monahan and me. Miss America was there with us, on the podium up there, so we came outside to take a picture. The chairman of the stock exchange, Richard Grasso, had his picture taken with the most beautiful woman in the most beautiful pickup truck. He asks what it would cost to replace it, and Craig said, 'Well, I don't know.' Mr. Grasso said, 'We're going to raise money. We're going to get you a new truck.' I don't think he would say it if he didn't mean it.

That truck has been such a hit. We'd be driving down the street, totally exhausted and beaten up from a day at Ground Zero, and here was our little war truck, coming back from battle. ■

I leave the firehouse and head for the site, thinking about the huge number of ruined fire trucks in the department. When I was talking with Sean O'Malley in 10&10 about the loss of Ladder 10's truck, which was completely wrecked and flattened in the collapse of the south tower, I saw a door, bent and U-shaped, in a dark corner of the firehouse. But I can still see the number on it clearly, the figure 10 in gold. This reminds me so of William Carlos Williams's seeing the figure 5 in gold, and I wonder what the poet would think of this image that has survived the complete ruination of the fire truck.

At the site I go up Vesey Street to Broadway. All along the street the light is casting shadows. I see bodies in these shadows, whole bodies, full and recognizable. Finally, I see they are cast by statues on the building next to me. But I have also begun to see human figures in the definition of cracks on the sidewalk.

I stop to sit for a while in St. Peter's Church, the oldest Catholic church in New York, built in 1785. It was here that Father Mychal Judge was laid upon the altar after he was carried out of the north tower.

There is truly a heavenly respite here from the incessant sound of motors in the streets. I am reminded of Italy by the fundamental simplicity of St. Peter's. A silent fan off to the side of the altar gives testament to the warmth of NYC this evening. Recent days have been either hot or cold, and nothing seems regular or expected.

There seems to be fewer firefighters at the site. The FD has scaled back its presence, though it is still in command.

Walking down Liberty Street I pass a man I have begun to call Lieutenant Mike. Mike Szczecinski works – volunteers – in the supply store, and has been here since Day 1 when he drove up from his retirement home in Maryland or Virginia or somewhere south of the Mason-Dixon Line. He is a former New York fire officer. I ask for a pair of leather gloves, for I left mine somewhere. In his commandeered store he has hundreds of shovels piled in a corner, rows and rows of gloves, flashlights, axes, saws, a kit to fix a flat on a golf cart. There is a sign from a woman named Rhonda: YOU NEED IT – WE'LL FIND IT. Stacked all around are oxyacetylene torches and cutting apparatus, power saws, Sawzalls, partner saw blades, ladders, ropes, picks, buckets, spray paint for IDing search points, safety goggles, masks, respirators, backboards, Stokes baskets, body bags – orange for MOS, black for civilians – garden tools, faucets and plastic connectors, and empty cigar containers.

He is a good and trusting man, and represents for me part of an ideal social order. We do what we can to help, always. Here we are at Ground Zero, with people of a common purpose.

The south cathedral wall is still standing, though it is now a little bent, and a small piece has been cut from its top.

In interviews over the years people have asked me to state my life's ambition, and the answer has really never changed – to get As on my report card, and always have time for my family and friends. But I have one addition now: I would like to remember in detail every minute I have spent these recent days with the greatest people I ever met.

Day 19, September 29

Tom Beattie was seriously injured on the job today at Ground Zero. He had been going up in a basket, at the end of a cable attached to a crane, to cut the windows of the cathedral wall.

'Two cranes,' he says, 'got hung up on each other,

437

and I was in the basket. The boom on the crane we were in – the 500 – started swinging to the right and the ear of my cable got hung up on what we call the rooster tail of the thousand tonner. By the time anyone sees it, it is too late. So it pulls us up like a pendulum, real high, and it suddenly uncatches itself and jumps off the thousand tonner. My basket now falls with full force on the end of this cable for about ninety feet, and crashes us into the wall.

'The guy with me, Jimmy, walks away from it okay, but I really crashed my shoulder, and it is too swollen to get an MRI, maybe it will be swollen for a few months. If I get hit in my head I'm always okay, but my shoulder is in a lot of pain. If things go bad, I guess I'll be looking for another line of work.'

The south cathedral wall is now half gone. The entire top of the wall has been removed, and while the Romanesque style is still evident in the top of the window frames, the Gothic height, the reaching-up-to-heaven quality of the wall, has now been reduced to the flat top of a row of windows.

Building 4 is now half the height it originally was, and a big ball on the end of a crane's cable is smashing against it.

The fire department presence has been cut. 'Guys are reeling,' Lieutenant McCaffery says to me. 'And no one has the slightest idea how it will end.'

Day 20, September 30

At mass at St. Jean's today, I can hardly keep my mind on the liturgy, for I am so focused on the anxiety that comes with being daily at Ground Zero and trying to accommodate the destruction there. I worry that that experience will come to every other idea that should be examined in this crisis – for example, trying to understand the people in the world who hate us so vehemently. I pray here in church that all of our citizens will be vigilant in making certain our rights are not eroded in the name of guarding liberty and progress. We must likewise protect people who do not belong to the majority. The differences of culture and religion among our citizens are substantive, and we in America simply integrate them in our view of a democracy. But by and large, we tend to underestimate how intensely felt those differences are in other parts of our world.

Vinnie Bollon, my friend who is general secretary-treasurer of the International Association of Fire Fighters, called today and voiced concern about the diminished presence of searching firefighters at the site. He told me that Chief John Fanning's wife was furious that some reporters had said that her husband was dead without a body to wake, and that Chief Ray Downey's wife had called Mayor Giuliani, saying, 'My husband is not dead. There is no body, and he is still alive. They will find him.' The mayor must have been stunned by this, coming from the wife of one of the most famous firefighters in history.

Vinnie also told me that the union's 9-11 Disaster Relief Fund had raised $14 million and pledges of $6 million more, though they expect that to grow even larger. They have already sent $10,000 to each fallen firefighter's family. The New York Police & Fire Widows' & Children's Benefit Fund has also sent a check for $10,000 to the families of each firefighter, police officer, and Port Authority officer. At least they will have some money to tide them over until death benefits, insurance, and the distribution of the various funds are resolved.

Getting ready to leave for Ground Zero, I place a call to Charlie McCarthy, a registered nurse, retired FDNY lieutenant, and friend since 1966. The 'big guy' wasn't at home, and so I chatted with his wife, Mary. 'As you know,' Mary said to me, 'our son Jim is a firefighter, and I told him, "As your mother, there are two things I want you to do, Jimmy – rest and pray." Jimmy is frustrated when he's not working now, but he needs his rest. He can't be on the job all the time.' That is just as I remember his father – always ready to serve.

I stop by Engine 33 on Great Jones Street, right off the Bowery, and though the companies are again out, this time I intend to stick around until they return. While waiting I meet a hoisting engineer, Ian Thomson, from Ontario. The firehouse is putting him up, because he came down to volunteer his services as an experienced crane operator. Although the operating engineers union would not allow volunteers at the site, the union was so impressed by his background and commitment for going to work at Ground Zero that the local put him on the payroll.

The firefighters of Engine 33 and Ladder 9 return from the run, and greetings are exchanged throughout the firehouse. Between these two companies, nine firefighters are missing, including Michael Boyle, David Arce, and one of the senior men, Bobby King, who was a master carpenter, and who had assembled a very sophisticated shop in the cellar. Someone takes me down to look at it, and I see a beautiful new desk that Bobby had almost completed, made for the housewatch area. Cut into it are the words *Dem Bums,* a reference to the firehouse moniker, 'Bowery Bums.' This desk will be completed and will be the memorial to the missing men. I see a big wooden locker with the name MICHAEL BOYLE written over its front door. The locker is open, and inside are Michael's clothes on hangers and hooks, and an overnight bag that has printed across its side: A PROUD PROFESSION, A BOLD UNION, A BRIGHTER FUTURE. Beneath these words is the insignia of the Uniformed Firefighters Association, the local union. At the side of Michael's locker are shelves with eighteen slots built specially for union forms, department forms, and medical and benefit information. Michael took his job as union delegate seriously, and he was always available for the men who had special administrative or union benefit problems. Whoever inherits this locker will have quite a legacy.

In the kitchen is posted a letter written by Lieutenant Warren Smith of Ladder 9 to the firefighters of this firehouse:

I don't really know how to begin this letter. But I will try. . . .

First, let me start by saying I love you all. I know you guys know that but before Tuesday most of us probably didn't say it enough. That has changed. Never again will I only shake a person's hand when greeting them. They will get a hug whether they want one or not.

Second, there are very few words to describe, explain or even fathom what other initial responders and I witnessed throughout that fateful day. But I will try. . . .

Ladder 9 responded somewhere between the 2nd and 5th alarm shortly after 0900 hours. While responding there were conflicting reports on the radio, that in addition to tower one, tower 2 was hit. As we entered tower one, dodging debris and bodies as we did, we were still unsure of all the facts. Were these buildings hit by accident? Was debris from one tower causing damage to the other? We simply didn't know. The direction of our response did not give us a good view of either impact. I don't believe any of us thought at that moment, that this was an intentional act. Even at the scene it was incomprehensible that a fellow human being would want to inflict such destruction on civilians of a country at peace. As I am realizing as each day goes by, these cowards are not fellow human beings we are talking about.

As we made our way into the lobby you could

sense you were about to embark on a mission that was unimaginable. As it turned out, my thoughts didn't even come close. After we stood fast for a minute or two, we were ordered to use stairway B and make our way up. No elevators were in operation. We were given no specific orders or instruction at this time. No floor. No specific job. No chief to report to. Nothing. It was then I realized how chaotic and desperate our situation was.

As we worked our way up we did our best to calm the evacuees, asking them to stay to the right and at the same time letting them know they were out of harm's way or would be shortly. Due to the fact that there were thousands of civilians on the same stairs as us, our ascent was very slow. Maybe 3 floors per minute in the early going. That soon became even slower. As we got higher into the building we heard numerous urgents and may-days from firefighters with chest pains, in need of oxygen or worse. I knew these firefighters were understandably trying to run up the stairs in full gear. I told my members to pace themselves. I didn't want to be useless when we arrived at our destination, still not knowing where that was.

On the way up, we met up with an FBI agent, who informed me that these planes were intentionally crashing into the towers and that there were still planes in the air, unaccounted for. We were incredulous. At this point, we stopped at the

31st floor to gather ourselves. A couple of my members were a few floors below and not doing very well. It was here that we met up with Engine 33. There were also other units and a battalion chief at the other end of the floor. We still had electricity and no smoke condition, but were hearing reports of jet fuel on the upper floors. I made my way to the chief for orders. Before I got to the chief to inform him of what the FBI agent had told me, our building was hit. Or so I thought. As I found out later, this was actually tower 2 collapsing. It shook our building like a rag doll. We all dove into the nearest stairwell. Once it stopped shaking I found myself next to the chief. I asked him 'What the hell are we doing now, Chief?!'

He said, 'I don't know, but I'll find out.' So I stayed with him, waiting for an order of what to do next. Probably in the next 20–30 seconds he received the order to evacuate. I passed this order on. We kept our masks and some tools, told the engine to drop their roll ups, and began an orderly descent. We picked up firefighters, unaware of the order to evacuate, on the way down. As we got lower, we lost electricity and things started to slow down, eventually bringing us to a halt at approximately the 11th floor. I found myself on a landing, by the door to this floor. At this point there was no panic, just a determination to get out and regroup. It was then a firefighter tapped me on the shoulder and said, 'Follow me, the stairs over here are clear.'

I turned to my members, Engine 33, and other companies within shouting distance and yelled to them to follow me because the other stairs were open. Many firefighters did. As dark as it was, it was impossible to tell just how many. As I found out later, it appears that Engine 33 and the members below me in the stairwell did hear my message, passed it on to Engine 7 and switched over on the 9th floor. At this point in the operation many of us thought we were being hit by numerous planes and still had no knowledge of tower 2's total collapse. I was thinking we'll all get down, regardless of which staircase we take, and regroup outside the building. As we got closer to the lobby, our stairs also became bogged down with firefighters. Again, there was no panic and we eventually made it down to the lobby. As we stepped into the lobby, I was trying to make sense of all the damage. Something wasn't right. I now felt we had to get away from this building. This was reinforced by a battalion chief directly outside Tower 1. He was frantically waving everyone north of the towers. It was here that I found Bert Springstead in the wave of firefighters evacuating the building. We wanted to wait for everyone from our company but we had no way of knowing if they were in front or behind us. The chief screamed at us to get away from the building and head north. At this point, he was still the only one of us who knew that Tower 2 had totally collapsed. We tried to contact Engine

33 and Ladder 9 members via radio, but we had no luck. As we headed up the West Side Highway, we saw Mike Wernick and Mike Maguire sitting exhausted and dazed on the highway median. Even though we were probably 2–3 blocks from Tower 1 already, something told me we weren't far enough away. I told them to keep moving (all the while still trying to contact everyone else). Thank God they did. In the next 20–30 seconds we heard a roar. We began to run. As we ran, I turned to see what it was. I saw a mountain of dust, rubble and debris chasing after us. At this point you had numerous choices. Dive into a car or building, stop and put your mask on or just run for your life. Once I realized it would catch us, no matter what we did, I figured my best choice was to keep my mask, pick a path and run. Once it caught me, it was utter darkness for some time. I eventually reached light. Almost immediately, I again hooked up with Bert. We tried reaching our other members. We contacted Mike Maguire, Mike Wernick and Don Casey either visually or via radio. Something I didn't realize at the time, but which struck me much later, the radios were eerily absent of chatter. Mike Wernick and Don Casey required medical assistance. After leaving Mike Maguire with them, Bert, myself and a probie from E-214 (who had lost his members) continued north looking for a command post or staging area to report to. As we discovered much later, there was no one readily

available to man one. We kept trying to contact Engine 33, or Ladder 9, or anyone, to no avail. We were covered head to toe in dust and debris. People were giving us water and cleaning us off. We were even taken into a small warehouse where they helped to clean out our eyes. They also allowed us to call our families who we knew were witnessing this on TV. Once we allayed their fears we decided to head for a firehouse.

Even at this point, we believed that it was just a matter of time before we hooked up with all our members.

We made our way to Ladder 8 for information. I wanted to get any information I could that would enable us to find the rest of our men. It was here I discovered how complete the devastation was. We heard there was a staging area at Ladder 20, so we headed there to await any word. I sent Bert up to Ladder 9's quarters to see if anyone had made their way there.

Eventually I went back to Ladder 9 as well, to await word on our brothers.

I will close on a more personal note. . . .

For those of us who survived there are no words to describe the feelings and thoughts we had once we realized the enormity of this unprecedented attack on our society. And will always have. None.

Every day I pray for a miracle.

Every day I think about what I can do to help

prevent this from ever happening to our job and our city again.

Every day I think about what I can do to help us all get through this in any way I can. I plan to love everyone, hug everyone I love any chance I get, and try to do something each day to make a difference in someone's life.

I love you all,

After reading this letter, I decide to visit Ladder 3 on 13th Street, where Paddy Brown was captain. Battalion 6 is also headquartered in this firehouse, whose walls are hung with old photos of the company working at some of the historic fires in New York, and of testimonial dinners held fifty and eighty years ago. Here is the Equitable fire, and one on lower 5th Avenue, and here is a portrait of eighteen men sitting around a long table like the figures in da Vinci's *Last Supper*. Ten minutes before the buildings collapsed, Paddy Brown got on the department radio to say there were many seriously burned victims on the thirty-sixth floor. Then he called 'Mayday, Mayday.' Chief Hayden tried to raise him in return: 'Paddy Brown, Paddy Brown!' But no one ever heard from him again. It was reported later that somebody warned him: 'We gotta get out, Paddy.' Paddy only replied, 'We gotta get these people out.'

No sooner do I walk into the kitchen than the firefighters invite me to dinner. Brendan Gillen of Ladder 3 will cook. Tall and bald, he is the father of four, with the kind of New York manner you could write a book about. He points to a newspaper article about the steel that is being taken illegally from Ground Zero. 'The fact,' he says, 'that this dust and steel is being taken by the Mafia, somehow, it makes it so New York, you know, and it's all right.' 'But,' Brendan continues, 'what are we gonna eat? I can heat up some stuff from last night. Some neighbor brought in a meat loaf.'

Someone says, 'Let's order in Chinese. Make it easy on yourself.'

'No,' Brendan insists. 'I wanna eat like an Irish American. I'll make some potato soup. I need ten pounds of potatoes and a bag of onions. I got some cheese here to spice it.'

Brendan looks around the kitchen and sighs. The tables are filled with pies, cakes, cookies, candy, meat loaf, a canned ham, rolls, and breads of all kinds – enough food to supply a birthday party at Camp Lejeune.

'I just wish people would stop bringing this stuff in,' Brendan says. 'We can get back to normal.'

I meet Sal Pinciatta, Ray Downey's nephew, and we begin talking about Paddy Brown.

'He was a modest man,' Sal says. 'He could do anything, and he was okay with himself, you know. When I first met him, I thought he was gonna be ten feet tall, and here he is, five feet eight inches. I worked with him for three years. At jobs people would go haywire, and he was calm, always calm. And then, he always had a "special girl." Every time we went out the door he could meet a "special girl." He was so cool. So contained.'

At the front of the firehouse Paddy Brown's photo is

at the top of the twelve photos of its missing firefighters. His face is black with smudge from some fire, which is an appropriate way to remember him, for he was always in the middle of a fire.

'I know he's happy up there,' Sal says. 'And, I know he's in charge. He doesn't need orders from anyone.'

At Ground Zero, the south cathedral has still not been dismantled, but now it's leaning toward the south, and the Statue of Liberty.

Shortly after I arrive I meet Eddie Tietjen, of Ladder 48, and almost don't recognize him. When I was studying for a master's degree at NYU in 1971, the department allowed me to work special hours for one semester at Ladder 48, which is in the Hunts Point section of the South Bronx. Eddie and I spent a springtime and summer together there, in the Seneca Avenue firehouse. 'Look at me, Dennis,' he says, 'still carrying the irons after thirty-three years.' To me, it is a great testimony about the congeniality of firehouse living that a man can stay in the same job for thirty-three years.

A cold rain begins to fall, and Lieutenant Mike, the supply guy, hands me a yellow slicker. I don't want to wear it, because it will look like I've joined an army of construction workers. But now it begins to rain hard, and the mud is splashing up to my knees. I'm at the base of the south cathedral wall on Liberty Street, waiting to get into a basket. They have dug out about 150 feet from the base of the Gothic wall, which is where the huge pyramid is standing, just to the east. We're standing on a precipice that is ground level with the street of Liberty

Street, and looking down into a hole – a crevasse in the Earth – that goes down 50 or 60 feet, filled with twisted steel and debris like the rest of the landscape. A small group of firefighters is lowered in the basket down into the crevasse, where they will face incredible and dangerous conditions. A huge piece of steel juts out precariously above them, and the rest of the steel is crisscrossed throughout like a pile of pickup sticks. The rain is making everything slippery and even more difficult to handle. I worry that the rain will loosen the dirt, which will in turn undermine the already precarious stability of the steel. The potential for serious injury is everywhere, yet the men go on with their work, because they've found a body, which they've identified as a Brooklyn firefighter.

All the firefighters remove their helmets and salute their fallen colleague. Everyone is waiting patiently for a decision to be made about how the transportation of the body will occur: Will it be taken on a golf cart, or by ambulance? It is very cold and the wind is merciless. I see that a firefighter from Ladder 111 is taking his place by the Stokes basket, and also members of Rescue 2, so I assume they'll take the fallen firefighter away.

A golf cart appears, and Battalion Chief John Norman of Battalion 4, an expert in rescue and collapse, steps onto the cart as two firefighters place the draped basket securely behind him. The cart begins to move slowly away, as it fights its way through the rain to the morgue. We take our helmets off again, about sixty of us, and salute until the cart travels a good distance away, a respectful distance. Then an officer says 'To,' and our hands drop.

The enactment of this extraordinary ritual in such terrible, howling weather has surely been embedded in the minds of every firefighter here, and it is precisely such historical memory that will enable men to face danger so bravely in the future.

Day 21, October 1

Frank McCourt calls first thing. He has just returned from a trip, and wants to know how things are going, what's changed, and the general view of the verities. I tell him how much I have been thinking about our need to better understand Muslims and the Islamic religion. He points out the fact that 42 percent of all Arab Americans are Catholic, while just 23 percent are Muslim. We then talk about the numbers of Muslims in the world, and he compares them to Christians. 'Remember,' he says, 'how the Romans tried everything to stop the Christians? The lions had indigestion with the amount of Christians they ate, yet the Christians kept coming. And don't forget our first saint, who preferred being stoned than renouncing Christianity.'

There is certainly a valuable lesson in this whimsy, for there are many historical antecedents of large numbers of people willing to give up their lives for their faith.

I am greatly encouraged by the number of remains that have been recovered today. Thirty victims were removed, of whom sixteen are believed to be firefighters. DNA analysis will have to be run before identifications can be made with certainty. Whenever we receive such news I say my silent prayers for Jimmy Boyle, Lee Ielpi, and Ed Schoales, hoping they will find their sons. Yet every recovered body will bring solace to some family. It is not closure, for there is so much practical and emotional work that has to be done after a death in any family. But now, at least, a family can say, Yes, that's right, this is him, and this is where he died, and he has been returned to us.

On the way home today I stop in to visit with the members of Ladder 16 and Engine 39 on 67th Street and 3rd Avenue. It is a historic building, having once served as the department's headquarters and then its training school. I climb up the wrought iron stairs to talk with Lieutenant Jim McGlynn. Lieutenant McGlynn had left the north tower when he realized one of his men was missing, and he and two firefighters reentered it to find him. Less than a minute later, the building came down, and Lieutenant McGlynn and the two firefighters from Engine 39 were trapped just below the men of Ladder 6, Port Authority Police Officer Dave Lim, and Chief Picciotto.

I have been looking forward to meeting this quiet, studious man, for not only did he live through the most nightmarish experience that anyone could imagine, having 110 stories falling in on him, but he was then in contact with Chief Prunty, who lay dying twenty feet below. His ability to endure all that, and then to make certain his men were all accounted for before he was willing to work his way out of the collapse, is an inspiration to me, and I believe will inspire firefighters for years to come. I don't know why the department does not have the ability to promote on the battlefield in the way the military can, but clearly there are many men who acted in such commendable ways on 9/11 that a way should be found to circumvent the civil service laws to bring them up another rank. The lieutenant and I are speaking when we are suddenly interrupted by an alarm, and Jim McGlynn goes rushing out the door. I sit quietly for just a moment before I head back down to the kitchen, and notice twelve clipboards hanging on a wall, each of which has a title page: Hydrant Inspections, Annual Evaluations, Wake and Funeral Arrangements, Proby Development Program, Standpipe Sprinkler Log, Officer Mutuals. It stuns me to see that 'Wake and Funeral Arrangements' have become one of the normal, day-to-day duties that a fire officer has become responsible for since September 11. I am in the kitchen when the firefighters return, and a few of them ask if I would have a meal with them. It is a welcome invitation, since firehouse food is as good as you can get anywhere in the world, and so, on this chilly fall New York night, I sit down to a plate of hot lasagna.

While getting ready for bed, I watch a television commentary on Mayor Giuliani's new proposal to have a special extension of his term for three months into

2002. While he has some justification in believing he has the best and most experienced team to get the city through the challenge of making itself whole again, it could also be argued that he is forcing the voters to recognize the dimity experience of the opposition candidate. Whatever the case, any incoming mayor who would agree to let his predecessor retain power for even a short amount of time should take a course in self-confidence.

Day 22, October 2

At Captain Terry Hatton's wake at the Frank E. Campbell Funeral Home, I run into a crowd of the mayor's oldest friends, all of whom I know because I was active with the mayor when he first ran for public office. Because Terry Hatton was married to the mayor's long-time assistant, Beth Patrone, all the city's important powers had come to know him well. My wife said to me tonight, 'I remember when we first met them together at Gracie Mansion, he was so, so handsome and she was so, so proud of him.'

Beth herself is an extraordinary woman. She has worked in Rudy's emergency office every day since Terry was reported missing in the attack. Focusing on the totality of the disaster and the multitudinous demands made on the mayor's office, I know, is a way for her to focus her energies, and to put off facing the singularity of the great loss with which she will have to learn to live. I have admired Beth from the first day I met her fifteen years ago, and today she inspires me, for it is evident that she will get through this in a positive way. She won't pine, at least not for a while. I know that Terry would realize how much there was to do in the mayor's office, and he would urge her, 'Just grab the irons and go to the roof, Beth, and see what you can do there.' Now is not the time to sit around. Not in New York.

Beth has a smile in her greeting, and she introduces me to those around her. It is never easy to find consequential words at a wake. Everyone says much the same thing, and like everyone I offer my sympathy. I know it is in the sincerity of the words that comfort is found, and not in the words themselves, but still, I wish there was something more meaningful I could say to Beth.

I hear the voice of Terry's father, Kenny, a retired deputy chief, with whom I worked when he was a firefighter in Ladder 19, and I offer him my condolences. He tells me, 'Dennis, you know the boys from 19 Truck came down today, and I saw them all.'

'I've been seeing a lot of them, Chief, and Engine 50, too, down at the site.'

'Yeah, well,' he says, 'there's still a lot of work to be done.'

The words Kenny Hatton left with the members of Rescue 1 are becoming famous: 'Don't let them throw you outta there. Take our boys home.'

And here, at least, is the body of his own son, a

captain, but even apart from his rank Terry was one of the most respected firefighters on the job. Now I try to imagine what it must feel like for Kenny to see his son lying in a coffin, to wait for a funeral service. I can only think of my own sons, and how I would pine if ever I lost one of them. Resting on top of the flag-covered casket are Terry's helmet and a small Halligan tool, the two things Terry brought into every burning building. I say a prayer and touch the casket. I did know him, but I know a lot about him, which is often enough to miss a person deeply.

Day 23, October 3

Signs of restoration are all over Ground Zero. Streets have been washed clear on all the surrounding blocks, and even the dirt roads through the site seem packed and tidy. Sixteen firefighters were found the day before yesterday, fifteen yesterday, and three civilians and one firefighter have been found today. The pyramid is down by about thirty feet, and a crane is now pulling steel out of it. Four WTC is half gone, but #5 is still a burned-out hulk with its threatening waterfall.

What kind of monument will they put up here? My idea of a suitable memorial is a simple, green grass mound with a bronze ring surrounding it, with every one of the victims' names cut into the bronze, so that it will last for centuries, like the green grass burial mounds of the American Indians in Ohio, or the mound graves in Ireland. Two construction workers tell me they've been cutting pieces out of the wall and storing them down the street, so the cathedral wall, this symbol of containment and control, already figures into whatever design will be chosen.

Day 24, October 4

At St. Patrick's Cathedral this morning, firefighters from all over the country are lined up on Fifth Avenue to bid farewell to Terry Hatton. I put my gold shield on and take my place on the line in the dignitary section behind the director of the Federal Emergency Management Administration, Joe Allbaugh. The governor, mayor, and fire commissioner are just off to Allbaugh's left. Behind us two ladder trucks have raised aerial ladders holding a large American flag, thirty feet long. The Emerald Society Pipes & Drums are here in almost full force, more than fifty of them, and they can be heard playing 'Going Home,' in a slow march. And then, there is just the slow, agonizing sound of muffled drums for several blocks until the band passes. They are followed by a spare rescue truck (the actual Rescue 1 rig was lost in the collapse), and then a fire engine carrying the body of Captain Hatton. The limousines behind the fire engine stop, and the Hatton family lines up with Beth and her family, and they wait for the firefighters of Rescue 1 to take the casket from the top of the fire engine. On the steps of the cathedral, a lone bagpiper plays 'Amazing Grace,' and Monsignor Alan Placa waits to bless Terry Hatton before he is carried into the church, followed by thousands of men and women filing past the great bronze doors of the cathedral.

In a blessing, a family friend, Monsignor Czajkowski, becomes confused over the governor's name, and there is an awkward moment as he searches simultaneously through his notes and his memory. But it also provides a bit of humor as Kenny Hatton wryly observes that it would take a missionary priest named Czajkowski to hang on the name Pataki. Another laugh comes when Lieutenant Penna reminds us that Terry loved to ask people if they knew how to get to heaven. The answer was: on a tall ladder. He also recalls that with Terry, something was 'either outstanding or wholly unacceptable.' Mayor Giuliani refers to Terry's good looks, saying, 'He was so handsome he could have been an actor, but if he was an actor he would be playing men like Terry Hatton.'

It is mentioned at the end of the service that Beth has recently learned she is going to have a baby, and the whole of the church echoes with applause.

It has been three beautiful hours of homage to Terry, to his ability and courage, and I wish that every fire-fighter could be sent off like this, with a tall ladder. At the end of the service every person in the cathedral begins to sing 'The Battle Hymn of the Republic,' and though I try to join in, I am stunned to silence, for I am paralyzed by the tears I feel dripping on my unyielding heart. I want my heart to open like a floodgate, but instead I feel that I am being made old as the shock of all this is permeating my extremities.

In the afternoon I meet Dan Potter and his wife, Jean, for lunch. They are such a sweet couple, and their love reinforces for me the tragedy of other couples who have been torn apart with the attack. I am certain there are hundreds if not thousands of poignant love stories that were buried in the collapse of the buildings. But now, I reprimand myself: I should and must not see everything in my life with respect to 9/11. Things can be taken as they are at the moment, and not in relation to the past. Now I want simply to have the pleasure of getting to know a couple like the Potters better.

A memorial service is also being held today for the seventy-four employees of the Port Authority of New York & New Jersey, and the thirty-seven police officers among them, who lost their lives at the World Trade Center. More than five thousand people gathered in Madison Square Garden, and even in so large a crowd there was utter silence as the photographs of all the men and women were flashed across the panoramic screens. And then, as the image of one man appeared and hung on the screen for a few moments, a small, thin voice was heard from among the throng, saying 'Daddy.'

At the site today there was a most unusual event, perhaps one of those things that 'can be kissed up to God,' as we used to say. In the middle of the rubble, a construction worker found a beam that had toppled and split, and then landed straight up in the shape of a perfectly symmetrical cross. It seems as if it is the only artifact that exists on the site, for all else has vanished – not a computer, or copy machine, or filing drawer has been found.

The loss to the city has been estimated at $105 billion, which includes $35 billion in property damage, and $60 billion in lost revenue. It has even been estimated that the city will lose $82 million in parking ticket revenue alone.

Paul Crotty is being honored tonight at a dinner for the Flax Trust, an organization in Northern Ireland that funds incubation businesses if they agree to hire equally from the Catholic and Protestant populations, the idea being that economic self-interest will bring people of the two faiths together in working partnerships.

During the dinner, Paul Crotty tells me that the Irish foreign minister, Brian Cowen, came to America with a mission from a woman he knows whose son is missing. He was told the boy was wearing a claddagh ring (a traditional Irish ring). A cop at the scene tells him, 'Minister, we have found more than two hundred claddagh rings so far.'

Day 25, October 5

A small, two-column story in today's *New York Times* reports that a man in Florida has come down with pulmonary anthrax. This leads me naturally to the thought that of all the weapons of mass destruction available to terrorists, the chemical and the biological are the most readily accessible. I am reminded of Chief Ray Downey's testimony in Congress in 1998 warning how unprepared we are as a nation for such an attack, and I wonder if anyone took his words to heart.

Today is the funeral of Thomas Cullen III at the church of St. Ignatius Loyola. I love this big, cavernous church for its music, but also because it has an extraordinary collection of old vestments, some of which date back to the seventeenth century.

Tom Cullen worked in Squad 41 in the South Bronx, a firehouse to which I was detailed many times. Tom was such a young man, only 31 – the same age I was when working in the South Bronx. Tom is the father of a 2-year-old son, and his wife has said that he was studying to be a lieutenant, and that he always accomplished what he set out to do. He comes from a family of lawyers, and he also took the LSAT exam in his last year of college. He aced the exam, but decided he wanted to work the fire department more.

My second stop today is the St. Frances de Chantal Church in the Bronx, and the funeral of Lieutenant Ray Murphy of Ladder 16. I call Jimmy Boyle from the car, who tells me that his family has just decided to have a memorial service for their son, Michael, at St. Patrick's Cathedral on November 5, and they will ask the Arce family to join with them. I smile as Jimmy speaks, because I know that Michael and David Arce had been best friends since grammar school, were next-door neighbors, and went on to the fire department together and to Engine 33 together. It is only right that they will be memorialized together. There has never been a time since 9/11 that I don't have a tear in my eye after speaking to Jim Boyle. It is the collective memory that so moves me – Jimmy has done so much for firefighting over the years, and it pains me deeply to see so good a man suffer through this.

I meet Chief Tom Kennedy standing at the end of a line of men preparing to receive the funeral cortege. 'I don't go into the church anymore,' he says to me, 'for I am tired of seeing these young girls standing on the steps with two or three children at their knees. I stand on the end, just to be here, that's all.'

Tom Kennedy had called last week to see if I could help with TV coverage of fire legislation. He is very active in a movement to enable fire chiefs to be more comprehensively trained, particularly in terrorism mitigation. Congress is talking $600 million when we really need $5 billion to bring the fire service to ready preparedness. I tell him that I contacted ABC for him, but have heard nothing yet. Tom spends so much of his time on this issue that I wish I could just snap my fingers and interest people in the media in it, however unsexy a subject it is for the nightly news.

I take my place at the side here, for I arrived a bit late, and the cortege has just moved to the front of the church. Eight priests have come to meet the casket and the family. A firefighter plays 'Amazing Grace' on the bagpipes, and once again it brings me back to the days I played the pipes with the band. We played on *The Ed Sullivan Show* one year, which up until then may have been the most exciting thing I had ever done – if you don't count the fires.

There are no seats free in the church, and I line up with scores of uniformed firefighters along the side. The mass is being said by Monsignor Jack O'Keefe, the principal of Stepinac High School in Westchester County, and my next-door neighbor growing up on 56th Street. His father was a firefighter, and was killed in the line of duty in 1966. Lieutenant Ray Murphy's parents also lived in our neighborhood, and were good friends of the O'Keefes, and as I stand leaning against a wall I begin to remember Ray from his days as a firefighter in Ladder 42, and this funeral begins to take on a very personal resonance for me. I cannot get Ann O'Keefe's face out of my mind, one of the kindest women ever on Earth's surface, and when her son in his homily begins to talk about his mother's dealing with the line-of-duty death of his father I begin to weep. The Murphys and the O'Keefes are me, really, for I am from the same culture, religion, and neighborhood, and now, here before me is Ray Murphy, lost at 46, with a wife and two children left behind. My memories turn to my own mother, who was such a good friend of Ann and Jack O'Keefe, and how hard it had been for her being left with no man in the household and two small babies, me and my brother. And how hard it had been for Ann O'Keefe and her five little ones when Jack was taken from them, and now for

Ray Murphy's Linda and her two sons, Sean and Ray, Jr., just entering adolescence.

For me, this day will never be forgotten, for it is a deeply moving coming together, of the family of the fire department, of the family of neighborhood, of the family of religion, and of the family of the Irish immigrant, all of us bound in this profound and unrelenting grief.

My shoulders are shaking, but I keep my head high and I face straight ahead, for I do not want someone to be concerned and offer me a seat. I can get through my own pain, and I know that it is an easy pain compared to what Linda Murphy will now carry with her every day. Finally, though, someone says something to make me laugh, for all firefighters, whatever their ethnic background, want to behave like the Irish at a wake – fatalists making the best of it in a bad time. Ray Murphy's cousin, Firefighter Jim Carney, is on the altar telling the story of how Ray hated listening to his endless jokes. 'He would say,' the cousin related, "Uh-oh, here comes Jimmy, and I have to get away from his jokes." And, so, last night at the wake, I figured I had a captive audience – so I told him one more joke.'

Day 26, October 6

Twenty-six funerals and memorial services are being held for members of the New York Fire Department today. I have chosen to attend that of Pete Vega of Ladder

118, which will be held at the Church of Our Lady of Peace, in a small, mostly Hispanic community in Brooklyn. I did not know Pete, but I wanted to come to this service to pay my respects to a brother firefighter, on a day when every firefighter must make an effort to find himself inside of a church. My mother grew up just a few blocks from here, and as I walk up Carroll Street I realize that this is one of the few neighborhoods left where the church steeple is still the highest structure in sight. It is a neighborhood of row houses and three-story tenements, and I am certain it has changed little from when my mother was a little girl on these streets during the First World War.

Normally thousands of firefighters gather when one of their own is laid to rest, but today I count just fifty-five firefighters and fourteen officers lined up in the street, and a crowd of parishioners. The Knights of Columbus #126 Bagpipe Band has volunteered to play, and they escort the fire truck as the single bell in the steeple tolls slowly. The vehicle is on loan from the Freeport volunteer fire company in Long Island, and on it are a few wreaths of flowers and a large color photo of Peter Anthony Vega.

On the morning of September 11, he left a phone message for his wife, Regan: 'Babe, it's a big job. Gotta go. I love you.' There were six men on Ladder 118 when the truck sped over the Brooklyn Bridge to the World Trade Center, and none of them returned.

The lone bagpiper plays 'Amazing Grace' and then rushes away for another service in Queens. There are only twenty rows of pews in the church, but they are brightly lit, and flowers cover the altar. The family enters with Regan, followed by a little girl, the firefighters, and neighbors, filling Our Lady of Peace as if it were Easter Sunday.

A redheaded woman sings 'Ave Maria' so beautifully that I think I would buy it were she to record it. The hymn ends with the patter of a steady rain on the metal roof. The mass is said in English by a priest with a Hispanic accent, and he introduces the speakers. The mayor's representative is Commissioner of Trade Waste Carey, and he delivers an intelligent and rousing celebration of firefighters, and how very much they are asked to do by their employers, the taxpayers.

I look over at Regan and her daughter, Ruby, and I see another young woman with a little one at her knees, the very image that so moved Tom Kennedy. Pete's mother is there, in the front pew, with her husband, who is in a wheelchair. She has a glowing, rosy face, and she looks, as my mother would say, 'like the map of Ireland.' She is married to a man named Rosenberg now, and her son, David Sean Rosenberg, is asked to say something about his half brother. David is a slight young man, a computer expert, and he is a little nervous, but he finds the rhythm for his eulogy:

■ You know, for some reason, everybody who was at Peter and Regan's wedding seems to remember the speech I gave as the best man. Even to this day everybody still remembers it. Well now I'm faced

467

with the task of not only having to top myself, but give a speech I never thought I'd have to give at this point in my life. Deciding where to start talking about Petey is not easy. I could start with how brave he was right until the end. I could start with what a loving person he was. I could start with what a great listener he was and was even greater at dispensing advice. I could start with any of those things, but instead I think I'll start with his big head.

Petey . . . had an amazingly large head. From several miles away, you could see it approaching like the sun rising. His head reached legendary status, like that of Paul Bunyan or Johnny Appleseed in American culture. It was said that his hair could brush itself in the morning, and a single head butt could knock down a wall. And woe unto the poor fool who would be on the receiving end of that head butt!

Many of us here have memories or stories about Petey. I will always remember him as big brother. The big brother who would torment me with tickling, knuckle cracking, and bear hugs, but if any of his friends would start with me he would come right to my aid. The big brother who would let me stay up late when Mom was out. The big brother who stayed with me overnight in the hospital when I had pneumonia. The brother who first showed me comic books, watched the Three Stooges and Abbott and Costello movies with me every Sunday, and tried to teach me to ride a bike. The brother who let me move in with him when I was still in high school, who let me go drinking when I was just barely of age. He took me to Puerto Rico for two weeks as a college graduation gift, and trusted me completely with the safety of his daughter. Essentially, he was *the* best, a fair person, without prejudice or bigotry. He believed in giving a person a fair chance and had no qualms with someone's background. He was always helping somebody, he was always giving someone a hand when it was needed.

The only people he had no tolerance for were the stupid and thoughtless. Pete was a survivor. He went through a lot in his life growing up, he got himself into a lot of trouble, had many personal problems as we all do. He also had dyslexia. And he overcame them all. In the air force, he got to travel all over the world, he made lifelong friends, and got some well-needed discipline and focus. Later, when he came home for good, he got himself a job as an ironworker for several years, planned to take the fireman's test, and he met his true love: the Nu Nile Pomade hair care product that he would comb into his hair to keep it looking good.

And then he met Regan. These two lunatics were retarded for each other and seemed to keep each other in check. I always said that their marriage gave me hope about love, and their time together was too brief. Petey was a thinker. He had a plan that by the time he was 35, he wanted to be married to a woman he loved, have a job he liked and was proud of, get his college diploma, have a home, and start a family.

And he did them all.

He did it and he did it well and that's why I'll always look up to him and admire him. During one of the several low points in my life, Pete told me something that I'll never forget. He said: 'Life isn't about being great, it's about being true.' And he was. All the things he went through, all the steps he took to better himself are something that we should all aspire to. Petey was funny. His laughter was contagious. His sense of humor was very much a product of our family, and Abbott and Costello, *The Honeymooners,* and Jerry Lewis. He loved it when his antics would eventually cause him to get that 'you idiot' look from others, and I loved it when he and I would team up and flank our mom or Regan with our weirdness. It was the best feeling for me. The two of us making our mom laugh hysterically was just magic. We made a great team.

Petey was brave. He had no fear of death. He took a lot of chances in his life, and pulled through. He had three careers that are considered some of the most dangerous. Serviceman, ironworker, fireman. Personally, I think he was just crazy. Part Irish, part Puerto Rican, no way he wasn't a little nuts.

He knew that when he went out on the 11th, that it was going to be bad. But he, his crew, the 342 other firemen who went in that day . . . they did what they had to do. Run in, get people out, save lives.

Petey was taken too soon. He had so much more to do. He had to watch Ruby grow up into a woman, he had to grow old with Regan, he had to be there when I eventually get married and have my own kids, we had to become cranky old men together, and he still had two more *Star Wars* movies to see.

It's not fair, and it's not right. I have a lot of anger about that. But putting aside that anger for those of us that are left behind, I think I speak for Petey when I say that he would want us all to get along, take care of each other, and love each other as much as we are able. To let go of any stupid, petty misgivings or problems we might have with one another, and just be happy that we're all alive and that we all know each other. Petey was a son, husband, daddy, best friend, hero, protector. My big brother. Petey was love. He loved his family no matter what. He loved his friends despite how aggravated he could get with them. He loved his wife immensely. And he loved his daughter more than anything else in the world. Petey, we all love you right back, we will always miss you, and I don't know what I'm going to do without you. But I'll try to make you proud, I'll tell Ruby everything about you – almost everything. And be there for Mom and for Regan.

And now when you think about Petey, picture him in the great beyond, looking down on us, smiling, and wearing the biggest freaking halo over that massive cranium of his. ∎

The rain stops at the exact moment Pete's family leaves the church, and the sun breaks over Carroll Street.

I am thinking that the origins of literature lie in the elegiac. In prehistory, someone died and something good was found to say about the deceased. At the next funeral, they found something good to say, not only about the deceased, but about his exploits as well. And then, the next time the exploits of the family would be orated, and then the tribe, and the tribe's history. Until, over a period of thousands of years, the society had a laced and woven tale of ancestors that leaped from the eulogy to the epic, and gave birth to the *Odyssey* or *Beowulf* – the stuff of mythology. But, here, we simply had David Rosenberg missing his big brother, and I will always remember that.

I stop at the firehouse of Engine 224 on Hicks Street to have a cup of tea and to change my clothes for Ground Zero. I meet a young woman there who was recently appointed to the job. She is one of the seven women who passed the difficult physical test out of the four thousand men and women who took it. She looks strong enough to keep up with the heavy pulling work of an engine company, and I wish her luck.

The corner of Liberty and Church streets is now leveled, and the pyramid has been reduced to almost half the size it had been, though it is still about fifty feet high. A permanent command office for the fire department has been built in front of the Green Tarp Café, and it looks like a summer cabin built into Ground Zero. Orange netting has been hung all around and covers a few of the buildings on Liberty Street, while a sheet of black netting runs halfway down the Bankers Trust Building.

The piles today seem crowded with construction workers, while firefighters are mostly huddled in groups off at the periphery. In a two-block area there are only a dozen or so firefighters, but I do see a few friends working just up West Street. From there, I climb to the south cathedral wall where the Cats are pushing with their long arms against the side of this structure, which is all that remains standing of the south tower. Lee Ielpi of Rescue 2 stands watching as one of the machines pushes until its wide steel tracks skid in place over the dirt, almost lifting it off the ground. Still, the structure does not budge. Lee walks away, unable to just stand around leaning on a shovel while his son Jonathan is missing. He goes toward the pyramid to see where he will next fall to his knees and clear a crevice with the little shovel he carries. Like John Vigiano, he is here every day, along with other fathers searching for sons. Captain Bill Butler digs each day with a small pick and shovel, hoping to find his son Thomas of Squad 1, and Lieutenant Dennis O'Berg searches for his firefighter son Dennis Junior.

Lee has two friends with him, both well-known firefighters. The first is Pete Bondy, also of Rescue 2, who jumps from beam to beam over the sharp and angled steel rods with a feline grace. He is a quiet, highly observant man who is a registered nurse, and who spent some years with the Peace Corps in South America. He occasionally follows Lee, but mostly is off by himself searching every shadow. Both men keep a sharp eye out for each other, and as soon as one finds something interesting, the other is there almost immediately.

The other man is Battalion Chief Marty McTigue, who ten years ago, as captain of Rescue 4, responded to a report of a burst boiler and steam explosion in the Con Edison plant on 40th Street. 'We didn't get good information from Con Edison,' he recalls, 'and I followed the workers up and down, in and out, and then they point up. There was a tremendous whistling, so loud it actually hurt to hear it. It is like in a ship's engine room, grated floors, and I have to get on the other side of a wall to get to the problem. I climb over the wall, and conditions get worse quickly. The steam rushes, and the temperature is up over a thousand. A fireman with me runs past, and jumps down the stairs. I jump, too, but my mask gets hung up on a banister, and I am dangling there. Three firemen run in for ten seconds at a time to try to dislodge me. I passed out and my face piece came off, searing my throat and lungs. My ears were burned off except for bits and pieces, and my nose. I was in the burn center for ninety days, through eighteen major operations. My eyes wouldn't close, they rebuilt my nose, parts of my ears, and they only got my neck to move with tissue balloons.

'Why be in this job unless you're going to give your most?

'Father Judge came in to see me three times a week. He would tell jokes for forty-five minutes, and then pray for fifteen. Lee came in and sat with me just about every day for ninety days. He was my proby. I was sitting at the housewatch desk in Engine 227 when Lee first walked in. I welcomed him, and we've been friends since.' I follow Lee, Pete, and Chief McTigue through the site, searching and spotting. There is nothing to dig, at least not now. As the big Cats get closer and closer to the center, they turn things over, and provide us new opportunities.

At home, Lieutenant Joe Daly of Ladder 42 called. He retired only recently after a bypass operation. He is a 6th Division firefighter as well, and we've been tight for many years. He tells me he went to the site, because he wanted to pray. 'I dug at a mound,' he explains, 'but since my bypass, I didn't want to use up another ambulance.'

Day 27, October 7

Only a year ago I was asked to deliver the keynote address at the annual ceremony of the National Fallen Firefighters Monument, in Emmitsburg, Maryland, where the Fallen Firefighters Foundation had assembled the families of 110 firefighters who had died in the line of duty in 1999. I remember how carefully I worked on that speech, a fourteen-minute address, and it may have been the most important talk I have ever given – important to me. I wept in the middle of it, recognizing the contribution of all those families from coast to coast, and the depth of their suffering never left me for a moment.

President Bush gave the keynote speech this year, a commitment he had made to the firefighters long before 9/11. Mrs. Laura Bush decided to attend as well, in acknowledgment that firefighting is at heart a family affair. When the president finished his address, he sat next to Hal Bruno, who thanked him for taking the time to honor the fallen firefighters of America.

'This is where I want to be,' the president answered. 'This is where we belong.'

It means so much to the firefighters to know that their president and the first lady have made room for them in their lives, especially, as we find out later in the day, that the air war against the Taliban and Al Qaeda has begun.

One of the president's key points is to make it clear to the world that the Afghan people are oppressed, and that our operation, Enduring Freedom, is not a war against the Islamic world.

At 7 P.M. at Ground Zero, there is still much activity around the site. The south cathedral wall is finally coming down. Only eight windows are left of the original twelve, and the high red crane continues to tug at it. The big Lomma crane in the corner of West and Liberty now has an orange fence around it to protect the passersby. There are now eleven Cats up on the piles, and the pyramid has been razed to the level of the rest of the piles, about forty-one feet high.

Because the piles have been reduced in places to street level, I can see the pit for the first time from the West Street side of the site. The golden globe is still standing.

To the north is the cross of steel that was found sticking up from the rubble, a perfectly formed Christian cross. The construction workers moved it out of harm's way to stand as a shared symbol overlooking the site. I feel confident that the Jewish and Muslim firefighters, POs, and construction workers accept this gesture, for any connection to God in this disaster can only be good – especially if a benevolent God.

At the bottom of the piles on West Street, I meet Mike Rowley, a battalion chief from the 14th Division. 'My grandfather Houlihan,' he says, 'came from Ireland, mud between his toes. Same with all these cab drivers and grocery store people from Egypt, Pakistan, Iran, Iraq. They are poor people looking for a chance, and here in New York they get it. We've always been a place for immigrants. Some of these guys on the piles don't even know we've gone to Afghanistan. They'll find out soon as they go get something to eat.'

A store near the river with a large sign in its window announcing VIP YACHT SALES now dispenses Salvation Army supplies – water, soda pop, coolers, buckets, and myriad items in unidentified boxes.

Another site memorial has been erected here, filled with photos of the victims and surrounded, as they are at firehouses around the city, by flowers, candles, and notes. This impromptu shrine stands within one block of the marble memorial space of the New York Police Department, which contains the names of all New York City police officers who have been killed in the line of duty etched into a fourteen-foot wall. The last name listed is Police Officer DZIERGOWSKI, Feb. 14, 1999. There is no indication of how he died, but no explanation will be necessary regarding the fates of the next twenty-three names that will be added to the wall.

I return to the makeshift memorial wall to study the photographs, as the wind rushes the blue plastic covering it. There are patches from police departments from all over the country – from Goodyear, Arizona; from Maryland; from Los Angeles – all saluting the twenty-three fallen police officers and thirty-seven Port Authority police officers. Photographs of firefighters have been tacked all around, and one, that of Firefighter James Pappageorge of Engine 23, has a letter from his sister attached to it:

> Jimmy, my big brother. I miss you so much. I can hardly wait to see you again. . . . I love you.
>
> Your little sister, Helen

Jimmy peers out with sharp, intelligent eyes that could be either blue or green, here in the odd amber light behind the World Financial Center.

Six entire large boards are filled with letters for Lieutenant Kenny Phelan of Battalion 32. I see Captain Bill O'Keefe and Rescue 1 firefighter Patrick O'Keefe, and I wonder if they are related. There is a photo of Gary Geidel of Rescue 1 with the president of the United States, George Bush, Sr., near a helicopter somewhere. There are flowers, dolls, and teddy bears everywhere – a diorama from schoolchildren of three court officers who were killed. I look at Peter Vega's photo, and I think of Ruby. When I see Terry Hatton in his captain's hat, smiling, I think of the little one who will be called Terry or Terri. Here is Ray Murphy, serious and handsome in his gray mustache; Pete Ganci, with his wide and confident smile; Geoffrey Guja, curly headed, laughing so heartily in the photo that I'm surprised it's in focus. I see Michael Boyle, Jonathan Ielpi, Tom Schoales, the Vigiano brothers, Joseph and John, and I think of their fathers. I see Joseph Angelini and Joseph Angelini, Jr., and I think of the man who is their brother and son, and the woman who is their wife and mother. I am drawn to Moira Smith, for her name is one we considered for our own daughters. I look at Moira, cool and self-possessed, a police officer who knows the drill. And I think of her husband, who is also a cop. Every face has left a loving someone behind.

Back on the piles, the smell of the dead carries on the autumn chill. I walk up a roadway built from 10&10 to just behind the pyramid, at the edge of the pit. A Cat is lifting up steel, and it doesn't seem as if anyone is spotting its work for remains. I watch every bite for an hour. Four firefighters from Rescue 2 come up, and they join me for a few moments. One points to the direction of West Street, saying, 'That's where our guys are, I think. We were just there and found two legs in police officer shoes – could be Port Authority, could be NYPD.'

I walk down into the pit for a ways but stop when I see blond hair, long, in curls. I can't get to it, but I signal to the Cat operator to stop, and I call a firefighter from Ladder 50 to help. He is already down deeper than I am, and digs around it and pulls. It is a wig. I take it to the chief, who says, 'If it was worn by a human it will have DNA in it' The Ladder 50 firefighter also turns up a bent and rusted can of Rheingold Beer, a company that went out of business the year the WTC was topped. Some construction worker apparently threw the can into a beam well, and I decide to keep it. It is junk, but the junk of memory, representing a person who built the tower, and it is the only object I have seen that has survived intact. I would not take a memento of someone who met his fate in the towers, for I consider that holy.

The construction workers have mostly moved to the perimeter, leaving groups of firefighters and police officers to work on the piles, one north and one south. The dust of this history beneath us is picked up by the river wind and smarts the eyes. Fifteen bodies are found.

Day 28, October 8

I call Engine 82 to collect on the meal that I was invited to a few weeks ago, but that was canceled. 'I'm only a proby,' the housewatchman who answers the phone explains, 'and I can't do the inviting. Let me give you to the senior man.'

The senior man says that they will be eating tonight at about 8 o'clock, and I tell him I will arrive at 7:30. But Engine 82 and Ladder 31 'catch a job' around 6:30 and forget to send someone to the store. They remember their lack of provisions at 8:00, and the fireman who is dispatched buys four-pound chickens that take an age to cook. We finally sit to dinner at 11:15, and I realize that nothing much has changed in the firehouse in all the years since I worked here. When we were doing forty runs a day during the blaze days of the sixties and seventies, dinner often merged in with breakfast, and I am guessing that the men of this house want to keep up that tradition.

But the delay is time well spent in this building where I experienced six of the best years one could ever ask for. Three firefighters who recently worked in La Casa Grande were lost at Ground Zero: Geoff Guja, Chris Sullivan, and Tom Haskell (Tom's brother Timmy was also lost). I study their photos and look at the small shrine created by the neighbors in front of the firehouse. Guja is laughing in this photo as well; Chris had been promoted and went on to be a lieutenant in Ladder 111; and Captain Haskell had just been promoted and was still covering in Division 15.

There are hundreds of letters from schoolchildren taped to the kitchen wall. One is a poster filled with little four-inch handprints in different colors, saying 'With these hands we pray for all of you.' One from Isabelle McCarthy, 1st Grade, Proctor Collegiate, Brooklyn, reads: 'Dear Firefighters, I hope you are not dead. Thank you for being so brave. Love, Isabelle.'

The evening is spent in typical firehouse kitchen talk – a lot about the job. Later this month four hundred probies are being appointed in a night and day shift at the proby school. Two firefighters died the night Timmy

Stackpole was burned over 70 percent of his body, but Timmy said he heard a voice, and it said to him, 'Not tonight Timmy.' About Squad 288, the hazmat squad, where eighteen firefighters were lost, someone says, 'They don't run much, but when they get to a job, everyone says, "You got it, hazmat." About heroism, someone says, 'Think about that stewardess who boiled water for a weapon, 'cause that's heroism.' We begin to talk about the times we are in. 'Our families are so great,' someone remarks, 'because we're never home. All I'm doing is working and going to funerals.'

The chicken, finally, is truly delicious, and worth the wait. I cannot stay for dessert, for it is after midnight, but I see the amount of the bill the store man has written and circled on the kitchen blackboard. This is the way it is done: Someone volunteers or is selected to go to the store, and he puts the price of the meal on the board. Someone else divides the number of firefighters eating the meal into that amount, and that is the price of the meal. Tonight's tab is $39.70, and there are ten men eating, including me, so the price is $4.00 a man, not bad for a meal that would have cost five times as much in any of the downtown restaurants, and not have been half as good.

I put two twenties down, because I would like to buy the meal, which is what firefighters do on special occasions – weddings, promotions, retirements, and the like. This is a special occasion for me, and someone writes 'Thank you Dennis' beneath the circled price.

Day 29, October 9

News reports today announced that military jets have pounded the Taliban for a second full night, which will show the commitment of the government to root out the terrorist mentality in Afghanistan. But to bring the focus of war home, it was also reported that the man who had anthrax poisoning in Florida has died, and that another man in his office was exposed to the disease. The FBI has sealed the offices of American Media, Inc., which publishes some of the biggest tabloid shock sheets in America, the *Globe,* the *Star,* and the *National Enquirer,* and hundreds of people who had worked or visited the company are taking prophylactic antibiotics.

It was announced today that 4815 people are missing at the WTC and 417 are confirmed dead. The number of missing declines every day, as people who are feared missing are found to be alive, and as lists of missing from the various companies and agencies involved are collated.

At the site, the FBI and the NYPD are closely monitoring the movement of the steel being transported out, since some of it has been found in Long Island junkyards. There are 78,000 tons of recyclable steel in the rubble, which will net tens of millions of dollars. There is a total of 308,900 tons of steel, 351,000 tons of concrete, and 1,200,000 tons of debris overall at Ground Zero.

Day 30, October 10

Only four windows now remain in the south cathedral wall. On the Vesey Street side, the north cathedral wall is still leaning against the Customs House, building 6. It will be months before that part of the site is cleared.

I walk past 10&10 to the food area to see if anyone I know is taking a break. David Bouley's Green Tarp Café has been closed, and only scattered chairs and paper wrappings are left in the storefront, the enterprise having ended as unceremoniously as it began. The contribution of these volunteers who served everyone food for twenty-nine days will never be forgotten by the thousands of men and women who sat to those fabulous meals on hard wooden chairs. When it comes time to give awards for meritorious service, Bouley should certainly be given one to place at the front of his New York restaurants. He has every reason to be proud. As do other restaurateurs. I know that Bruce Schneider of the '21 Club' wouldn't let an opportunity to be helpful pass. Many of the employees of '21' are giving up their days off to keep the food supply going at Pier 61, which has been established as the temporary administration offices for the World Trade Center disaster. Tomorrow, '21' is donating 10 percent of the day's receipts to the families of the employees of Windows on the World. For the first few weeks after the attack, Buzzy O'Keefe of the famous River Café fed the police and city officials working at the east side morgue.

The Red Cross, too, has done much commendable

work during the second, third, and fourth weeks in providing food and respite in the boat it rented, *The Spirit of New York,* berthed in the Hugh L. Carey Battery Park City Marina, the North Cove at the shore of the Hudson. These respite centers have received very little press attention because news people were not allowed into them, the fear being that little respite would be available if reporters were conducting interviews.

The huge piece of the south tower that had pierced the skin of the Bankers Trust Building and hung like a modern-day, gargantuan sword of Damocles has been knocked away and now lies at the curbside of Liberty Street. It must have raised an incredible cloud and made quite a crashing sound when it fell. Next to it, at 90 West Street, the scaffolding on the ornate Gothic mansard roof reminds me of the construction of a great twenty-two-story church. The building had been undergoing major refurbishing before 9/11, and I wonder about the holding strength of the scaffolding after the shock of the two collapses, but I know that if there is any concern the chief in charge of the site will be very aware of it. It continues to surprise me how much can change at the site in only two days. The West Street side of the collapse area is now flat and level with the street. Gone is any trace of the mounds of debris and twisted steel we climbed those first few days, passing buckets from one to another. No one seems to be spotting the steel as the Cats eat at it in every direction, and I wonder how many souls are being carried away unnoticed.

On West Street, I see three officials of some kind in

blue nylon jackets, and two guests, riding around on a golf cart. One of the guests is busy taking photographs as the cart moves along the street. A fire lieutenant walks in the middle of the dirt road that is West Street, and puts his hand up, saying to the man with the camera, 'Are you down here officially, or do you think you're in Venice?'

The man puts his camera down behind him and apologizes. As they leave, the lieutenant comments, 'Mayor Giuliani says that the curious should not come down here, but if you have friends . . .'

I am standing on the edge of the south slurry wall, a long, concrete barricade about two feet wide and seventy feet deep that runs around much of the perimeter of the site in the shape of a rectangle.

A conveyor belt has been set up to pour sand between the slurry wall and the base of the pyramid, for engineers have found a crack in the wall on the Liberty Street side. A break in the wall would bring a cascade of water into the site, so truckloads of sand are being emptied against the side of the wall that has been exposed, which extends down around twenty feet, to guard against a break. No one can be certain if it will hold, but I have to believe that the engineers will save it, for the alternative is another disaster. If this deep wall cracked open anywhere the whole of the Hudson would come pouring in, which would not only create a very unsafe condition here, but would set the cleanup and recovery efforts back for months.

I meet Lieutenant Jim Cooney, whose dad died in the line of duty in 1971 at a job with Engine 321. I try to remember if I attended his funeral, but there have been so many deaths in the department over the years. About 6 firefighters die in the line of duty annually, which would mean 228 had been killed since I became a firefighter in 1963. While this is a large number, when measured against the 343 who were lost on a single day, the full horror of the WTC attacks is placed in striking perspective.

Jim has two sons, one of whom is a probationary firefighter in Engine 282 in Brooklyn, while the other is a college graduate who, as he says, had been 'talking about going to Wall Street and building a fortune. But now, after this, he's decided to go into the department. He's on the list and will go into the probationary school on October 28. It's in the family blood, and we are three generations in the department. We can't say that about Wall Street.'

I decide to walk over to the police department's crime scene to find Sergeant Orlik, to thank him for having driven me through the storm on his golf cart the other night. The cops at the site are so cooperative and friendly that I don't want to miss any opportunity to reciprocate. I suppose they consider our loss to be so enormous that we need every bit of support we can get, and in that they are right.

For the police department to have lost twenty-three of their own finest in a single incident would under normal circumstances be front-page news all over the world. But in comparison with the fire department casualties, their own have come to be regarded as almost anecdotal, an attitude I deeply regret. Here we are six weeks after the disaster, and we still have not recovered the remains of a single police officer, although we have found more than forty firefighters. I feel for each of these cops who is in mourning just as we are, the special mourning that comes with the badge and with waiting for some resolution, any resolution, to these twenty-three missing.

As I turn onto West Street I see a number of police officers and a few firefighters from Engine 92 fall suddenly to their knees. I have learned that any sudden activity here signifies that something important is about to unfold, and rush over to see what it is they are doing. As I approach them someone hands me a shovel, and a firefighter commands, 'Start digging here.' The very heavy smell of a cadaver fills my nostrils. I once heard someone describe life as being what lies between the stink of the diaper and the stench of the shroud, an inelegant reference, but somehow accurate. The odor is so powerful that my head jerks backward involuntarily, as if someone had just thrown a punch at my nose that somehow ended up in the far reaches of my nostrils. I pull up my filter mask, but I know it is too late. Later on tonight I will soap my fingers into a lather and shove them up my nose halfway to the cerebellum, hoping to clean it out. I know, though, that once this smell gets into your head it will stay there for days, and maybe a lifetime.

There is a body lying in the powder and concrete bits before me. I kneel down, and together with two firefighters from Engine 92 start digging around what is obviously a person's arms and legs. The body is intact, but headless, and it appears, because it is small and the legs are shapely, to be a woman. Before we have a chance to open a body bag we come upon another. Now we have four arms and four legs, and parts of two mid-sections. I cannot tell the sex of either, even though each body is without clothes, for their backs are to me. They do not appear burned. The smell of the cadavers comes through the mask filters, and it is choking. Still, each fire-fighter takes a part of the first body, hands underneath from heels to shoulders, and places it in unison into the black plastic body bag that has been spread out on the long, narrow, thick plastic of an orange Stokes basket. We do the same with the second body, and then I lay my shovel off to the side, while a firefighter tags both sets of remains.

With the others we pick up the Stokes baskets and carry them off to the temporary morgue, which has recently been housed in a tent on the west side of West Street. Everyone stops doing whatever he is doing and removes his helmet or hard hat as we pass. These are not uniformed personnel, but civilians, who have become a part of a ritual that is performed as often as twenty times a day. I feel as if I am an anointed visitor doing God's work, and I thank God to be a part of this valiant group that is performing a task that is so fundamentally human.

After we place the Stokes baskets on the ground in front of the temporary morgue a young chaplain comes to us and says, 'You have a biohazard on your hands. You will have to discard your gloves and your clothes. Don't take the chance of touching your eyes. Wash up immediately.'

I decide to take the chaplain's advice and walk to the wash station on Church Street. It is consoling to walk alone through this area, for it forces a new kind of contemplation, one that is completely other directed. I see many engineers holding clipboards as I walk and I think how easy, relatively speaking, their job must be. Their responsibilities are all mathematical. If it doesn't add up, it doesn't exist. For the firefighters, every body, whether complete or incomplete, and every human part, however small, represents a person, and is treated with the same reverence as the whole. While it may not meet the definitions of physics, it is the right way to think.

In the evening I am asked to attend a Police Athletic League dinner at the Rainbow Room. I go because of my conviction that we must support our police officers as much as we can during this period, though I leave early. The odor of the afternoon is still with me, and I have a headache. One man, though, makes a memorable remark: He lives downtown, he explains, and has always had the twin towers in view from his apartment. The twin towers are gone, but now he sees what stands behind them – the Statue of Liberty.

Day 31, October 11

One month has passed since the attack. We are still in great pain, and all of us in America are trying to mend ourselves, and evaluate our sense of the country.

Patriotism is as strong as it has ever been in my lifetime, and for that I am grateful. Commitment to the future of one's country is not something that can be measured empirically, and for many years I have worried about our ability as a nation to muster the kind of human resources and dedication it would take to fight a serious war with an avowed enemy.

Now, however, I am convinced that young people in America would go to war to defend our way of life. Respect for our first responders has had an effect, and the heroism they have shown has given American youth a standard to idolize.

But we are still battered, and a particular concern is getting New York back on its financial feet. The stock market is in a depressed state, and the budget surpluses that the city, state, and federal governments have all projected for five years into the future have now been revised downward.

A friend of mine who is chairman of one of the largest real estate leasing firms in the city called today and said that some of the major firms that had offices downtown will not return there, including Blue Cross, Marsh McClennon, Lehman Brothers, as well as many other smaller firms. Much of the office space in the five buildings of the World Financial Center and the surrounding buildings will be unoccupied and available. I mention that Mr. Silverstein has announced that he intends to put another building on the WTC site, of which he is the leaseholder, and the chairman says, 'No bank is going to put up a billion dollars to construct a building that will be empty. The World Trade Center sat empty for many years, particularly from '71 to '75. It devastated the financial center's real estate value, but it was able to last through the bad times because it was funded to begin with by a government, or a quasi-government, agency. They will have to rent the World Financial Center first, because there will not be financing for another World Trade Center until those four buildings across the street are fully rented.'

Rudy Giuliani has let the idea of his elongating his term of office by three months quietly vanish, but he did make news today by returning a check for $10 million given to The Twin Towers Fund by Prince Alwaeed bin Talal, the sixth richest man in the world. The prince had issued a press release simultaneously with making the contribution, saying that 'at times like this one, we must address some of the issues that led to such a criminal attack. I believe that the government of the United States should reexamine its policies in the Middle East and adopt a more balanced stance toward the Palestinian cause.' Mayor Giuliani has the right to take umbrage with this statement, particularly if he wants the Jewish vote in New York, should he run for office again. But in fairness, it should be noted that many Americans might agree with the prince's conclusions. Of course, his $10 million would have meant about another $25,000 for the spouses and children of our 403 fallen first responders, and I wonder if those men and women ought to have been consulted about rejecting such a large sum of money. Lee Ielpi, though, told me he thought it was 'absolutely correct to return the money, because so much of the terrorist thinking originates in Saudi Arabia.' While the prince is not a figure like Qaddafi or Arafat, a government leader who has a known record of supporting terrorism, returning the check does send a strong message to Saudi Arabia that Americans feel its government has not done enough to monitor and control the activity of terrorist groups in their country.

Today it was announced that 4776 people were reported missing at the WTC, and 442 are confirmed dead.

Day 32, October 12

As of today there are 1050 construction workers at the site, 160 firefighters, 90 police officers, and perhaps another 1500 service workers, volunteers, and soldiers in the national guard. Thus far 290,000 tons of debris have been removed, and now as it leaves the site each truck is driven through a pair of high-pressure hoses manned by national guardsmen. All passenger cars are hosed down as well, so that the dust and dirt of Ground Zero does not leave a trail through the city.

Yesenia, Jonathan Ielpi's wife, and her in-laws Ann and Lee Ielpi, have decided to have a memorial mass for Jonathan on October 27. This must have been a difficult decision for them to come to, for I know they have been ever hopeful of recovering their son's remains. Everyone at the site has seen Lee there every day, digging, searching, covering every changing square foot. With a wake and a mass, at least, their friends and relatives will no longer ask what steps they are taking to try to bring closure to this tragedy within their family.

Anthrax was found in the office of Tom Brokaw's secretary at NBC today, which is a major development, for it establishes that the infection of the employees of *The Sun* is not an isolated case, and that the national media is being targeted. I call my son Sean at home and suggest that he inspect every envelope that comes to him at ABC.

'Where are you going today, Poppy?' he says. No one in the world calls me this but Sean.

'To Brooklyn, home of your ancestors.'

'No, they're your ancestors,' he replies, teasingly, with the big city pride of Manhattan.

It is an Indian summer day, calm, with not a breeze in the Brooklyn air. I check my watch, see it is 10 A.M. on the nose, and take my place on the line of firefighters outside of St. Patrick's Church, on Brooklyn's 4th Avenue. The firefighters of Rescue 2 line up on the steps of the entryway, Lee Ielpi and Pete Bondy among them.

I see Jack Pritchard, who is now a battalion chief, but who was once an officer in Rescue 2, and who might be the most decorated man in the history of the fire department. Bob Mosier, a deputy chief and another alumnus of Rescue 2, is also here today. To me, being here in the presence of these men is thrilling, because I am aware of the road they have had to travel to develop their reputations as legends.

Billy Lake, whose funeral we are attending today, whose life we are celebrating, was something of a legend in the job, too. He had a little bit of the wild man in him, but he molded that wildness into the courage that it takes to be up front in every fire, and to give everyone he ever met that wide, eye-winking, I've-got-a-secret smile.

There are twenty-nine HOGS (Harley-Davidson owners group members) in the funeral procession, and they are serious and reverent, even though they appear in their Saturday-night-bash leathers. They were a part of Billy Lake, and Billy Lake was a part of them, and so I love them for the bigness of their hearts.

Captain Mike Laffey, a cousin of Billy's, delivers the eulogy.

'Billy loved just three things,' Captain Laffey says. 'The fire department and Rescue 2, his motorcycle, and the New York Yankees. Now it's easy to understand the fire department, everybody in it loves it, and easy to see why he loved the freedom and excitement that came with the HOGS. But none of us in the family can understand why a Bay Ridge, Brooklyn, boy, coming from a family that has a long tradition of loving the Dodgers and hating the Yankees, can go to Yankee Stadium. . . .'

His words remind me of my own family, and its Brooklyn roots. It is said that when my grandfather got off the boat from Ireland in 1921, he did not have enough carfare to take the subway out to Brooklyn. He did have the number of a candy store, however, and someone was sent from there up the street to fetch his cousin, who then searched the neighborhood for someone with a Model A to drive to the pier where my grandfather stood, a hopeful immigrant, with his wife and six children. Captain Laffey's portrait of his cousin also reminds me of Alice McDermott's book *Charming Billy,* and how that title so aptly describes Billy Lake. 'Billy was a fire buff with his brother when they were kids. The men in the firehouse put up with them hanging around the firehouse, until they memorized the boxes and began to outrun the fire engines, and beat the firefighters to the box. His brother, Brian, is the captain of Ladder 156 today.

'Three days were important to Billy: September ninth was his mother's seventy-fifth birthday, and the whole family got together to celebrate. It was a beautiful gathering, all the little ones you see here today, and the cousins, uncles, aunts. Then on September tenth Billy had his twentieth anniversary of coming on the job. His pension time was complete, and so he bought the meal in the firehouse that evening. A lot of the guys came in just to say hello, to pay homage to the senior man in the company.'

Phil Ruvolo, the captain of Rescue 2, steps up to the altar and says just one thing — 'Billy was the toughest man I ever met' — and the church fills with applause.

This observation gets to me, for it is the toughness combined with the heart I know these firefighters to have that makes them unique. In my years in the job I have always found that it is the toughest who have the biggest, the strongest hearts.

The priest says, 'Blessed are the grieving for they shall be soothed.' I look at Billy's family in their sadness, and I think, *I hope and pray this is so.*

A young, burly priest with a blond crewcut, Father Sauer, begins, a cappella, the introductory lines of 'God Bless America.'

While the storm clouds gather far across the sea,
Let us swear allegiance to a land that's free,
Let us all be grateful for a land so fair,
As we raise our voices in a solemn prayer.

He sings in the most beautiful, strong, and untutored voice, a combination of gusto and poetic sensitivity. As he finishes the introduction, the entire church joins him in singing 'God Bless America,' or at least we try to sing it, between tears. There have been only a few times in my life, and always before in the theater, when I have witnessed anything so perfectly dramatic.

Outside, the rescue rig goes by, and a fire engine pulls up, slowly. A single bell begins to toll, seven seconds between each striking. A trumpeter plays taps. The coffin is lifted by the rescue firefighters to the top of the fire engine, which then pulls away slowly. The motorcycles pass, twenty-nine of them, their exhausts very loud and illegal. One of the riders has a cigarette dangling from the side of his mouth. It was placed there deliberately, I think, to show that even in a funeral cortege, one should be a little bit different, a little bit irreverent. Like Billy.

All the rescue men line up, many of them with rows of medals across their chests. It is not an easy thing to be awarded a medal in the fire department, for the standards are high. You have to come pretty close to buying it, or at least put yourself in situations in which no sane and thinking person would be by their own choice. I am reminded of Tom Neary, another of the few in our job who wear their medals in rows. I worked closely with Tom in 82 and 31, and know him very well. He was someone I respected greatly, so I was especially pleased, in those days when I ran *Firehouse* magazine, when in 1988 Tom won the highest award in our heroism awards program. I never played any role in choosing the winners, a responsibility that was given to fire chiefs in different parts of the country. A check for five thousand dollars came with the award, and I asked then-Mayor Koch to present them in the City Hall Blue Room. The mayor agreed, and when Tom Neary showed up he was wearing a uniform with several rows of ribbons.

'That's very impressive,' Mayor Koch said, pointing to one multicolored ribbon, 'and what is that for?'

Neary laughed, and said, 'I don't know, Mayor, this is

not my jacket. I was late and grabbed the first jacket I came across in the firehouse.' Of course, I knew that Tom's own jacket was even more highly decorated.

Day 34, October 14

Anthrax is coming through the mail. More than 680 million pieces of mail move through the post offices of America every day, and in response to this new threat the government has begun an awareness program. It urges us to be prudent, to be aware of the potential hazard, and to inspect each envelope carefully. This is as good a subject to think about in church as any.

What worries me is not that I know next to nothing about the agents of biological terrorism, but that most firefighters and other first responders also know very little. The federal government should immediately begin a major training program, and not wait for the states to develop syllabuses and protocols. We are at war, and it is the responsibility of the Senate and the House to get together to protect us through our first responders.

At the site twenty-four wells have been dug all along the Liberty Street side of the slurry wall. Pumps will be sunk into these wells, with the goal of draining the area and lowering the water table. The engineers must be getting more worried about the stability of the slurry wall.

A revolver was dug up at the site tonight, on the pyramid side, where the south tower went down. It is the first police officer found, and there are only police officers on the pile, digging with steely concentration. They soon unearth a second revolver, which makes them even more optimistic as they dig further. Soon, they come to the body of Tommy Langone, and thereafter that of Paul Talty. The police officers are gratified that, after thirty-four days, they have finally found two of their own. A large police emergency van pulls up, and the emergency service officers gather to take the men out. Everyone at the site seems to have gathered, and they are all at attention, saluting. The van leaves Ground Zero, and goes up 1st Avenue, the lead vehicle in a caravan, followed by the police commissioner, and then by Joe Dunne and the other staff officers. All along the route, from the site to the morgue on 29th Street, police officers pull their squad cars over, get out of the cars, stand at attention, and salute.

As of today, there are 4688 missing and 450 confirmed dead.

Day 35, October 15

My son Sean calls and informs me that they have found anthrax in the offices of ABC News. He is a producer there, and works with Peter Jennings. On September 28, a Friday, an associate and friend of Sean's came into the office to visit, and to introduce her co-workers to her baby boy. The offices of ABC are busy under normal circumstances, but this is one of the greatest news periods in history.

Knowing Sean's love for children, his friend came out of her office, which is right next to his, and plopped her 7-month-old son in blue pants and a bunny-etched shirt onto his lap. Sean is always rushing about New York with his own three children, and he loved entertaining the little one.

When his friend returned home that evening, though, the baby became ill, and on Sunday his parents took him to the hospital, where he remained for two weeks. The doctors thought he might have been bitten by a spider, but that was just a guess, for they could not find anything specific to which they could attribute the symptoms of fever, vomiting, and constant crying. And then on Friday, October 12, the story broke about anthrax's being found in an office at NBC. The child grew worse, and on Sunday suffered renal failure, but at the same time, his parents and others at ABC had followed the news at NBC and made the connection. The child was immediately tested for cutaneous anthrax, and was found positive. Now, he is on Cipro and progressing well. And so the terrible news at NBC helped a little boy from ABC and probably saved his life.

Sean is now dispossessed from his office, and waiting for the health department to inspect his equipment. He has been tested for anthrax, along with anyone else who came in contact with mother and child that Friday. And so now my son has been drawn into my story in a way that I could never have anticipated.

Could I be any closer to these events, I ask myself, having been so close to the disaster at Ground Zero — losing friends, seeing other friends weep for their losses, having a neighbor and friend in my own building lose her son, having my sister-in-law's nephew saved by an exchange of tours, having the next-door neighbor from my youth saying the funeral and memorial masses, and now having this beautiful baby who is front-page news, laughing on the lap of my son Sean? But being close to these things is not suffering the way Jimmy Boyle and Lee Ielpi have suffered, and I realize that every waking moment.

I am a trustee of the New York Fire Foundation, and today I attend a meeting at the office of the president of the John Jay College of Criminal Justice, Gerry Lynch. I learn that because *The New York Times* has included us since 9/11 on their Neediest Cases list we have raised $1.4 million, a substantial amount of money for this small organization. We decide to give two thousand dollars to each beneficiary of the 343 missing firefighters, and to use the rest of the money in helping with long-term rehabilitation, therapy, and crisis intervention for the families. The thought goes through my mind that perhaps we should distribute every penny of it, and I look over to the presidents of both the firefighters' and the fire officers' unions, who are present. But they are comfortable with holding some money back for longer term care, and that is how we vote.

This will be an issue that will have to be confronted by The Twin Towers Fund, and also by our own Police &

Fire Widows' & Children's Benefit Fund – to distribute all the money, or to hold back some for future use, which simultaneously makes the fund more stable. There are many factors to consider in determining the administration and distribution of such large amounts of money. Our fund, for instance, has been doing this work for fifteen years. Since 9/11 we have so far raised more than $30 million. We have an ongoing mandate, responsibility, and promise to the widows of our firefighters and police officers, and so it would be appropriate to continue our policies as before, with an increase in the amount of benefit determined by the amount of money we raise. The Twin Towers Fund, which has raised upward of $50 million, might better serve the intentions of the givers with a complete distribution. The state attorney general, Elliott Spitzer, wants a central board to oversee how all the generosity is coming to New York, how it is registered, banked, and distributed, but some of the bigger charities are reluctant to share their information with others. The United Way and the American Red Cross have already raised several hundred million dollars each since 9/11, and they have not yet indicated how they will disburse these contributions.

Day 36, October 16

Marty Geller calls to offer his firm's hockey tickets to the firefighters, which is a touching act of consideration. In conversation, he mentions that four firefighters from the local firehouse came to his synagogue, the Jewish Center on West 86th Street, to participate with the congregation in saying the Yizkor. This is one of the holiest of Hebrew prayers, and it is said just four times a year. 'There was deafening applause for the firefighters,' Marty said. It is a sweet thought to imagine these men in the local temple, especially at a time when people are conscious of the importance of coming together as one American society.

In the meantime, it has been found that the anthrax sent to Senator Daschle's office is pure and highly refined, consisting of particles so tiny they could spread through the air without detection.

Day 37, October 17

Anthrax was found in the office of Governor George Pataki on 3rd Avenue in New York. The governor and his staff immediately evacuated the premises and began to take antibiotics.

In Washington, a preliminary investigation has shown that thirty-one workers on Capitol Hill were exposed to anthrax spores from the letter sent to Senator Daschle. All six House and Senate office buildings have been closed for further screening.

At the same time the government closed its congressional buildings, it announced that it was negotiating with four drug companies to buy three hundred million doses of the smallpox vaccine – a sufficient quantity to protect every American.

All of this information makes me wonder if we are creating an environment of fear, or one of prudence. It is a little like looking at a glass of water as half full or half empty.

Day 38, October 18

At Ground Zero, the south cathedral wall is almost down to street level. All its arches are gone, and so it has been robbed of its Gothic quality, and its base, eleven windows long from west to east, a section about forty feet wide, is now a crude wall containing a longer pile of debris, still twenty-five feet high. I have come to feel that this wall was keeping the rest of the world removed from the fire-fighters' realm, and that the piles beyond it were a sort of sanctuary. Now that the wall is down, it is like just another ruined choir of the sort I remember seeing in photographs of Dresden. Our separateness from the rest of the world is not quite as defined without it.

The West Street piles are completely gone, and the street has been widened again, and covered with fresh brown dirt. The construction companies running these operations certainly know how to provide orderliness and cleanliness to a construction site. Elsewhere, the piles have been inverted, and are now pits, and each day as the workers go deeper they will add to the width as well as to the depth of the pit. The north cathedral wall is now clear in the distance, like a wall of Chartres seen across a brown field. It has fifteen high Gothic arches, with another ten window frames wrapping around the east side of the pile. This is the only wall that remains of the north tower, and as I look at it I cannot help remembering that its windows were all shattered when the building was shaken by the impact of the plane. There was so much broken glass that the arriving firefighters and police officers had to step through inches of it to get into the auditorium-like lobby.

A huge drilling machine is at work on the slurry wall. The machine is forty feet long, and it is hammering lengths of twenty-inch pipe through the wall, to tie them to the back of the wall. 'Like a butterfly nut going through plasterboard to hang a picture?' I ask one of the workers. 'Yeah, exactly,' he says, 'only big time, very big time.' It is hoped that these stabilizing pipes will strengthen the slurry wall.

I see something that is as surprising to me as if I had come upon the Brooklyn Bridge up on 34th Street. On Liberty Street, just south of the pyramid, stands the huge golden globe that until now had remained in the middle of the plaza area. It was moved because the ground there was no longer level. To me the globe has become like an old and proven soldier. I walk around it and place my hand on it. Its top has been sheared, and I can look up through it to the clear blue evening sky. It is perhaps four times as high as I am tall. Inside the globe is the rubble of 9/11, the papers, the dust soft and sandy, some scattered concrete. It brings me back to the day itself, thirty-eight days ago when I first saw it in its surviving pride. It is beautiful, this sculpture, its gold patina glowing with the majesty of experience, and I almost want to say a prayer before it. It holds in its inner core the sacred dust of our fellow human beings, and conveys to me, more than anything I have ever seen, the awesome power of the human spirit.

On one side of it, someone has spray painted: PORT AUTHORITY POLICE. Is it evidence? Is it being saved for a museum or a monument? Or will it be examined and then sent to its end in some Staten Island junkyard?

I walk up a new road that runs from Liberty Street into the middle of the pit and meet Chief Hayden, with whom I exchange greetings. It is almost as if we were on a country road, and we were neighbors saying hello as we go about our business. Farther along I meet Mike Leddy, who has a small camera with him and is documenting the site for his own archives. I am always sensitive to the fact that this is hallowed ground, that beneath us and around us are the remains of several thousand people, and that if a photograph is taken it should be taken for a reason that respects its holiness. Cameras are not allowed in the site area, but many firefighters carry small pocket cameras in the normal course of duty, in case they are in a situation that needs to be photographed. Rich Ratazzi, for instance, took the last known photo of Lieutenant Ray Murphy, simply because he had a throwaway camera in his pocket.

The top of the road ends at the beginning of a steep

incline of rubble that drops down for about seventy feet. Just below me, about thirty feet down, I see Lee Ielpi working an inch-and-a-half hose line. There is quite a bit of smoke in this area of the pile, and he is cooling whatever he can. It is a perilous climb down, and I begin with prudent trepidation. Mike gives me a hand the first few steps, and then Nick Liso, standing just above Lee, comes up to give me a hand as well. Nick and I spent a great deal of time together, 'humping hose' in burning buildings, when he worked in La Casa Grande. His nephew is on the job, working in Engine 73, a good 6th Division company, and I think for a moment of how alike he is to his uncle. Nick himself is off the job now, having suffered a stroke, from which he recovered completely. He would never be able to balance himself in this mess if he hadn't.

I give Lee a hug, as much as you can hug a man with a hose line in his hand. I see he is lifting the full weight of the line, and get behind him to lighten the backward pressure a bit. Several firefighters are on the pile searching for voids and indications of a possible void.

I can see that Lee is pleased that I am backing him up on the hose, and I ask him, 'You wouldn't want me to relieve you on the knob?'

This is an inside joke among firefighters, for no one ever relinquishes the nozzle until he has taken so much of a beating that he can't continue. This is particularly true in companies like Rescue 2 and Engine 82, where we come from.

'Not a chance, brother,' Lee laughs. 'Not a chance.'

It is not a full laugh, but a shared natural feel for the

tenor of firehouse humor. I wonder if Lee will ever fully laugh again, or for that matter, if anyone at Ground Zero will. I feel a deep admiration for Lee, for he is given the kind of respect that does not come easily to a man, the kind that is earned only through the experience of a lifetime. Yet Lee is fundamentally a simple man and can be defined in just three words – straightforward, honest, and hardworking.

It is now about 7 P.M. I notice that Mike has taken a photograph of us. Lee is covered with the dust, dirt, and grime from his day's work here, but, as the saying goes, 'Where there's muck there's brass.'

A chief signals us from below that we should move the hose over about fifty feet, and so I yell to a nearby firefighter to shut the line down at the check valve of the 'Y.' With that done it is much easier to move the hose line, which runs down the pile to the check valve and then for a few hundred feet to a fire engine. We are now standing above a small precipice in the middle of the pile, a safe, level place about three or four feet square.

I turn again to the firefighter down below and tell him to restart the line, shouting, 'Charge it.'

After wetting down the area Lee chooses, he indicates that he wants to move the line back to where we were originally standing. With one hand on the line I motion to the firefighter to shut it down once more, but instead of pushing the control valve left, he pushes it right, and a great and powerful surge of water is forced into the line. It surprises us both, and we have to momentarily fight to regain control of the hose, which shoves us backward several feet, and almost lifts us off the ground. I know that Lee will have a firm grip on the hose, so I loosen my grip on it and just grab him as tightly as I can around his waist, and push my shoulder into his back. Still, the force of the surge is so great that we continue to move backward.

The firefighters around us, Liso and the others, begin to yell, 'Shut down, shut down!'

But Lee puts all his strength into the nozzle control handle and finally adjusts it to relieve the pressure.

I didn't consider how precarious a position we were in, until the hose stopped moving, when I looked down and saw how close we had come to falling off the edge. Below is an acre of jagged steel, iron pikes, and cracked concrete. Tripping on such a landscape might lead to a small injury, but to fall into it from above would have serious consequences.

'Good save,' a firefighter says.

And, as I am looking down, I say, 'It's all right, I had another two inches.' I turn to Lee and see that experienced, knowing smile of his. He knows the incident is not worth commenting on.

A large Cat now comes toward the pile. These machines look small from a distance, especially if seen next to one of the megasized cranes. But now I think, *Steven Spielberg himself could not have directed this machine to move more ominously.* The Cat seems to slide as it approaches, its four-foot-wide tracks grinding heavily into the earth. It progresses slowly, inch by inch, until its huge and forceful teeth go squarely into the mound just next to us, tearing at the pile. It seems almost pre-sumptuous for this Cat to come so near us as we work, and I fear it is gestures like this that add to the overall tension that has begun to grow at Ground Zero. The construction workers have a mandate to get the job done, while the firefighters want to make certain that they do not miss a single opportunity to find a body or a body part. Most of the families of the missing fire-fighters are Catholic, and the Catholic Church will not provide a funeral mass without a body or a body part.

I have heard of one family that was given just a scalp. It was determined by DNA testing to have belonged to a certain person, and that person's family was deter-mined to bury it with a funeral mass. A coffin was purchased, a grave site was selected, and the person was buried. The amount of the body, the percentage of bones and skin and organic matter that exists, does not matter, for the soul exists in any amount, in any percentage. This family has a grave site, but most important, they have the confirmed knowledge that, yes, their loved one is dead, and here is the grave, the coffin, and the person for history to corroborate. I say a silent and separate prayer, as I have done many times, that Lee will soon find the remains of his son Jonathan.

Later, Mike, Nick, and I go over to Red Cross respite station 3 in the Marriott Hotel. I wanted Mike to take a photograph of Nick having his shoes shined, for there is a boot-cleaning station located outside the hotel, part of an effort to control the amount of dust brought into the area. The dust is pervasive, and no one really knows the consequences, long and short term, of regularly breathing it. A security official is checking boots as we enter, and sends back a construction worker whose boots are still muddy. This is one of three respite centers the Red Cross is running to serve Ground Zero, and at its entrance is a big sign that reads: AMERICAN RED CROSS DIRECTORY, NO PHOTOS ALLOWED. The carpeting has been covered with thick plastic, which gives a definite temporary air to the place. Below the sign are listed the services of the respite center: first aid, mental health, rest rooms, cafeteria, oasis, TV, entertainment, relax, supplies, massage therapy, chiropractic care, friends of Bill W. (I had to ask what that was, and someone politely explained it was for recovering alcoholics), pay phones, showers. So the Red Cross has pretty much everything that an exhausted rescue worker might want, although I know that Ielpi, McTigue, and Bondy have never stopped in here for anything more than a quick snack. There are other workers here, though, who do not have nearby homes to go to, but hotel rooms, and so the center might be a godsend for them. David Bouley has evidently contracted with the Red Cross to move his catering operation over to this center, and the food is still excellent.

I meet a steelworker on the food line, and ask, 'How ya doing?'

His answer is simple. 'I want to go home,' he says, 'and I want winning lotto numbers.'

On the third floor is a large, high-ceilinged rest room, which seems to have been a small ballroom. Three rows of eight lounge chairs are lined up across the floor, big leather lounge chairs that could be pushed out and almost serve as beds. A sheet covers each one of them. Every chair is occupied, and six different televisions have the ball game on, the Yankees playing Seattle.

There is also a row of six computers, each one being used by a rescue worker catching up on e-mails. Farther back there is a whole row of pillows and blankets, as well as six tables at which guys are sitting around playing cards. Smaller tables hold coffee and all kinds of drinks and snacks. Outside this room is an area for massage, with nine massage tables and three smaller back-massage tables. The masseuses have to commit to working twelve-hour days, or the Red Cross won't accept them. Sleeping accommodations are available at the respite center north, which is at the St. John's University building on Chambers Street.

As I sit with Mike Leddy and Nick Liso in the dining area, I meet an attractive Mormon woman from Boise, Michele Gamblin. Michele is a mental health therapist, having studied at Brigham Young University. I ask her to tell me about her experiences being a professional mental health worker for the Red Cross.

■ A lot of the workers down at Ground Zero really aren't used to seeing what they are seeing, mutilated bloody parts, very graphic and gross things. There is the pressure of digging to find body parts, anything for the DNA, and of course some of the rescuers have family members here and they are digging for them, so finding remains isn't as joyous as you might think.

The construction workers deal with machinery and build buildings, and here they have a lot more to deal with. I had a crane operator tell me that he was picking up something with his big shovel, and he didn't recognize what it could be, and it was hanging human skin. They had to stop. Whenever they find anything they have to stop, and he had to call them over to take the skin off his shovel. How do you deal with the psyche when that happens? And a lot of them were having nightmares.

I would say that out of those that came in to eat a very small percentage stayed around for the other services available. The policemen and firefighters down here expect to see trauma, and so they would come to eat and then go back to the piles. Maybe they don't need the rest. And of course, they have their own brotherhood, and they have a system already in place to talk about these experiences. But these construction workers, and ironworkers, and truck drivers that are hauling this stuff away that aren't part of a brotherhood, these are the people I found I do a lot more counseling with, because they don't have any place else to turn to. Plus many of them live far away, and they don't go home. One guy told me he hasn't seen his wife in four weeks.

I hung out in the area by the massage, where they might be waiting, and I would chat with them, and probe a little. They would ask about me first, and I would tell them that I am a marriage counselor back in Idaho, and I am here just to talk. And basically, when they find out they have an opportunity to talk with a professional person, they might just begin to talk, and I might refer them to long-term counseling, or to take some time and have a long chat. Some we just gave stuffed animals, and though it might seem funny giving a large, tough man [a toy], remember – it's something to hold. And I would advocate for them by asking, Are you getting the rest you need? Are you taking a day off? Are you just staying in the grind and wearing yourself out? If you are, you might be making yourself unsafe out there with all the machinery.

Each day, maybe I have two serious conversations where I can say it is genuine counseling, where I can go into the feelings of a person. But it is all therapeutic. I have a friend here who is assigned to the family service center, and she is totally exhausted, because she is seeing three or four hundred people a day, and they all have issues. She never has a break, because there they do the processing for relief money, but they also do counseling.

Counselors, and nurses and doctors, have to agree to come for a minimum of two weeks, and other volunteers have to agree to three weeks. The Red Cross pays for our airfare and hotel, and they give us money each day for our food and expenses. When I was in Mississippi for the tornado there, they gave me sixteen dollars a day, and here it is just a little more because the city is expensive. I am staying at the Marriott on Times Square, and I am sure the Marriott gave the Red Cross a good deal, because that is an expensive place.

All of us are hurting by this. Because of what's gone on here probably [went on] all over the country, the same guys are feeling the same hurt. It might be a little closer to home here, but firefighters feel the hurt all over. They are coming here in droves from cities all across the country, just to help us out, to clean up firehouses, to cook for us, to go to bat for us wherever they can. I think we would do the same thing for them. ■

I am finished with whatever work I was able to do here today, and though it is near midnight, Ground Zero is lit up like a stadium night game. As I look around, I see lights on in the interior of the World Financial Center buildings, and also in some of the homes around Battery Park City. It is reassuring to know that these buildings that have been so beat up, especially those of the World Financial Center, with its Winter Garden pavilion that was nearly destroyed, still have life in them. People are in them, getting them ready for a renewal. Work will be done here again, and soon.

Day 39, October 19

I spoke with Vinny Dunne today. When I first met Vinny he was, at 40, the youngest deputy chief I had ever met, and one of the youngest in the department's history, and I remember being very impressed then with both his

516

boyish qualities and his authority. Now, he has recently retired and is devoting his time to writing. It could be said that Vinny is a noted expert on building collapse. He has written several books on the subject and is published regularly in *Firehouse* magazine. I asked him to explain why the World Trade towers had an internal flaw that was largely unknown to professionals in the fire service.

'History,' Chief Dunne stated. 'You have to look at the history. Before 1938 all steel in New York City buildings had to be covered with two inches of brick and concrete. Every bit of vertical steel in the Empire State Building is covered with two inches of brick and concrete, and every stairwell, every elevator shaft, is surrounded by walls of brick and concrete at least two inches thick. The floors in the Empire State Building are made of concrete that is eight inches thick. In contrast, look at the floors in the World Trade Center. They are corrugated steel with just two or three inches of concrete, and the steel had a sprayed fire retardant. The stairs and the elevator shafts were boxed in with one-inch drywall, Sheetrock, for goodness' sake. Everyone knows that fire resistance is directly related to mass, and the architects and engineers have been chipping away at mass since 1938. Everything gets thinner and thinner, everything is more steel and less concrete.'

'What about the construction type?' I ask.

'The World Trade Center buildings have a bearing wall construction, and they are the only buildings in NYC to have this type of construction.' The buildings had an interior support system that consisted of forty-seven vertical beams that were placed around the elevators and stairwells, which held 60 percent of the floor weight. The support of the vertical steel running up the sides, and some trussed steel beneath the floors, held the other 40 percent of the floor weight. 'In the whole country,' Vinny continues, 'I think the Sears Tower in Chicago might be the only other example. All the other high-rises in the city are of skeleton steel construction, which means that there are vertical steel beams in the interior, between floors, as well as on the outside.'

'Would you have predicted that these buildings would have come down, Chief?'

'No,' Chief Dunne replied. 'I would have anticipated that some of the building might collapse because of the high heat load, but I would have stayed inside fighting the fire. Look, the architects who built these buildings were interviewed, and said that the buildings would stay up even if they were hit by two airplanes. But they were wrong.'

Day 40, October 20

The summer seems reluctant to fade into fall on this beautiful Saturday morning. I have come down to the Church of St. Stephen, the first martyr, on East 29th Street, to participate in the thirty-fifth anniversary mass of the 23rd Street fire. On that October night in 1966 a fire in the basement of a drugstore collapsed the flooring and killed twelve firefighters. It was the worst disaster in the history of the New York Fire Department, and there I was working in Engine 82 in the Bronx and not even on the assignment card for a fifth alarm. Several firefighters and I asked Chief Bill Burke if we could take the spare chief's car and go down to the collapse to help. He let three of us go, as I remember, and we worked throughout the night, moving stones and bricks and mortar by hand, until seven the next morning, when the last body was brought out.

It was one of the saddest days in my life. I had never anticipated that there could be a fire in which so many firefighters would die, and I was grief-stricken for months afterward. The funeral for these men ran the length of a mile down Fifth Avenue – one firefighter was brought into Temple Emmanu-El on 65th Street, two were brought into St. Thomas's on 53rd Street, and the rest were brought into St. Patrick's Cathedral. Thousands of firefighters lined up at attention along the avenue, and the band from Power Memorial High School played a dirge as the cortege passed. I never got into the church, but stood in line for more than two hours until the ceremony was done and the coffins carried away on the tops of the fire engines.

There was a great outpouring of generosity from New York's citizens, and a fund was established that ultimately gave each family affected several hundred thousand dollars. I felt grateful that those families would receive such assistance, but something had happened shortly before that tragedy that made me see how a turn of fate can occasionally be unjust.

My next-door neighbor during my childhood, Jack O'Keefe, died in the line of duty just two months prior to the 23rd Street fire, when he was drafting water out of the East River, and four of the twelve men who died at 23rd Street were pallbearers at Jack's funeral. His wife, Ann, had to then learn to make do on Jack's pension for herself and five children, and I always felt disappointed that the O'Keefes were not included in the 23rd Street fund.

My own mother would often say that most of life depended on the luck of the draw, but here was an inequity unwittingly caused by virtuous citizens being moved by the tragedy of the 23rd Street fire to make a contribution to a fund. There was no fund for the O'Keefes, but Ann was never one to make a fuss about things she could not control. She raised her children, and did her best with a family that stayed close and loving through the decades. And it was a good Irish family that had a cop and a priest, too.

If you are going to die in the line of duty, I once heard a firefighter say, it's better if there are others who die with you.

Serving the mass today is Monsignor Jack O'Keefe, and it is great to see him and to take time for a chat afterward. Here also is most of Jack's family, for whom it has become an annual ritual, as Jack has been saying this mass every year since he became a priest more than twenty years ago. I meet his aunt Florence, whom I haven't talked to in thirty years, and I remember that she was the maid of honor at the wedding of Ray Murphy's father, Lieutenant Murphy of Ladder 16. Outside the church I run into Lieutenant Mickey Kross, who is feeling no residual problems from the collapse, and we walk together to a reception that Engine 16 and Ladder 7 are giving to honor the 23rd Street heroes.

While there I chat with Rosemary Cain, whom I have known for thirty years, since she was a neighbor of my brother's in Sunnyside. She has lost her son George in the collapse, and I can see she has been through a lot. They have not found George's remains. I think as I hold her hand of the poem by Patrick Pearse that has a mother saying, 'Lord, you are hard on mothers, you pain us in their coming and their going.'

Rosemary tells me that she just received a letter from a family in New Zealand. 'They were all so sad in New Zealand,' she says. 'They love the U.S., and they will pray for George.

'We are going to keep George's memory alive. He was so athletic, and was going to run in the marathon. And so a friend in Massapequa is running in his place. Also, we are starting a foundation, the George Cain Memorial Foundation, and we hope to bring the children of firefighters who need some restful time to Connecticut where my daughter Erin has friends who have a horse farm. We will bring kids up there, and there are accommodations for several days at a time, and we will teach them all about horses, and how to ride. Erin is a very experienced horsewoman.

'All I can think of is that there are so many missing, that some will lose their identity – no one will ever know who they are. So I was motivated to make sure people will know about George. I went on a television show with my daughters. Anything to let people know what a wonderful man he was. I don't know what else to do. I didn't know where to go. Common sense tells me that George isn't coming out of there, and I have to sit here and wait and wait and wait.

'As long as my son is there I don't want anyone telling me I can't go there, or go to someplace in Brooklyn to get on a boat to take us over there. There should be somewhere I can go to look for my son. How can they ask me to go to Ground Zero just to pick up some stupid little urn? [The mayor has organized a day of mourning at the site, at which the city will present a wooden urn with dust from Ground Zero to the next of kin of each victim.] They will have six hundred families, and it will be chaos.

'My son is there, and I should have the right to go there whenever I want or need to. I feel so close to God when I'm there. I couldn't feel closer to God if I was in the Vatican. It is the holiest of places, acutely holy.

'Our birthdays fall on May 13 and May 16, and so we always share the day. We'd all be together at his sister Nancy's. He loves solitude, could go anywhere by himself, and make friends with everyone. He was so kind and generous — he just drew people to him. And he could laugh at himself, too, at his own expense.'

Rosemary's daughter Nancy is married to a detective in Suffolk County, and Erin, her daughter who does makeup and wardrobe for the television and movie industry, comes over to join her mother. Erin is a slender, attractive young woman, two years younger than her brother George. She lives in Deep River, Connecticut, where she keeps horses. 'I always go to the firehouse when I am in the city,' she says, 'and I usually park my car there. On Monday the tenth we were shooting a commercial for a portable defibrillator down by the World Trade Center, so I went to the firehouse first. George showed me on the map how I should ask a taxi driver to take me there, and so I went down and shot exteriors near West Street. He called my cell and asked me to come to the firehouse for dinner, and when I got there he lifted me up completely and put me on his shoulders, like you would do a baby. George is always funny and sweet like that, a very close person.

'So they had cheese and crackers, and prosciutto out on the table, and we just had a great time in the kitchen. Richard Muldowney was cooking shrimp scampi and Bob Foti was doing the pasta – I am so honored I was with them for this last meal they prepared in the firehouse. George was a marathon runner and he was always encouraging me to quit smoking, and I had just quit the week before. And so Mickey Kross came into the kitchen and said that he has no sympathy for ex-smokers because they are all self-righteous, and Mickey and a few others began throwing cigarettes at me, saying "C'mon, c'mon, forget it, have one." They are such great teasers. It was a wonderful meal, and time, and I left about 11:00 for uptown where I was staying at a girlfriend's apartment. George called a firehouse up there, and asked them if I could park the car by their firehouse. Of course they said I could.

'But George didn't want me to leave. He never wanted me to leave the firehouse, and he kept me standing there in front of the firehouse, just talking, for the longest time.'

Later in the day I drove down to Ground Zero, arriving there about 2 o'clock. Someone tells me at the site today that building 7 is just about completely cleared away, and I walk up Church and down Vesey to see what the level ground looks like. This is the building I watched fall to the ground, taking the offices of the Secret Service and the Securities and Exchange Commission with it, and burying $200 million in gold and silver. It was a sleek, modern, soulless building that still seemed sleek and modern after twenty-four years, but I cannot any longer think of any architecture down here as being soulless. The very air is filled with souls, so many to continue to pray for.

The Indian summer day lingers, and I take off my sweatshirt and wrap it around my waist. The battalion chief who is directing the men is suggesting that they don their face masks. He himself is wearing a mask, and at first I cannot tell who it is. 'I see the guys from OSHA,' the chief explains through the layers of cotton and charcoal in his mask, 'and we should do what they want.' I put my own mask up so that he sees I am complying with the rest of the firefighters there, four or five of them handling a hose. The smoke is still rising on the piles of building 6, just on the other side of this flat field of pressed dirt on which we are standing. The chief turns, and I see his name written on the back of his coat: BATTALION CHIEF SCHOALES.

Ed Schoales is a man I know well and with whom I worked in La Casa Grande. He was in Engine Co. 85, in those days, with Harry Meyers. Even then, Ed was a sweetheart of a guy, and respected by every firefighter I knew. I feel my heart rise to my throat when I see him, for I know that Ed has lost his son here in the piles.

Tom Schoales had almost two years on the job, but because he was a police officer for three years before that, he was nearly at full salary. He was the youngest of four boys and two girls, and worked in rotation at Engine Co. 4, a company that was on the first alarm, and that lost all of its firefighters except the chauffeur of the pumper. Tom's rotation was just ending, and he was so looking forward to returning to his own company, Engine 83 in the South Bronx. But now Tom is still missing somewhere within the piles of the north tower, just a short distance from here, on the other side of tower 6. I offer Ed my condolences, and we talk for a while. He is working in the 15th Battalion in the Bronx and has been detailed down here for the day. It is bittersweet that he is working at a job he knows so well and loves, while feeling the agony of having lost his son in the same job at Ground Zero.

I am again amazed at how clean the outhouses lined up along Church Street have been kept. Still, the graffiti writers will have their way. In the one I have stopped in an arrow points down to the urinal, with the caption:

TALIBAN HELMET. And, on the other side, a bit of doggerel:

> *If you're brave*
> *Come outta your cave*
> *And we'll put you in your grave*
> *Mr. Bin Laden*

As the piles go down, other structures arise. The Environmental Protection Agency has erected a huge, tentlike structure of some three hundred thousand square feet, which covers almost half a block just north of Vesey Street. It is being used for biohazard cleaning and decontamination, though at least half will be available for other purposes.

One very positive development can be seen across West Street: The Winter Garden is being rebuilt. All of the broken glass has been removed from its roof, and the debris has been completely cleared from its interior. The tops of the palm trees growing within it are visible through the skeletal steel, and I can only think of how amazing it is that they have survived the atmosphere down here.

A benefit concert is held tonight at Madison Square Garden, featuring performers like Mick Jagger, Elton John, Billy Joel, as well as many other celebrities and politicians. The firefighters break into wild cheering for Mayor Giuliani, and then, peculiarly, they boo several others of whom they do not approve. It is a strange gesture, for the concert is being internationally televised, and millions and millions of people will be watching, but the firefighters seem to be asserting some other heretofore unstated complaint about what they are experiencing. They are angry, and I don't think anyone can blame them or begrudge them that. They have lost so many of their best friends, and they haven't figured out yet where the blame should lie.

There is just one person who has been consistently blamed since Day 1, and the mood of the evening was best articulated when a firefighter who lost his brother in the attack, Michael Moran, got up on the stage and, taking off his cap, said, 'Hey, bin Laden, take a look at my face, I live in Rockaway, Queens, and you can kiss my royal Irish ass.' Not surprisingly, Madison Square Garden shook with the approval of thousands of police, firefighters, their spouses and supporters.

It is such a happy sight to see Elton John at the piano wearing a firefighter's cap, singing a duet with Billy Joel, wearing the cap of a police officer. The whole country seems to be saying, 'We love New York, and we especially love New York firefighters and New York police officers.'

The last thing I do tonight is return a call to my good friend Hal Bruno. Hal was the chief political correspondent for ABC, but now he is known by all firefighters for his important role as chairman of the National Fallen Firefighters Monument in Maryland.

'I was going through my desk,' Hal tells me, 'and I came across a letter from Michael Boyle. He was helping me on getting a list of firefighters to invite here to the

Fallen Firefighters Monument, and he was just such a terrific, helpful young man. . . . And it makes me just sick to think what his family has had to go through because of this.'

I think of all the families I know who have had to make adjustments in their lives. Yesenia Ielpi can't rely on Jonathan to take the boys to the park anymore, and Jonathan's mother, Anne, misses the telephone calls. Lee Ielpi does not take a day off from the steady determination of his digging across the piles. Donna McLoughlin had the men from Port Authority remodel her house so that it is now wheelchair ready. She has hardly seen her children since Day 1, because she has spent most of her time at the hospital with John. Jean and Dan Potter are looking for an apartment where they can open a window and breathe air that is not thick, and hear something, anything, but the constant clanging of heavy-duty machinery. The losses are different for everyone, but as real as a setting sun.

Day 41, October 21

I stop on the way to the site to read a plaque that has been placed on the side of a modern building on the corner of 23rd Street and Broadway, in the shadow of the Flatiron Building. It is a simple plaque that lists the names of the twelve firefighters who perished on this spot in 1966, and I wonder why there is not a monumental bronze statue here. But then I remember that the midsixties was a period during which the city was deteriorating rapidly, and the power brokers of New York were not of a mindset to build great commemorative statuary. Still, even this modest plaque has the power to inspire a thought and a prayer.

I meet up with Lee Ielpi and Chief Marty McTigue at the site, and we are shortly working in the middle of a pit, twenty-five feet wide and twenty feet deep, not far to the north of where the golden globe stood.

John Salka is in charge of the recovery operation, which suddenly changes from ordinary to intense when multiple bodies are discovered. I am on the slope going down into the pit, and Chief Salka hands me a big, yellow bucket for the body parts, which are handed to me one by one: a scalp, an ear, an arm. I wrap them in a yellow plastic bag.

The firefighters are cutting elevator cable, and Chief Salka throws down a pair of goggles for the man using the partner saw. Another man, though, intercepts them and puts them on. I find it humorous, because if he was actually the backup man, the goggles would actually be useful. As it is, the partner saw operator is throwing up a steady stream of sparks and he has nothing but steady blue eyes to protect him.

It is reassuring to observe Chief Salka at work. He doesn't miss a single trick, watching over and under the Cat and every firefighter in the pit. The objective is to remove a misshapen stairwell that is stuck like a building's foundation in the debris, for they have found four, five, six firefighters in and around it.

Lee Ielpi sees that he can get under the stairwell platform only if a crane lifts it a little. Lee wraps a thick, woven metal cable around the stairs, as a crane line to which to tie it is guided down. The crane pulls, and struggles so much that it seems if it snaps, all eternity will let loose. The platform is finally lifted a little, maybe nine inches, but it is enough for Lee to slide under the stairs and get in position to pull a body out. Chief Fellini is above me watching the operation with Chief Pete Hayden and about fifty firefighters. The time passes slowly, and I stand there, holding my footing on this slope, as the bodies are carried out on Stokes baskets, flags draped over the remains. One after another the bodies come up out of the pit, and I reach for each Stokes basket, along with firefighters from Rescue 4 and Rescue 2, and enginemen from the boroughs. We pass the body from one to the other like a bucket brigade from the bottom of the pit to the flat of the mound above. The process takes an hour of passing and waiting, passing and waiting.

'Make certain,' Chief Salka yells, 'that only men in fire helmets carry the body – no hard hats.'

Salka is ensuring that if there are long-distance cameras trained on us, they will see only the unique helmets of the firefighters carrying out their fallen brothers, crossing the expanse of West Street, up Vesey to the morgue, rows of men and women all the way, saluting. We have taken out as many as nine firefighters, though by the middle of the recovery I can no longer count. The number does not matter, for each is important, and there are so many more yet to find. It is simply progress.

Lee Ielpi tells me that 'it could be twelve, at least Chief Hayden told me it could be twelve because of all the body parts we found. It was a good day.'

On the way out of Ground Zero, past the roar of the machines, and the hustle of hundreds of men and women, firefighters, cops, ironworkers, teamsters, construction workers of all trades, I am thinking now only of the fire-fighters. There are thousands of victims and I think about them all the time. But firefighting is what I know, and so I reflect on these 343 men who were killed in these piles, who were prepared to go into the twin towers strengthened by their many previous confrontations with the unknown. It is not easy to walk or crawl into a situation where nothing definite can be anticipated, apart from the fact that there is something wrong, something seriously out of the ordinary. There is a fire. It could be large, it could be small. It could be sucking desperately for oxygen and waiting to leap over you if you open a door. It could be an innocuous run-of-the-mill fire in a chair or on a stove. Or it could be eating away at the very floor where you stand. Any building can have a crack in the ceiling. Any building can fall down.

As I lie in my bed, my scratchy eyes red and tearing, I relive in my mind what I have experienced today.

Where's the bucket for body parts? Listen up, I need a bucket for body parts and another bucket for tools.

Where do you want them? Right there? I'll send them up to you. What I want to do is, I want to get all three of the

*Stokes and bring them up together. Is that all right with you?
Two feet over that way. That way. I want to get three Stokes
baskets, and then we'll bring them out together, okay?*

We're waiting for a bag here.

*Dennis Smith — here. Body parts. Will you hold that for
me, please?*

*Give him a bag. . . . I got a bag right here. Give me that bag.
This guy had a radio. He had a radio?*

The garbage bag has a radio in it.

*There were three radios, and three men here. I want to
know where those two radios are. Who walked away with them?
Where is that third radio? And that says what on it? Does it
say a company number? Whose was it?*

*Where is that tool bucket? We'll bag the bodies, and then
we'll bring them all out together. They must be able to tag them.
Lee, be careful there.*

Sammy. Sammy. Is that tools? Is that bucket tools?

No, it's body parts.

We're going to have to open this up to get this guy out.

*Why don't we make a hole and just keep on going? No,
let's tie it up and lift it. Get the crane wire down.*

*What is in there? Anything else visible? Radios, helmets,
anything?*

Anything in there? Any body?

*Do we have the chaplain there? Give me that Stokes bas-
ket.*

*Right here, down below. Right here. Dennis, grab the other
end of that.*

*Hey, that's just body parts, you know. That's just body
parts.*

Doesn't matter. After we get him out, we'll keep all this with him, the radio, everything. Where's his helmet? Is his helmet there? You guys can use a tool.

That bag's got to be closed.

I know. I know. Just take it down.

Hey, Sammy, where are you? Zipper this up.

We're bringing in the crane, all right, fellas?

We got to get some of these guys out now.

He's gonna bring them out that way. There's two of them.

We have a chaplain here already? He can stand there, but careful. I'll cover him back there.

I need another bucket, another bucket. We have all this . . . body parts.

Hey, guys, nice and neat. The other one's nice and neat, only fire helmets, no construction helmets. Same with the other ones, do you hear me?

Chief Salka, you wanna wait for the third, we're bringin' two out now.

No, bring those two out and and get 'em going.

Get him down. . . . Hey, listen. . . .

Comin' down, comin' down.

I want a lot of movement here. Clear this area, and watch the cable. All right, we're gonna do this quick and then we'll come back to this area, all right? I don't want anybody hurt. Clear the area.

What do we have there?

Teeth.

What else? It's equipment?

DNA on it, probably.

Put the equipment in the equipment buckets.

Chief Salka?

All right, everybody out of the hole, please, I want everybody . . . Watch the beams.

Everybody that don't have to be here, get out of the way.

Hey, fellas. I need you to get up onto stable ground in case you got to move, in case it snaps, all right?

Coming down. Coming down. Watch your back. Coming down.

Attach it right there. You know how it works?

Yeah, they do. They do.

Trouble. Trouble.

Hey, you got a bag over there?

Here's a shovel.

Big man. We need a chain here.

Hold on. Boy oh boy.

Take the stairs up then. Slowly.

Chief, it's getting smoky here.

Right. Do we have another body bag there?

Uh, up on top. Body bag?

Right there. Hey, Chief Salka.

We need another one.

Hey, we need another one, too.

He's going up. Everybody's going up. Zipper that bag.

We'll continue. This guy's out. Everybody's out. You do your service up there with the chaplain, and then come on back down. Take that body with you.

Everybody on the hill stand still.

We'll get this guy up there. We'll tag him. Then everybody will come up all nice and easy and we'll all go out together. We'll do the line and everything. Once we get out there, then we'll talk. Okay, where's the equipment bucket?

It's up here. What've you got?

A helmet. Take it, please. It should go with that body.

Yeah, but you know, it doesn't matter, because I talked to the guys at the morgue, if it's not on the body, even if we found it next to it, they don't consider it part of the body anyway. So you know what? It's not important that we know who it is. They'll figure it out with DNA.

Okay, let's move them.

Hey, tell Chief Hayden, at least one of these guys is from 10, Engine 10. They're right there on Liberty Street. Let's get somebody over here from Engine 10. Without using the radio.

The fire is coming through, Chief.

Everybody here that is not working, I want four or five guys to give these guys a send-off.

Everybody help right now. Everybody.

Okay, we're going to take three more remains out right now. Anybody that's not part of it is in the way.

John. John. John. It's ten, the guy is ten.

Okay, so we'll keep him last. We'll get the ten guys.

Let's go. Everybody on the edge of the hole, move up.

I need guys, I need people as carriers.

Want to wait until Engine 10 gets here?

Chief Sulka, a radio message, any of your guys that you can take from your operation to send them over to the north tower? It would be greatly appreciated.

Chief, we got one right here.

They need more body bags. Hey, Chief, we have some flags here.

I want a line in here to knock down that fire.

Yeah, I see it, Chief.

That would be great.

Okay, yeah. Okay, guys, we got to pull the next three out now.

Lee, you have it.

Get the flags ready.

Day 42, October 22

There are just two words that come naturally before all others when thinking of firefighters – *duty* and *sacrifice* – for one cannot exist without the other. When I was a young firefighter, standing in Madison Square Park in October 1966, and listening to Chief John O'Hagan talk of our worst tragedy, I felt deep in my heart the sacrifice that those twelve men made. I prayed I would never again have to dig through piles of broken stone and concrete, and never again see another day like it. How young I was, but hardening even then, knowing that my prayers were a natural response, but that they wouldn't work. In the fire department, when duty is the motivating force, sacrifice will always follow. There must have been hundreds of firefighters at Ground Zero who were at the 23rd Street fire. Where was Joe Angelini on that fall night on 23rd Street? He must have been there, too.

I head over for a visit to the firehouse of Rescue 4 and Engine 292. It was here I was working the day President Kennedy was buried, and just a week before that I faced my first significant house fire in a small home in Maspeth, Queens. Captain Finnerty taught me to keep my nose down to the floor until I got into the back bedroom, shifting the hose little by little, letting the men behind do most of the pushing and pulling. Just get to the door, he kept saying, that's all. Get there and open the nozzle. That's it, that's it.

I go back to the kitchen and don't recognize it at all, for it is twice as big as I remember – they have annexed some square footage from the apparatus floor. No one in the firehouse has heard yet how many men we actually took out yesterday. I remember having seen some of the men from Rescue 4 on the piles, and they probably came home and reported it had been a good day. Every member of Rescue 4 who responded to Ground Zero on Day 1 was lost – the captain, the lieutenant, and five firefighters. Captain Hickey was one of them, and just a few months before he had been blown out of a hardware store, in Astoria, Queens, on Father's Day. Lee Ielpi and his son Jonathan heard about the fire on their radio and responded, assisting with the removal of the injured, and taking Harry Ford to the hospital. They then returned to the scene and helped Brian Hickey, whose knee had been smashed fairly severely. 'How is Harry?' Captain Hickey kept asking, but Lee and Jonathan didn't want to say. In the end, three firefighters were killed, John Downing of Ladder 163, and Brian Fahey and Harry Ford of Rescue 4.

Captain Hickey had been on medical leave with injuries from that fire, and had just gone back to full duty. He and his wife, Donna, had dinner with Jean and Dan Potter a week or so before, during which the captain talked about how much he missed the job, being out on leave. He was still limping, but he made up a story for the doctors at the medical office, who put him back to full duty. His second tour back to work was September 11.

I offer my best wishes to everybody in the kitchen, and continue on to Squad 288, the hazmat unit, to say hello again to those men in that firehouse where eighteen were lost. This is Jonathan Ielpi's firehouse, and Phil McArdle's, and Captain Dennis Murphy's, located on Grand Avenue at the junction of the Long Island Expressway.

The kitchen has been turned into something of a war room. Three tables have been set up, at which people are now working, and boxes of information files and folders are spread all around the room. Nineteen envelopes are taped to the room's walls, on each of which is written the name of one of the company's fallen firefighters. Every piece of information about a particular firefighter is placed into his envelope – whether his family or a friend calls, someone comes to visit, a telephone call is received from the department looking for a date of birth, anything. The contents are checked every day, for there are tasks to see to, people to be called, relatives to be consulted. The system has been designed for one purpose – to assist the families.

Captain Dennis Murphy is sitting in the kitchen with his men. He missed being at Ground Zero on 9/11 because of the Father's Day fire, in which he had been blown against a wall and broke his leg. After spending a month in the hospital he was released, but was still on crutches on September 11, and he watched the entire day's events on television. He knew his men would be responding, so he called the firehouse, but they had already left. On crutches he would be useless at the scene of an emergency, so he went down to headquarters, and began answering phones, coordinating information, anything to help out. He would not be rendered helpless on this terrible day.

He told the story of Brian Fahey, a firefighter from Rescue 4 who was killed at the Father's Day fire. Another firefighter, Bronco Pearsall, the drummer in the pipe band, had become the best friend of and helper in the Fahey family, assisting them through the wake and funeral, shopping for food, taking Mom to the dentist, taking the kids to the movies and the Mets games, planning their summer. It is very common after a line-of-duty death for a member of the firehouse to step up and attach himself to a grieving family. Bronco, himself the surrogate son in the household of Harry Meyers, did all he could for the Faheys. The family became dependent on Bronco, loved him, and then on 9/11, just eighty-six days later, Bronco perished at the World Trade Center with all the others from Rescue 4, and the Fahey family was left disconsolate.

Phil McArdle, a large man, comes into the kitchen in hazmat coveralls. He is a nationally recognized hazardous materials expert, and for the last fifteen years has gone regularly to work with the San Francisco bomb squad. Phil, a man who has the reputation of being willing to give you the last dollar in his pocket, is all business. When there are nineteen firefighters lost in your firehouse, there is no time to be anything but all business.

■ Two things we looked at in the beginning. We have the guys who are missing in the piles, and also we needed to focus on the guys who were living. The number-one issue is the families that were left behind. So we asked, 'If something happened to me, what would I want the guys to do for my family?'

The answer is simple – everything that we can do to support the family we will do. So the first thing was information. Information is what the families were starving for right from the beginning: Is he okay? Where is he? We divided the company into groups, and we assigned people to call the families every two hours, just to see how they were doing. We called the families every two hours up until midnight and then we started the next morning at 8 o'clock. For the first twenty-four hours, we called them just to make sure the kids were okay. That would be important to me, if something should happen to me. I would want to know if my kids were okay, and then, what my wife needed immediately, and stuff like that.

Initially, there were three of us calling all the families for the first twenty-four hours, so that they wouldn't be inundated with calls. After the first twenty-four hours we assigned a specific man to each family – the man who would [be responsible] for supplying them whatever information they needed. Within the first week, we had all of the families come to the firehouse here and we had an informational meeting. We made a PowerPoint presentation to show the families just what was going on and what to expect in the future. Usually the fire department can be relied on to be helpful to the families of men lost in the line of duty. But the department is overwhelmed. We told them that we weren't going to give up hope, and based on the meeting we had, we [put together] a survey form [that asked]: Are we doing enough, or are we not doing enough? We had never done anything like this before, so we didn't know.

So we got some feedback from the families and then, based on that, we created a book, a two-inch binder that has every form – city, state, and federal – every resource for a family to get through all the administrative options and requirements when a firefighter dies in the line of duty. And before we gave it to the families, we required the firefighters who were going to be their representatives to sit and read the book themselves. If the family had questions, he would then be able to answer them. If not, then the questions would be directed back to me. It [was a] divide and conquer [strategy] – to be able to give every family individual attention, and provide them contact information with phone numbers, and counseling information, because I thought that was critical for families. Counseling for the kids was going to be very important, because if the kids weren't okay, the wives weren't going to be okay. It was important that we give a lot of attention to this. Then we have financial information, and information about how to conduct a funeral or ceremonies. [We explained] what the department would provide, and also, what the department cannot provide – if the job couldn't provide a bugler, for example, we will make sure [one is found]. If they can't provide a caisson, we make sure it gets done. If it can't provide transportation for the family, we make certain we have transportation for the family. Essentially, we told the families that we didn't want to see them [responsible for] any out-of-pocket expenses. One of the things that we did very early on in this firehouse [was consult] the union delegate manual, which said that I'm supposed to open an account. So I had one of the guys open an account, and the very first check that the families received was from us. The money came from the community at large.

We actually lost eighteen from this firehouse, but we took Chief John Fanning and included him with us, because he was so much a part of our team, and because he was a headquarters chief, there was no team to care for his family, as there would be in a firehouse.

Of the guys who survived, there were five of us who were down there when the event occurred: myself, Jeff Polkowski, Herman Karolian, Bob

Hunter, and Anthony Castagno. We were all in different areas. My situation was that I was getting off work that morning, early, because I was going to a meeting with Chief Fanning, to discuss a hazardous materials training program. So the guy who relieved me to go home, he's missing. I was late to the meeting with Chief Fanning . . . and I went to Ground Zero to meet up with him . . . ■

It is a long day at Ground Zero today. The pumps along the slurry wall are in place and working, and entire sections of the site have been taken over by the construction companies, with trailers or small buildings to house their offices. There are no longer any tables along the periphery offering flashlights, gloves, and refreshments, and some of the tents have given way to storefronts, including ones taken by the Salvation Army on Liberty Street, near the river, and by the medical supply center.

A small crowd is waiting to have their boots cleaned at the Marriott wash station, and so I fall in with them and then head upstairs for a meal. I see a few police officers I had met the night Sergeant Orlik drove me out of the rain and take a seat with them. Detective 1st Grade Hal Sherman explains, 'This is the biggest crime scene in our history, about four thousand murders. We need to document and classify every piece of evidence we can.'

The crime scene unit has gone in recent years from investigating two thousand homicides a year in New York to fewer than eight hundred. But, now, there will be about four thousand added to the seven hundred or so that already occurred in 2001.

At Ground Zero the CSU is responsible for photographing the site, recovering physical evidence, documenting body parts and any other physical evidence like weapons or wallet, manning the temporary morgue at the site (as well as the city morgue up on 28th Street), inspecting the debris that leaves the site, and inspecting the debris as it gets sifted out at the Staten Island landfill.

'We are stretched out to the max with our regular homicide investigations, for we only have fifty-five detectives in all, including six sergeants, one lieutenant, and one inspector. All evidence is documented – airplane parts were essential to the beginning investigation, but now they look for any little thing that fits into our categories of evidence – fingerprints, blood, hair, fibers, glass particles, semen, ballistics (weapons, bullets, discharged shells, trajectory). We document everything using photos, diagrams, 3D drawings, and aerial photos. We ID every part. Pillars and beams are swiped for hair tissue and blood, evaporated body evidence. We have two police officers with mortuary degrees, and they are either in the medical examiner's office or the police lab, because you must be a sworn police officer to take evidence.

'If you step on a fly ten times there will be nothing left. And here we have no couches, no computers, no chairs, no glass. Any small trace of anything is evidence. Anything to bring closure to the families. Human body part, clothing, jewelry, equipment and tools, anything. If there is ever a trial, we will be prepared.

'We've been there from Day 1, and we'll be there well after the regular police officers go home, when everyone is packed up and gone.'

Mayor Giuliani today swore in 307 new probationary firefighters. 'The reality is,' the mayor said, 'that we're mournful, we're crying and we always will, but we are stronger than we were before.'

There are 4207 missing, and 473 confirmed dead at the World Trade Center.

Day 43, October 23

As of today there are 4129 missing and 478 confirmed dead.

Day 44, October 24

This morning's *New York Times* reported that its unofficial tally of the missing and dead at Ground Zero, including the passengers on the two planes, comes to about 2950. This figure is 1657 less than the number of victims that had been reported in the previous day's official tally. The police department, charged with calculating the official count, has said that its goal was not to be prompt, but accurate. However, *The New York Times, USA Today,* and the Associated Press each had a figure of between 2600 and 2950.

Surely there were many bystanders, visitors, and undocumented workers who would not show up immediately on such lists, but we are now six weeks from September 11, and the consensus among the news organizations is that the number of casualties will be fewer lost than first expected. Of course, we have seen the official number revised downward as the reports of the missing and the dead are refined. It will ultimately take old-fashioned police work to arrive at as near to a definitive number as possible, by sending detectives to interview next of kin, and rooting out mistakes and possible frauds. Even though the final count will be lower than anticipated, it would be difficult to take this as a victory of sorts, because the sadness that exists in our department, our city, and our country will remain overwhelming.

In the late afternoon Harry Meyers responded to a report of the collapse of fourteen floors of scaffolding on a building on lower Park Avenue. Five men were killed, and ten were severely injured. Firefighters and police emergency squads worked for hours to cut away the piping to free all the laborers. In all, thirty-six were injured, of whom twenty-nine were accounted for and seven ran away, presumably because they had no documentation permitting them to work in our country. And so, the work of our first responders continues.

Day 45, October 25

At St. Patrick's Cathedral my thoughts are on Chief Bill Burke as I watch the fire truck approaching with the photo of his son Billy, whom we will memorialize today. Behind me are two ladder trucks with crossed aerial ladders, one from Hempstead, and the other from Manhasset. FDNY doesn't have enough trucks to send around to the services.

When I was first assigned to Engine Co. 82 in June of 1966, I had to call Bill Burke, because I needed a day off. My son Dennis had just been born, and it would have been easier for us if I could stay home with Brendan, the other little one. Chief Burke didn't give me a yes or no, because I suspect he did not have the official authority to grant my request, but he simply said, 'You just take care of your family, Dennis.'

William F. Burke, Jr., was just 11 years old when my son was born, and like his father he grew up to be a firefighter, but he was also a professional lifeguard, which perhaps suited his unmarried life. A friend of his said, 'To know Billy was to make you want to be a better person.' I thought this same thing of his father, and I am sure they both had pure hearts. I remember how the chief introduced me to his son when he first came on the job, not making anything special of it, just a casual meeting between firefighters in a crowd.

Mayor Giuliani spoke at the service and said that 'Bill Burke died for freedom, to protect our way of life.' I have never thought of this horrendous tragedy in such patriotic terms until now, but of course, the mayor is right. Billy Burke was doing his job, just as every honest, straightforward, hardworking person in our country protects our way of life.

Another friend said, 'People don't die if they are remembered, so keep telling those Billy Burke stories.'

There are 3958 missing, and 506 confirmed dead at the World Trade Center.

Day 46, October 26

Forty-five days after the attacks, the soft weather of summer has finally begun to give way to the anticipation of a hard and cold winter. I push my collar up against the wind as I cross the wide expanse of blacktop, and into the back entrance of the large firehouse of the Vigilant Engine and Hook and Ladder Company in Great Neck, Long Island. Jonathan Ielpi was a firefighter here with these volunteers for more than twelve years, since he turned seventeen and became old enough to be a member. It was his life's dream to be a firefighter, a dream he could hardly avoid having in the Ielpi family, which lived and breathed fire departments, both the paid and the volunteer.

The meeting hall, an expansive, low-ceilinged room, is filled with several hundred people, most in uniform, and as I pass a friend named Joe Daley, a lieutenant from Rescue 3, he tells me, 'They are just closing the reception line, Dennis.' I rush down the center aisle, for I want to make certain that Lee knows that I have come to share this farewell to Jonathan with his family. Lee is at the front row, before a hard, wooden church bench. He gives me a hug around the neck, and says, 'Thanks for coming, Dennis.'

I think, as I always think whenever I see Lee, that this is a full man. He is at one with himself, and knows that he has done his best in every situation in which he has found himself. The eight rows of medals on his uniform are testament to what everyone who knows him understands – just how much he can contribute when people are in trouble. It is a great thing to be so admired, but Lee is a humble man, and I realize as he introduces me to his family, to his wife, Anne, and his daughter-in-law, Yesenia, that his family has inherited his gift of humility. What moves Lee, I know, has only to do with the shared heart. No one can know what it might be like to lose a son, especially one who has become everything you wanted him to be and who is so very young. Not unless it has happened to you. Though Lee keeps humor and determination in his eyes, he cannot hide the grief and the sense of utter loss. To be with him even in these terrible circumstances has more meaning for me than being anywhere else in my life. I see in Lee the true and fundamental texture of being human. And I see how his faith keeps him strong for those around him. The memorial service begins, and I stand off to the side of the room, against a paneled wall. Lee has his arm around Anne's neck, just beneath her graying hair, and she is holding his hand, rubbing the back of it against her cheek from time to time. His grandson Austin, Jonathan's son, has found a perch beneath the altar placed before us at the front of the room. The altar, a kind of bier, is covered with a blanket of deep red roses, upon which rest two fire helmets — one is Jonathan's from the Vigilant Engine and Hook and Ladder Company, the other his helmet from Squad 288 of the New York Fire Department.

An old-timer from the volunteer company, Warren Hecht, delivers one of several eulogies. When he isn't fighting fires, he is a dentist. Dr. Hecht has known Jonathan for a lifetime, and said, 'Jonathan is with us. I will not refer to him as gone. I will tell stories about him. I will tell how when he was a boy of nine, his father in Brooklyn working in the firehouse, his mother called and asked me to come over to bury a family cat that had died. I dug a trench in the backyard for the cat, and when I looked up I saw Jonathan with his nose pressed against a window, his cheeks wet with the tears that were streaming from his eyes. This is the way I see Jonathan today, as a boy, and as a man, who loves and respects life.'

Lee, Anne, and Yesenia all share a small, knowing smile. Yesenia is very beautiful, with large brown eyes beneath a thick wave of black hair. She is mindful of her children, who are climbing on and off her lap. If the children grow impatient during the ceremony, no one could possibly mind.

A large photo of Jonathan stands at the rear of the room, and even at this distance I can see what a handsome fellow he was. I am sure everyone in the room has a similar thought this evening – that God would have blessed this family, and all the families affected by their loved one's being missing, by allowing the fire department to recover Jonathan's remains. Lee held back from setting a date for a memorial service during all those many weeks since the attacks, during which he climbed the piles and shoveled out every crevice that looked as if it had potential for a void, for it is in the voids that bodies are being found.

Several chaplains speak and afterward the firefighters begin to approach the shrine of roses and helmets. They come to attention, and bring their hands to their caps in the slow-motion salute of remembrance and respect. Austin is back on his perch below the helmets and the blanket of roses. He is sitting directly in front of the firefighters and waves his little hand up at them as they salute. He does this again and again, through a hundred firefighters, until his big brother, Andrew, appears in the line, accompanied by a firefighter from Jonathan's company. Andrew is wearing a fire department shirt and his father's dress cap, Jonathan's chrome badge glistening below at the peak. He performs a slow-motion salute, and from under the bier Austin waves up at him. But Andrew does not acknowledge his little brother. He remains at attention, then snaps his hand to his side in military fashion and turns with the firefighter to march back up the middle aisle. I look over at Lee, and I can see how proud he is of these two boys. They have Ielpi blood and the Ielpi character. *Get out there. Do the job.*

Get it done. And, if you can, add a little good humor as well.

After the ceremony, people congregate. Jonathan's wife has love and depth in her eyes, and she makes a connection with every person she meets. The boys are around her now, and I see how Yesenia's good looks and Jonathan's have been passed on to them. But I can also see how the hearts in this family are what will be forever impressed in my memory.

Joe Daley rejoins me. 'How many times,' he asks, 'can we do this? This sadness is just endless.' I calculate in my mind, if there were as many as five services each day, it would take more than two months to honor the firefighters. And, then, there are the police and the Port Authority police, and the thousands of others.

Before us on a stand is a montage of family photographs, and one, in black and white, captures my eye. It is of Lee and Anne, and the four children, not one taller than their parents' hips. It was taken on the steps of City Hall one Medal Day, and Lee is wearing a ribbon and medal around his neck, one of the many he has received. It is hard to take my eyes from the photo, and as I am standing there contemplatively, Joe Casaburi, the former chief of department of the New York Fire Department, approaches and gives me a bear hug, which is an appropriate gesture for a man who is known in the department as being a warm and considerate leader. When I was a young firefighter I wrote a training film for our Saf-T-Harness, a lifesaving belt that every firefighter wears, and Chief Casaburi was the head of the training division at the time. We became good friends, a friendship that lasted through the years and his rise through the ranks to become the highest-ranked uniformed firefighter in the largest fire department of the world.

Others come to greet the chief and I notice that he gives everyone an equally affectionate hug. He has obviously made a commitment to himself not to miss any opportunity to express to every firefighter just how he feels. I remember Lieutenant Warren Smith's notice alerting every firefighter that he was going to get a hug from him whether he wanted it or not, and I can see that that sentiment has spread like the gospel through the department. The chief and I talk about Lee and his son, and I point to the black-and-white photo. Suddenly I feel a great burst of sadness within, and I try to control my breathing, but my voice cracks completely as I begin to say, 'What a beautiful family Lee Ielpi has.' And now, I cannot speak any longer, as I feel the tears clouding my eyes. Chief Casaburi begins to rub my back.

'It's such a difficult time for us, Dennis,' he says.

Day 47, October 27

Vinny Dunne calls to tell me that John T. O'Hagan, one of our most famous fire chiefs, was very much opposed to lightweight construction, and to large, wide-open spaces in high-rises, and to spray-on fireproofing of steel beams. Yet even during Commissioner O'Hagan's tenure through the late sixties and early seventies, buildings with these features were built. He was against the building of the World Trade Center, but his objections had no effect, for the project was being constructed by the Port Authority of New York & New Jersey, which by law could circumvent the established fire codes of New York.

Day 48, October 28

A ceremony is held at the site today, at which a rabbi, a Muslim cleric, a minister, and Edward Cardinal Egan speak. It has been designated as a day of mourning, and the city will give a wooden urn containing the dust of Ground Zero to each participating family that has suffered the loss of a loved one. Cardinal Egan prays for the victims, and laments the 'futures snuffed out by villains filled with hate.' Andrea Bocelli sings 'Ave Maria,' his salt and pepper hair waving in the wind. Behind him four hose lines are keeping the smoke of the piles momentarily suppressed. David Rodriguez, a New York City police officer who should be singing at the Metropolitan Opera, or at least on tour with Andrea Bocelli, sings the national anthem. I am moved and proud to see someone so talented wearing the blue uniform of the police department.

Day 49, October 29

I am thinking today that a reformation and a renaissance are needed in the Middle East. The culture and the scientific contributions of the Islamic countries have been so great in the past, one wonders why they have allowed themselves to be intellectually marginalized. Arabia, Persia, and Syria were once, during the period of our Middle Ages, the scientific center of the world.

When the prophet Muhammad went from Mecca to Medina in 622, he constructed in his mind a world and life's view that resulted in the holy book the Koran, or Quran. This book gave rise to an enlightenment period in Islam because of its teaching to 'seek knowledge.' And so the precursor to our modern university system was first developed in that part of the world, and discoveries were made that changed the world and the world's thinking. They translated and ingested Greek knowledge before the West had any sense that it was important. The Arabic-speaking people discovered the concept of zero; they developed non-Euclidian geometry, and algebra. Our system of numbers and numbering is Arabic. The precise position of the stars was calculated and mapped at the Samarkand Observatory one hundred years before Galileo. They were the first to develop empiricism, and it was their knowledge that fed the Italian Renaissance and our Western scientific revolution.

The formalized study of the history of Islamic science is a field that has not yet been created in our

Western universities, yet Islamic thought was at the very forefront of scientific thinking for five hundred years.

I realize that every statement made about the history of Islam requires chapters of reference and supportive documentation, but I think about this subject only because I know that it concerns in some way every shovel full of Ground Zero debris that Lee Ielpi moves. It is the backward thinking of large segments of the Islamic world that led to the attack on our country, and to me this is more regrettable and indictable because the Islamic people were once the intellectual leaders of the world. Science and open thinking are encouraged by Islamic teachings, but the progressive nature of openness has been effectively skewered by the repressive ideas and practices of bin Laden, Al Qaeda, the Taliban, and the whole bunch of them.

It is reported that 3958 people are missing and 506 are confirmed dead.

Day 50, October 30

Steve Dannhauser and I are sitting in a mahogany-walled private dining room at the top of a Park Avenue building. The views of the cityscape all around us are extraordinary, and they can be viewed either as brash expressions of power or as an inspiration for poetry. I wish Robert Frost were here. We have been invited here for breakfast by Jimmy Lee, the vice chairman of

J. P. Morgan Chase, because the bank is developing a special leadership program in the wake of Ground Zero. Two women appear, and we all take seats at the table and make polite conversation until the telephone rings, and Mr. Lee explains he has to cancel his appointment with us because of some urgent company development. Of course, J. P. Morgan Chase is one of the largest banks in the world, with ninety-five thousand employees, so this is entirely understandable.

I am given a moment here among this canyon of eighty-story buildings, however, to reflect on how advanced we are as a society. Our growth in America through the industrial, electronic, and computer revolutions has been based on our openness and our ability to think in new ways, which is precisely what Mr. Lee wishes to impart to his executives. How does one think with courage, and how do our firefighters and police officers think in ways that might manifest courage? The answer probably falls between the first responder's ability to honestly assess a situation, without a hidden agenda or prejudice, and his ability to act in a forthright manner based on his belief in himself. Of course, if a responder does not have these qualities, he might get himself killed, and place others in jeopardy as well. Executives have a little more latitude, and it would be worthwhile for them to recognize that they are privileged in this respect.

Day 51, October 31

There is a phenomenon called the rumor mill, and it is not an insignificant phenomenon, for it can have terrible consequences. At Ground Zero I heard, at various times, that U-Haul trucks were missing in great numbers, that bin Laden got all his money from a mine he owns in Colorado, that a man rode an I beam eighty-six stories down a collapsing tower and survived, that Joe Dunne had been lost in the collapse, and that there would be another catastrophe today, October 31. Such rumors, which can be divisive and destructive, inevitably develop after every newsworthy event, whether it is a storm, flood, fire, or earthquake. Today's major rumor supposedly originated with a friend of a friend who was engaged to a Muslim man, who left her without explanation. Before he disappeared, however, he did warn her not to travel on an airplane on September 11. He also urged her to keep away from any shopping mall on October 31. I know there are a great many people avoiding shopping malls today, and it is a shame that their time has been wasted by this idle gossip. Also, a great deal of money was undoubtedly lost by businesses based in malls.

I will wait to see if this particular rumor can be associated with the prophecies of Nostradamus, but in the meantime I have been alarmed by a featured item in the morning paper. It was announced today that the city is reducing the presence at Ground Zero of the fire department, police department, and Port Authority police to 25 personnel from each uniformed service for each eight-hour shift, down from between 60 and 80. It also stated that the uniformed men and women would be placed on a stand-by basis until a crane or a Cat or an excavator exposed a body, and then the uniformed group would be summoned in to remove it. There would be no more continual hand digging, and no more on-the-site spotting. The first thing that strikes me as I read this piece is that there are still 250 firefighters unaccounted for out of 343 reported missing. I remember George Cain's mother, and her admonition to me: 'I want people to remember my son Georgie, and I don't want him to be part of a number.' More than a thousand children of firefighters have been left fatherless, and there is a valley of tears among them. This decision to reduce personnel is so impersonal and mathematical that I am certain there will be a strong reaction from the firefighters and police officers.

Also, more than $200 million in gold and silver was recovered today from a vault deep in the basement area that was leased by the Bank of Nova Scotia. This represents a significant treasure – 379,036 ounces of gold, and 29,942,619 ounces of silver – and more than twice the 14,000 pounds of gold that was said to be buried beneath building 7. The bank is fortunate that the bullion survived, and that it didn't end up like First Deputy Police Commissioner Joe Dunne's car and sunglasses. It is found money, and my mother always taught that found money has a special luck attached to it if you give it away. I allow myself to imagine for a moment that this gold and silver might be donated to appropriate charities, charities caring for the families of those who died here. The firefighters might then not feel that the city places more value on recovering precious metal than it does on recovering the bodies of its heroes.

In the evening I meet Vina Drennan at a coffee shop at 86th Street and 1st Avenue. I thought of Vina Drennan as soon as I heard that Paddy Brown was among the fallen brothers, for Paddy had always been there for Vina through her long suffering with her husband, John. The whole job knew that when John Drennan died, he was leaving a legacy of love behind, and that Paddy Brown was there every moment he was needed, in the hospital, at the wake, at the funeral, everywhere.

John Drennan was the captain of Ladder 5, down on Houston Street, and he loved Irish music. He knew that I played the bodhran with an Irish band on Thursday nights up at Paddy Reilly's on Second Avenue, and used to stop in every once in a while with a few firefighters. On the night of March 28, 1994, he went into a fire, an extraordinarily hot one. Captain Drennan was searching above the fire, and the heat increased so rapidly that he and two of his firefighters were suddenly on the floor, unable to move. Jim Young was dead when they brought them out, and Chris Siedenberg died the following day. But John Drennan, a big and robust man, lived through major burns for forty days and forty nights, fighting every inch of the way, until one clear and beautiful May afternoon his heart gave out, and he closed his eyes forever.

Vina was also a good, close, and longtime friend of

the department chaplain, Father Mychal Judge, and I can tell that she loves to share her memories of the Franciscan. She is now smiling, relating how people were always trying to do something nice for Mychal, because he was always doing something nice for others. "One fruit basket is lovely,' he said to me once, 'but fifteen baskets is a nightmare – have you ever seen fifteen baskets of rotting fruit?"

■ Mychal would get overwhelmed, hundred of people asking him for something. Can you come to my promotion party? Baptize my child? Marry me and the girl of my dreams?

He would spend so much time with me, and after that there were twenty-eight more widows to follow up with . . . until 9/11. He flew to Ithaca to marry my daughter. He was comfortable with the mayor, Dr. Ruth, the Clintons. He was at the White House, and Hillary sat next to him. She talked about it at his funeral.

We all gave a dollar to the homeless, but Mychal knew their names, and whenever I was with him, I felt his touch, and with it came God's presence.

I felt that his sense of Christian faith was the best I knew. I am a Lutheran, and he was always aware of my faith. At the hospital, he once said to me, 'I love Lutherans.' Two other widows close to him were Lutherans, and of course his two sisters were Episcopalians, so he was surrounded by Protestants.

After forty days of being with John Drennan we

were at the funeral. My family, Paddy Brown, Mychal. And, there in the church, a butterfly came and sat on Mychal – the symbol of resurrection. It was a beautiful moment.

Mychal became a brother figure in my life, a constant. He'd call at 11:30 at night, wake me up, and say, 'Did I wake you up?'

'Yes, Mychal, you woke me up.'

"Oh, I'm sorry. Do you want to talk?"

Mychal would never criticize anyone, but he was fed up with the deaths. He felt God was using John's death through me to make firefighters safe. I keep writing these diaries about it.

'Don't ever stop writing,' he told me.

The last conversation I had with Father Judge was about the Father's Day Fire. Two teenagers had started the fire in the back of a hardware store, and then three firefighters were killed. Why weren't those kids prosecuted? I kept asking him. Made an example. We have no juvenile fire-setter programs. I kept complaining. Michael said I should have written it for the *Times.* If those two kids threw a brick down into a car on the highway with three cops, and the cops were killed, they surely would have been prosecuted.

Father Judge believed God was using me for this safety program. He was proud of me, I think, I hope. I did my best. I never could picture this man dying.

'I'll always be there for you' is sometimes an empty promise. And Mychal thought it was better not to promise at all, if it couldn't be kept. Firefighters often can't remember their own mothers' birthdays, and why should we expect them to remember a widow's birthday?

Now that the phone doesn't ring any longer, I have to think of what I have to remember about him. He would laugh if he realized he was going to be in the middle of a building collapse. 'If you want to make God laugh,' he once said, 'tell him your plans for tomorrow.'

'Today is what is important,' Mychal advised me. 'Your job might be simply to sow the seeds.'

After John died Mychal opened me up to possibilities. I could have been a bitter old biddy. And no one would have criticized, for we had been through so much. But he gave me a sense of an obligation to sow seeds.

I feel still I know what Mychal wants me to do. Mychal was sixty-seven. 'I should be retired,' he used to say. 'Guys my age are all sitting on golf courses.'

He was a twin but the sister was born two days later. He said his mother never forgave his sister for lingering. He loved his two sisters very much, and was always talking about them. Born in Brooklyn, he could sing all the Irish songs, the ones you never heard.

Mychal went to stern Catholic schools. He always sided with the underdog – a natural follower of St. Francis. He believed in keeping everything simple, the Campbell's soup syndrome. Soup is easy. People just want simple. Violence and intolerance are so complicated, and soup is so simple.

He always wanted to be a fireman and a priest. Now he is saying, with a proud chuckle, 'There I was, serving my firefighters – who would believe it that I'm a line-of-duty death?' ■

It is great to see Vina, for I have heard her many times delivering inspirational speeches on fire prevention and firefighter safety, and have believed, like Father Judge, that she had been given a mission. I say goodnight and tell her to keep writing. My last thoughts of the day are about Vina Drennan, and losing two of her very good friends. 'If my head could know it,' she said in reference to the number of firefighters gone, 'I could convince my heart to accept this grief. But, this seems like an illness, and I don't know the cause.

'Six weeks after a death,' Vina explained, 'grief turns to anger, unless these emotions are channeled into something positive. In New York City, the grief is turning to anger in our firehouses.'

Day 52, November 1

It hit all the papers today that the city was reducing the work force of firefighters at the site of Ground Zero to twenty-five. I know that firefighters have been anticipating this, and it will not be received with an acquiescence of any kind.

Day 53, November 2

More than a thousand firefighters have gathered at the site to protest their diminished numbers on the piles, and they are very upset. The city has cited safety advice from experts as the reason for the reduction, a rationale completely rejected by the firefighters. There have been only two significant accidents at the site since Day 1 – Tom Beattie, the ironworker of Local 40 who did damage to his shoulder, and another ironworker who broke his leg when a taut piece of steel snapped back at him after it was cut. There has not been a single firefighter injury, apart from what they are calling the World Trade Center cough. Who would know better about safety concerns than the very people who are charged with protecting the public safety – our firefighters? No, safety may be the stated reason for this decision, but everyone recognizes that the city understandably wants to make the area viable for business, and the slow process of looking for bodies is delaying that.

Until now, whenever a spotter or a digger found a body part – even one as small as a finger – all the machines in the area would stop for a comprehensive dig by hand. It seems the city now wants to continue lifting and dumping at all times (scoop and dump, as the firefighters say), except when an actual intact body is seen, while hoping that any parts of bodies that are in the mix will be separated out at the landfill area in Staten Island. I have heard firefighters ask, Why would one even discuss the passing of time when there are still so many bodies beneath us? And why are we thinking of businesses and the return to business as usual when so many of our brothers are still unrecovered?

Firefighters have complained, 'Now that they have the gold and silver out, they can take the rest to the dump.' 'They,' in this case, are the politicians, and this observation seems to be the expression of a frustrated sense of powerlessness. The firefighters have no advocate other than their union leaders and the fire commissioner, but even the commissioner is an important member of the mayor's team.

And so the union leaders took it upon themselves to organize a rally at Ground Zero today. The police stopped the march through the site, which led to some shoving and some resistance. A metal barrier was lifted and shoved against the police, bruising several of them. It lasted just a few moments, this pitting of firefighters against the police, and I believe that everyone quickly regretted it. As soon as I saw the video I wondered what was going through Captain John Vigiano's mind as he saw it. To have lost two sons, one a firefighter the other a police officer, makes him the person with the most authority to have a view on this subject, and I know this situation must be breaking his heart.

As I wait for a train in the subway, on my way to write all day, as I did yesterday and last night, a young Asian man comes into the station with a small stool and a one-stringed instrument in his hand, an Asian fiddle. It has the most unique and potent sound, a fierce vibrato, mournful and ardent. He begins to play 'Hey Jude' in the most yearning way. I have never heard it so beautifully interpreted, and as I recall its lyrics I begin to weep. I cannot control it, and I am by myself. I lean against a wall by a newsstand to gather my thoughts. 'Take a sad song, and make it better.' These words say it all about the firefighters at the site. They are living the sad song. But people do not realize the true depth of their sadness, and their inconsolability, and know only they are heroes, and to be admired for their heroism. I wish I could make it better, but all I can really do is lean against a subway post and hope for the best.

For the first time, the mayor has acknowledged that specialists have told him that because of the great force of the collapse, many if not most of the bodies at Ground Zero have been disintegrated, or atomized. Certainly, every firefighter who has been at the site knows that there is not a piece of glass or marble to be seen anywhere, not a desk, a sink, or a doorknob. It is the virtually indestructible bunker clothes of the firefighters that have preserved their remains, coupled with the fact that most of them were in stairwells that had odd configurations made from heavy poured concrete, which created crevices. Of the 542 confirmed deaths at the World Trade Center, about 100 have been firefighters.

News of Ground Zero is everywhere today, on television, on the radio, and in the afternoon editions of the tabloids. Nearly a dozen firefighters have been arrested, because, I believe, the police commissioner and Mayor

Giuliani have taken a surprisingly hard line. 'We cannot permit our police officers to be hit,' the mayor has explained. Yet the firefighters were arrested for trespassing on Ground Zero, and not for felonious assault. The firefighters themselves are furious with Commissioner Von Essen and Mayor Giuliani, and have called for their resignations. In fact, both are concentrated, hardworking men, men who have made their best efforts to attend every wake, funeral, and memorial service, men who have a sincere dedication to honoring the city's fallen heroes. It seems they have been ill-advised by people who are paid well to advise them correctly, and that they have not reliably taken the pulse of the firehouses, nor gauged the support that the firefighters have among the population.

If Vina Drennan is right about grief turning to anger after six weeks, the anger will quickly turn to rage if the mayor doesn't settle this dispute quickly. During the confrontation, the firefighters – who were chanting, 'Bring our brothers home!' – were told by the police brass to turn back at the checkpoint for Ground Zero.

Retired Captain Bill Butler spoke to the crowd. 'My son Tommy of Squad 1 is still not home. We haven't gotten to him yet. Don't abandon him.'

The firefighters then began to chant, 'Bring Tommy home!'

The march was stopped at the metal barriers, and with the confrontation now face-to-face, a top police chief ordered a group of officers to handcuff several firefighters who were at the front of the line. Pete Bondy was one of these firefighters, and Lee Ielpi, his wife, Anne, and his daughter-in-law Yesenia were off just to the side, with fifteen or so wives of missing firefighters.

'I'm sorry I have to do this,' the officer said as he grabbed Pete's wrist. Pete, retired after twenty-one years in Rescue 2, and one of the legends of the fire department, did not try to resist. But at that moment, another chief down the line ordered another policeman to hand-cuff additional firefighters, Captain Jack Ginty of the Fire Officers Association among them. 'We were being arrested for no discernible reason,' Captain Ginty said. 'It was then that things heated up, and the barrier was raised and used to push back the police officers.' With everyone's attention diverted, the police officer took the handcuff off Pete's wrist and said, 'Beat it.'

Later, an officer who was driving the handcuffed fire Captain Jack Ginty to the police station became so upset over the senselessness of the event that he had to pull his car over to the side of the road to bring his sobbing under control.

Kevin Gallagher, the president of the firefighters' union, called to the surrounding police officers, 'I know you have brothers in there, too, so come join us.'

In the end, the firefighters were let past the barriers. The crowd stopped in front of the piles on West Street, and someone began to say the Lord's Prayer, and soon everyone was praying aloud, firefighters, wives, children, cops, and construction workers, until they reached the end, '. . . and lead us not into temptation, but deliver us from evil.'

And then, they went home.

Captain Ginty told me the sergeants, lieutenants, and captains of the police department could not have been nicer or more considerate. The big bosses, however, must have had different orders because they would not permit the firefighters to see lawyers, and at one point they were taken out of the holding cells to be photographed. 'How can you photograph me,' Captain Ginty said, 'when you are arresting me for trespassing?'

'"We have been told to do it by the mayor's office," Captain O'Connell of the police department told me, and they took five Polaroid photos of me. They tried then to jack up the charge to rioting, but the district attorney told them there was no evidence of rioting. They wanted then to charge me with conspiracy to riot, but I asked them who I was conspiring with since I was by myself and had not said a word to anyone. So after the charge of trespassing was made, the DA threw the charge out for lack of evidence. But just remember this, they had video of everything, and not one firefighter was seen to punch a police officer.'

That short prayer in front of the piles should have been allowed to happen to begin with. The firefighters should have been allowed to walk anywhere they wanted to, for they had earned that right. The fact that these men were arrested and portrayed by the city officials as some sort of hooligans is a black mark on an effort that has been distinguished until now only with the gold and silver ink of the heroism of firefighters and police officers.

Pete Bielfeld is being memorialized today. I wait on Madison Avenue for Dan Potter and his pickup truck for a lift to St. Frances de Chantal in the Bronx, the same church where Ray Murphy's funeral was held several weeks ago.

There is much coverage in the newspapers of yesterday's altercation between the firefighters and the police. The *Times* termed it a 'scuffle,' and I suppose that is a legitimate characterization, though it lasted far less time than the 'scuffles' I have known. It was definitely not a 'melee' or a 'brawl,' as the tabloids have suggested.

Before the memorial service, Dan Potter and I stop in at the funeral home to offer our condolences to the family. Twenty or so men from Ladder 42 are standing at the entrance, and I greet them. This is also the fire commissioner's old fire company, and one I worked with hundreds of times at fires in the 6th Division. I go to the shrine at the front of the room to say a prayer, and there I see Pete's worn-out fire helmet, a cigar, a deck of cards, a pair of white gloves, and a football – the symbols of his life. He loved playing cards as much as he did chewing on a cigar. He won a football scholarship to Wesley College, in Delaware, where they recently retired his jersey and the number 42, and the white gloves symbolize the purity of his affection for the fire department. Off to the right is the small, wooden urn that was given by the city to the family, filled with the dust collected from Ground Zero. Another memento on the shrine is

571

an old Purple Heart in a purple velvet box, and I ask Pete's father about it. 'Someone just stopped in and told us that a man in his family won it in World War Two. He said it more properly belonged with Peter now, and he gave it to us anonymously.'

Dan stays back, and I know he feels a special connection to Pete, for on September 11 they began to enter the lobby of the south tower together when Dan said, 'I'd better go to 10&10 first for some tools, Pete. I'll be right behind you.' I am sure he wants just a few minutes alone with his thoughts of Pete, and his memories of that day.

Pete played football in college with a friend who had become an officer in the U.S. Marine Corps, a man who is now serving in the war in Afghanistan. A photograph that this friend sent to the Bielfeld family has been framed and stands plainly in sight. It features several military men standing next to an eight-foot bomb that is going to be used in Afghanistan. Printed in big letters across the side of the bomb are the words: TO THE TALIBAN FROM PETE BIELFELD — WHAT GOES AROUND COMES AROUND.

Pete was an extraordinary firefighter. He was severely burned when he worked a flash fire a few years ago, one in which Captain John McDonald died, and after he recovered he took up boxing, just to keep his body limber and to toughen up the damaged skin. On the Friday before 9/11, Pete responded to a third-alarm fire at which he got pretty beaten up. He was taken to the hospital, and then placed on medical leave. He was supposed to go to the medical office on Monday, but got permission to postpone that visit so that he could attend, with the fire commissioner and the mayor, the re-dedication of the newly refurbished quarters of Ladder 42, on Prospect Avenue in the South Bronx. He was at the medical office the next day, when, with several other firefighters, he responded to the plane crash. He went to 10&10, where he borrowed some fire clothes and tools, and must have had some sort of premonition, for he left a note in a firefighter's locker saying, 'This is Pete Bielfeld, Ladder 41. Tell my mother, father, sister, brother and Patti that I love them.'

It begins to rain heavily, and I have to make a run for it to the church steps. In the transept of St. Frances de Chantal I run into Captain Tom Armstrong, who is working with the Ceremonial Unit of the fire department. He has the bearing of an actor, lean, with a perfectly trimmed white mustache. Tom was a firefighter in Ladder 19, one of the companies I had written much about in my first book.

'This is our 200th ceremony,' he tells me, 'either a funeral or memorial service.' He doesn't have to say it, but I can see he is thinking that there are still 143 to be arranged in the future. I meet many friends in the back of the church, including Billy Knapp, the famous nob-man I wrote about in *Report from Engine Co. 82,* and Colleen Roche, the former press secretary to Mayor Giuliani, with whom I take a seat. She is a friend of the Bielfeld family and she whispers the names of everyone from her Throgs Neck neighborhood as they pass down the center aisle. It is reassuring to me to realize that in New York, finally, we are all neighborhood people.

The priest and members of Pete's family speak. Mr. Bielfeld has rewritten 'The Charge of the Light Brigade' so that it reads, 'Into the towers of death rode the three hundred and forty-three.' It gets a standing ovation.

The mayor has gone with the fire commissioner to Phoenix for the sixth game of the World Series, in which the Yankees are playing, and he has sent Deputy Mayor Harding in his place. The mayor has developed something of a public relations problem with firefighters since the confrontation at Ground Zero, and I wonder how the firefighters and the Bielfeld family will respond to the presence of his representative today. In this very church only a few weeks ago several hundred firefighters refused to rise to their feet when Monsignor O'Keefe asked them to give a standing ovation to the fire commissioner.

Mr. Harding, though, is smart enough at the end of his short talk to call for a standing ovation for Pete Bielfeld, something the mayor usually does in his eulogies, and as he leaves the altar the crowd applauses politely but, I sense, a bit dutifully. Next, a lieutenant from Ladder 42 presents Pete's helmet to his father, and when Mr. Bielfeld holds the helmet above his head, like the trophy of a winning athlete who has died young, the entire church jumps to its feet. The roar of clapping, shouting, whistling, and foot stamping can be heard for blocks. This is an expression of honest applause, if I ever heard it.

In the back of the church I see Lieutenant Joe Daley of Rescue 3. He has designed a memorial plaque, a raised

574

relief of the rubble before the cathedral wall of Ground Zero, with a firefighter's helmet lying upside down in the foreground. I think it is a great idea for raising money for the New York Police & Fire Widows' and Children's Benefit Fund, and tell him so. It is heart-rending to see how each of these firefighters is, in his own way, trying to lighten the load of this disaster.

The rain has stopped, and the sun is shining its approval outside of the church. The pipers begin to play, and among them I see Jacky Clark, with whom I played the bagpipes on the *Ed Sullivan Show* in 1970. I feel as close to these bagpipers as to any firefighter I know. They are a legion of heroes within the world of sacrifice. All firefighters are trying to attend these services and funerals, and to show up at as many wakes as they can, but the bag-pipers have made a personal commitment to do so. Of the two hundred services Captain Armstrong spoke of, the pipers have been at each and every one, either as the entire band or in small groups. They vowed on Day 1 that they would give every firefighter's family the respect of hearing the traditional music of a fallen firefighter, the martial music of a battle-slain hero.

Day 56, November 5

I arrived at Ground Zero at 7:30 this morning, two hours before Michael Boyle's memorial service, as I wanted to come down to this site to bid a personal farewell to him. This is where Michael rests.

I feel an overwhelming reverence as I look around today. I recognize that I am privileged to be here, because of the power of my badge and fire department ID card, and I wish I could bring others to stand here with me, the many people who cannot pass the check stations and who would like to join me here in prayer — the family members, friends, and well-wishers of these thousands resting before me.

The south cathedral wall is gone, and the rest of the site has now become a pit. About 60 percent of the north cathedral wall is standing, but the piles before it are gone, and the ruins of building 7 have for the most part been carted away. I am standing on West Street and looking down into the pit where the digging has gone down in some places twenty and thirty feet. But everything within the confines of the slurry wall is now ten feet below the street level. There are eight Cats chewing at the steel and concrete as they have been doing continuously for fifty-five days. There is still sixty feet of debris down to the lobby area, and thousands and thousands of tons to yet dig through and to cart out.

Harry Meyers, the assistant chief of department, said something to me a few weeks ago. We were talking about the gargantuan size of the job to get to the bottom of the piles to recover the missing firefighters. It is a slow job, tedious and sometimes frustrating work, but necessary. 'They came down to the World Trade Center in fire trucks,' as Harry said, 'and we should not let them leave in dump trucks.'

It is a harsh image, and we continue to hope that this does not come to pass. I close my eyes and say an Our Father. It is the daily prayer of the living, but I have always believed it is the one that God hears the most clearly, and so I say it devoutly, with the names of the missing sons of my friends planted firmly in my mind. I know they are here somewhere, whether as dust or bones or flesh.

At the wooden shack housing the firefighters' rest zone is posted a sign on a piece of plane wood, one by two feet, and written in Magic Marker. It says simply:

> Vinny Princiotta
> Ladder 7 – Engine 16
> Rest in Peace – Please pray

I love this wooden, makeshift sign, which conveys so forcefully the independence of our firefighters. I imagine that some firefighters, probably from Ladder 7 and Engine 16, decided on the spur of the moment to formally dedicate this quickly constructed wooden cottage, and now the Princiotta family knows that a building on the dirt of Ground Zero is named in memory of its loved one. It is the first happy thought I have had since arriving here this morning.

In the shack I find a group of nine firefighters, a sullen group, slouched back in their wire mesh chairs. I sense a mood of great disappointment, and then a profound sadness, linked to an anger that I have never seen before. The firefighters are doing nothing, for they are waiting for someone to call them, as if it were downtime at the firehouse. Here at the site where there are brothers buried along with thousands, they are sitting in downtime, restless and furious. 'If this was an African slave burial ground,' a firefighter says to me, 'they would hold up the construction for months if not years, and they would be right to do so. This is holy ground here. And look what they are doing to it.'

'They're carting our brothers,' another says, 'like garbage.'

'Why are we here?' asks yet another who is standing and pointing to a letter-filled wall. 'I'll tell you why, just read the letter. Of the hundreds of letters here, this is the one that gives me a reason to be here.'

The letter to which he refers is in color and includes a photo of a young woman within the text. Her name is Colleen Ann Meehan Barkow, and she is 26 years old. The letter is from her parents to the firefighters, and it reads, 'To all those who worked to return Colleen's remains to us. Having her means so much to our family. Thank you from our hearts.'

Captain Barry Meade of Ladder 162 walks with me to West Street, where we stop to look around. So many construction trailers are lined up all around us that it looks as if there are enough companies and local unions to construct the world's largest dam.

In the middle of West Street stands a large, blue tank truck with three-foot lettering on its side that reads WATER. On top of the truck is a water cannon, called a stang nozzle, and it is shooting a heavy stream into the smoke rising from the Liberty Street side of the pit. Just beyond it are two construction workers wearing yellow helmets, each of whom has a hose in his hand, also watering the rising smoke.

What immediately comes to mind is the fierce pride that is taken by the nozzleman at the fire. He will not give up that nozzle before the fire is out unless he is ordered to, or hurt and carried out tied to a Stokes basket. The stang nozzle on top of the water truck is operated by a remote. The two men operating the hoses here are construction workers, not firefighters.

I am puzzled by this. In the theatrical union, actors are not allowed even to move a three-legged stool from one part of the stage to another without a specific written agreement, but here at Ground Zero there are union men in the construction trades doing firefighters' jobs.

'They are calling it 'dust control,'' Captain Meade explains to me.

'That is an out-and-out unadulterated excuse,' I answer. 'Maybe they could get away with that if that smoke was steam coming from the Little Engine That Could.'

The captain laughs. He knows that the smoke is the product of deep-seated fires that have been burning here for the past fifty-six days. And they will burn for another two months, or until every last brick is turned.

'The men are overtaken with anger because of what's happening down here,' the captain continues. 'This is not a bunch of union rabble-rousers yelling and screaming at a union meeting, when firefighters complain that they are left behind.'

I agree with the captain. The change in policy down here is an attack on the very nature of what firefighters stand for, what we have been dying for, for 150 years in this city. The fire department and police department tradition of bringing out our dead is being abandoned. What civilized army in history left its dead on the battlefield?

I take one final look around the site. Firefighters and police officers stand in small groups here and there, but they are not working. And it dawns on me. Ground Zero not only looks like a construction site, it *is* a construction site.

As I look over this rough field I think back on my early days of firefighting on the streets of the South Bronx. How young and optimistic we were, Ed Schoales, Harry Meyers, and I working together in that Intervale Avenue firehouse. I suppose among the three of us we have been in as many burning buildings as any three firefighters that might have found themselves together anywhere. We went into those burning buildings with a great deal of self-reliance, I am convinced, because we believed that we knew what we were doing, based on our training and experience. And perhaps more important than this, we fought those fires because we had confidence in our colleagues, confidence that they would get us out of harm's way. If our time came and we were lost in the line of duty, as many of our friends were, we knew one thing for certain: Our bodies would not be left for a moment longer than necessary in the rubble of a fire.

At 9:30 I am standing before St. Patrick's Cathedral as the pumper pulls up. On the back are two flower arrangements, like two big floral trays, each of which has a fire helmet centered on it. Michael Boyle and David Arce will be memorialized together in the church that is the center of New York Catholicism. Jimmy Boyle and his wife, Barbara, step out of a limousine and take their place behind the firefighters who are carrying the helmets. Walking beside the Boyles is the family of David Arce. Michael was known as a stand-up kind of guy among his friends, and indeed he was the first to stand up in someone's face if he thought a principle was involved. It was this quality that made him a conse-quential union delegate in his company, and why everyone who knew him believed he would have gone on to be a great labor leader – just like his dad. He had that same wonderful soft spot in his heart that came from his parents. The family took in the homeless at Thanksgiving, visited every sick firefighter who ended up in a hospital, and worked tirelessly to take care of the widows and children who depended on their support. Michael worked with his father, and David worked with Michael, and it was their chain of good works that everyone in the fire department came to know.

The crowds of firefighters and dignitaries stretch for blocks up Fifth Avenue, a tribute not only to Michael and David, but also to the idea of continuity in the fire department. I look at Jimmy and reflect that fathers and sons have been serving in New York's fire and police departments since the city first incorporated them in the middle of the nineteenth century.

I have loved Jimmy Boyle since the day I met him. I can make that statement with as much certainty as stating that I love my children. When you become acquainted with someone who is fundamentally other-directed in his worldview, it is as if the sun bursts through a rainstorm for just a moment, and you remember it forever. St. Francis of Assisi changed the world with his simplicity, sanctity, and gospel of love, and I think that our own world could benefit from learning from the example set by Jimmy Boyle, Lee Ielpi, John Vigiano, and these many firefighters I have been watching these past hard weeks. Inspiration can come on the wings of angels, but it is more direct and more immediately appreciated when it comes from men like these.

By now you might expect the mourners to be spiritually and physically exhausted after having attended so many memorials and funerals. There are four fire-fighter memorial services today and one funeral. There are three memorials tomorrow and another funeral. This week thirty services have been scheduled for thirty fire-fighters: Boyle, Arce, Mahoney, Bedigian, Chipura, Margiotta, Russo, Pearsall, Bracken, Anaya, Fischer, Watson, Farino, Tirado, Bucca, Crawford, O'Callaghan, Halloran, Cordice, Allen, Barnes, Phelan, DeRubbio, Gullickson, Mercado, Princiotta, Spear, Powell, Haskell, and Brown. Brown. Paddy Brown. It is especially hard for me to come to grips with the passing of Paddy, the fireman's fireman, the object of affection of all who knew him, men and women. Like John Vigiano, Lee Ielpi, and Jack Pritchard, Paddy had a chestful of medals and he wore a modest heart. There would be no lasting pages of history in our department without men like these. And now Paddy won't be there any longer, first due at a fire and first to arrive at a burning door. It is as if Park Avenue had been lifted right out of the city, and though you could still travel uptown and downtown, the trip wouldn't be as memorable without it. Our future fires will be fought by good men, but somehow it won't be the same without Paddy.

The mayor speaks, followed by the fire commissioner. The biggest ovation comes when it is said of Jimmy Boyle that nobody knows how to treat firefighters in need the way he does. Everyone in the cathedral jumps to his or her feet in great, serious applause. In his speech the cardinal suggests that the firefighters climbed the hill of Calvary, like Christ, to their deaths. And, like Christ, they cared for others more than they cared for themselves.

Our former governor Hugh Carey spoke for Governor Pataki. Governor Carey just lost a son for the third time in his life, and I can see how he empathizes with these parents before him. 'Death is not an ending, but a continuation of the spirit,' he says, for he himself sees his departed sons in the lives of his eleven remaining children and many grandchildren, just as Michael Boyle will be remembered every time a man or a woman puts on a fire helmet to answer an alarm.

Whenever firefighters respond, whether to an auto-mobile accident, a flood, a hurricane, a suicide, a collapse, a terrorist explosion, or a fire in the largest city or smallest town, they will do honor to the spirits of all those who ended their earthly days at Ground Zero on September 11, 2001. I know that every time I hear an alarm or a siren for the rest of my days, I will think of the three thousand civilians who were killed, and then I will think always of every name that is found within these pages. I will say an Our Father, and I will thank God for the time I have been granted in the presence of heroes.

I am standing at the edge of the crowd outside the cathedral when Jimmy Boyle passes. I think back to fifty-five days ago when I embraced him on Vesey Street in the smoke and dust of the collapse. I wish he could have found Michael that day in that terrible haze of des-truction, but it was not to be. Now, we will go forward, but there is no end to this chapter of our history, at least for those of us living today.

Day 63, November 12

Marian Fontana recently founded the September 11th Widows' and Victims' Families Association, and she and Lee Ielpi and more than seventy-five families met tonight with the mayor and the fire commissioner. 'The meeting,' Marian said, 'was very, very difficult. The firefighters and the widows are so, so angry, and I feel sorry for Mayor Giuliani and Thomas Von Essen, who were called so many names. This anger is so great. I didn't realize it is so great.'

I remember Vina Drennan's comment that the fire commissioner ought to have a bodyguard, which is borne out by Marian's observation.

'You know, Marian,' I say, 'we go to so many memorial services and funerals, and we all say the Lord's Prayer with reverence, but it seems in the firehouses that the firefighters are not yet ready to forgive trespasses.'

I know that Marian identifies closely with this sentiment, for her husband, David Fontana, of Squad 1 in Brooklyn Heights, has already been memorialized, and I know that she, like all the widows, still yearns and prays that her husband's remains will be found.

'No,' Marian says. 'I don't think so.'

One woman said to the mayor, 'Last week my husband was memorialized as a hero, and this week he's thought of as landfill?'

The protest rally on November 2 resulted in the arrest of eighteen firefighters, which unfortunately has only served to increase the tensions between the firefighters and the city. The International Association of Fire Fighters canceled a huge memorial service that was planned for November 17, and although the mayor relented and announced an increase in the number of firefighters at the site, the resentment was not stemmed. And then when Marian Fontana started this association of families and victims, she removed the issue completely from the labor-management arena and placed it squarely as a matter between the mayor and the families. As such, the mayor's view was suddenly perceived as opposed to that of the widows.

To Mayor Giuliani's credit, he initiated this meeting at the Sheraton Hotel tonight to discuss the city policy at Ground Zero with the families of the missing fire-fighters. Jimmy Boyle is also in attendance, along with Jonathan Ielpi's widow, Yesenia, and his father, Lee. It is in the same room of the Sheraton where the reception after the memorial service for Michael Boyle was held.

I know that Mayor Giuliani does not believe that his policies represent anything but sensitivity to the families of firefighters, and that he decided to reduce the number of firefighters at the site of Ground Zero for safety reasons, but the families do not see it that way, and do not share his concerns about the timetable for clearing the site. Jimmy Boyle reported that the wives were 'chain-smoking mad,' and demanded that the same number of firefighters be present at the site as had been there since Day 1 of the recovery effort – case closed.

The mayor and the fire commissioner said little, but let the families vent their feelings and anger.

Lee Ielpi happens to be one of the most decorated firefighters in department history, as well as a family member, and has been adamant about changing the city's policy – now derisively referred to as 'scoop and dump.' He stood before the mayor and began to speak. He said that his son Jonathan was still buried in the piles, in the area of the Vista Hotel. His other son, Brendan, is also a firefighter. Jonathan has left a young widow, Yesenia, and two little boys. Lee told the mayor that his family was devastated, and that his wife has been robbed of a son. His daughter-in-law, Yesenia, was robbed of a husband. Lee told of how he has been digging with his friends almost every day since September 11. He was optimistic at first that there would be many rescues, and was saddened that there were so few. He was disappointed when the mission was changed from search and rescue to search and recovery, but recognized that the recovery of the remains of firefighters and police officers and other victims was vitally important.

The families had been told bluntly that many of the bodies had disintegrated in the disaster. Lee did not challenge that information, but as he explained, the issue was the remains, for there were still remains to be found. 'I believe in safety, and do not want to see another firefighter lose his life,' he said. 'But I know you can put more fire-fighters there, and you can run a better operation.'

Lee suggested that the mayor remove the huge red crane that stands in the middle of the site, gobbling debris in such large bites that it is impossible to spot the area for remains. It is too big and too indiscriminate in its function.

Mayor Giuliani was moved by the force and in-telligence of Lee's speech, and agreed to send Lee to the site with the fire commissioner to determine the best way to recover the remains. He also agreed to increase the force of firefighters from twenty-five to seventy-five and also to remove Big Red.

Later Lee told me that he wished the meeting with the families could have taken place two weeks earlier, for a lot of pain could have been avoided. I hope with all my heart that all's well that ends well. One cannot find a way to understand this issue from the point of view of the firefighters, the families, the unions, or the mayor without acknowledging Lee's sentiments: 'It was a subject so monstrous,' he said, 'we didn't know how to deal with it.'

Day 68, November 17

Kevin Shea of Ladder 35 appeared on the *Today* show this morning. I had asked Kevin to do the New York Police & Fire Widows' & Children's Benefit Fund a favor by appearing on the program and telling his own extraordinary story, as well as presenting the Beanie Baby dogs 'Rescue' and 'Courage' to the television audience. It was gratifying to see so many of these little puppies in the hands of the tourists who stand outside of the NBC studios when the *Today* show airs. Kevin gave the show a rationale to present the commemorative toys, and because of his goodwill we will no doubt receive many contributions in our effort to sell them.

Kevin Shea's story might be the most blessed of all. A young man, just 34, Kevin is the only surviving member of the entire 9th Battalion, which responded to the World Trade Center on Day 1 – of the thirty-three who died, eleven were from Kevin's firehouse, Ladder 35 and Engine 40. When he was pulled out of the wreckage from the collapse of the south tower, he was barely conscious, with three fractures in his neck, and a severed thumb. He was discovered lying between two burning pillars, and had no memory of the collapse. He remembers only going to the fire and then waking up in a New Jersey hospital bed with his family in the room. In any national defense emergency situation there is a civil defense plan in place that suggests a response for every contingency. The plan for the WTC was to take wounded emergency workers, police, and military personnel to pre-selected hospitals on the New Jersey side of the Hudson. This would not only prevent overloading at local hospitals but would ensure immediate attention for wounded first responders. Kevin was taken by police launch across the river and then by ambulance to the Newark Medical Center. There, the doctors worked feverishly to repair the broken vertebrae in his neck with a donated patella bone from someone's knee.

Kevin appreciates his fortune, and has dedicated himself to do anything that helps the brothers and their families. But this sensitivity is not merely a response to the events of September 11. A year ago, with his brother Brian, Kevin founded a Web site called fallenbrothers.com, which reports on line-of-duty deaths of firefighters all over the country, and sells T-shirts that commemorate them. All the profits from these efforts go to the widows and the children of the firefighters.

When the sky grows dark at the end of the day, and

as each week and month closes, September, October, and now November, I have to ask myself how, I, personally can come to terms with this worst attack in our history.

I think naturally of the question that most people, when discussing this tragedy, ask: What makes people run into buildings that everyone else has run out of? The answer can be found only in something called character, and that takes a lifetime to develop. I cannot speak for the police officers' institutional culture, but I did live in a firehouse for eighteen years, and I am familiar with the elements that build the courage and determination that enable men to step into danger. There is a shared historical memory that is in the atmosphere of every firehouse in the city, and it both preserves and renews the culture of the firefighter. The fires and then the funerals of brothers who have been killed in the line of duty are discussed over meals in the kitchen, along with the softball, hockey, and handball games, all part of the good-natured rivalry that exists between neighboring fire companies. That culture also includes Christmas parties, where the children of the firefighters come for presents and cookies, the way they might go to their grandparents or other relatives. It includes the annual company picnic, where the kids play with children of other firefighters, the way they would with cousins at a family picnic. It includes the annual company dance, a formal dress-up dinner, where the firefighters and their spouses celebrate their joy and humor and friendship, in a spirit much like that of a family wedding.

The firehouse is, fundamentally, a family environment,

and as in families, there is a code of behavior that its members are bound to honor. A firefighter is expected to help with chores like cleaning and tool maintenance, just as he is expected to know his duties on the fire floor or above the fire floor in a fire. He knows where he is expected to be, and he knows the consequences if he isn't there.

The firehouse is also about brotherhood and a strong, unquestioning relationship between the men (and women, in a few firehouses). If a firefighter falls into danger in a burning building, there is only the man next to him who will save his life, and that dependence is the unwritten code that binds them.

From his first day as a probationary firefighter, he understands that it is the older guys, the veterans, the first whips, who keep the job focused and unified, who serve as role models. It is easy to spot them: They are always the first to put themselves forward. They are what the young firefighter hopes to become. I cannot think of a more humbling experience than to stand among men like Lee Ielpi, Marty McTigue, and Pete Bondy, for they are not only the vessels of our historical memory, but are also our leaders. Knowing this firehouse culture, I understand fully why they rushed into the towers. Some went in with obvious trepidation. Pete Bielfeld left a note for his family. Terry Hatton is said to have met a friend in the lobby of the north tower, hugged him, and said, 'Take care of yourself, brother. I don't know if I'll see you again.' Apprehensive or not, 343 firefighters, 23 police officers, and 37 Port Authority police did not question their duty.

What is more difficult to understand is why they did not come out.

The media offers its predictable bromides, and has been telling us daily that we are changed forever. These assertions remind me of William Butler Yeats, who when he saw what he believed to be a catastrophic event in 1916, proclaimed that 'all changed, changed utterly,' and that 'a terrible beauty is born.' But this provides little comfort in coming to terms with the profound atrocity we have lived through – the first attack in the modern era on American soil.

Even in the best of times, we change a little from day to day. Here, at Ground Zero, the scale of frightfulness is what we must ponder, for it is the greatest warning the world has ever seen, a warning about the desperate measures that terrorism is willing to take.

Months later, people still remain incredulous that September 11 occurred. Lee Ielpi has told me that he simply doesn't believe it. He doesn't believe he will never see his son again, or that so many people have perished, or that so many firefighters are buried in the piles. 'Even a year from now,' he says, 'I won't believe it.'

Perhaps others have already settled the matter in their minds, and take comfort in the closure, but I still struggle to find a settled breath. If I could come to understand why this happened and why these men were lost, then perhaps all those I have spoken with during this time could do so, as well. But it is so hard. Finally, we all have to think and speak to ourselves, and as for me, I believe that the answers can come only in the little visits

I have from time to time with my spiritual connections.

Americans do not wear their religion on their sleeves. The reason for this lies, I think, in our fiercely independent nature, but it is also a product of our founding fathers' belief that a society is better off when church and state are separated.

In my local church recently the responsorial psalm was: 'I rejoiced when I heard them say, "Let us go to the House of the Lord."' I do rejoice when I see people going to a house of worship, any house. I do not, however, advise my children, friends, or anyone else to follow my example. People must do what they feel in their hearts, and what their education and intelligence prescribe. For myself, I believe that having a spiritual texture to life elevates my view of life and the people around me. It gives me patience in a demanding world, and it helps me to seek peace. I have often thought that my work as a firefighter has exposed me to so much of death and misery that it has tempered my anger. Instead of lashing out in sarcasm and ire, as once I did, my first inclination now is to try to understand the darkness of stupidity, and to guide it toward some kind of light. I also try to recognize that sometimes, the stupid and intransigent one is I. Understanding is a necessary first step in trying to end any misery. I think all firefighters, and all first responders, would gladly suffer with someone if it would end his suffering. It is in our nature, not only as first responders, but as Americans.

Our Judeo-Christian philosophy urges us to protect ourselves, our families, and our freedoms, while at the same time encouraging us to love our enemies. As W. H. Auden said in his poem 'September 1, 1939':

We must love one another or die.

The heroic souls listed in the dedication of this book were living lives of bravery because they believed deeply that they were here on Earth to help others. To me that is a truth based on the teaching that we must do unto others as we would have them do unto us, but though all religions expose some version of this ideal in a theological sense, it often doesn't filter into our day-to-day living. I know only this, that if the philosophy of 'doing unto others' truly illuminated all religions, we would have a better world.

All writers hope their work to be of some consequence, whether it is fame, or wealth, or social change. With this book, I simply wanted to record the efforts of great men and great women. Books will survive the movies, television programs, and magazine articles written about our world today. There are so many stories of courage and duty among the 343 firefighters, 23 police officers, and 37 Port Authority police officers – 403 such stories. And there are also stories of civilian workers in the towers going dangerously out of their way to help others. Even if it were possible to tell all of them, I do not think I have it in me to relate them, and in that I have failed. I knew so few of these nearly three thousand victims, yet I feel as if part of my family has sailed off on a ship, never to return. But I have tried to present an honest and direct account of this awful time in a sensitive way, and I hope that all the lives that have been lost will be honored by the stories found here. And that the wives and husbands, parents and children of the families left behind will find in these pages the heroism of their loved ones who no longer share their dreams.

Though he did not mention it specifically, I believe Seamus Heaney had our collective future in mind when he wrote:

> . . . No poem or play or song
> Can fully right a wrong
> Inflicted and endured . . .
>
> History says, don't hope
> On this side of the grave.
> But then, once in a lifetime
> The longed-for tidal wave
> Of justice can rise up,
> And hope and history rhyme.
>
> So hope for a great sea-change
> On the far side of revenge.
> Believe that a further shore
> Is reachable from here.
> Believe in miracles
> And cures and healing wells. . . .

I think of these many days that have passed since September 11, and I think of all I have seen and heard.

This awful experience has brought me closer to everyone I know, for the wonders of friendship and love have become so much more appreciable now. I hear the voice of love and friendship, and it is the voice of Ground Zero. It is the voice of kindness and generosity, determination and strength. It is the voice of tragedy, sadness, and grief, and the voice of inspiration. It is, finally, the voice of America.

And that voice continues to be heard loud and clear throughout the world.

Epilogue

On December 10, the 91st Day, my wife, Katina, and I invite a group of friends over for dinner, and the first to arrive is Michael Angelini. I am glad he has come before the others, for it gives me an opportunity to talk with him about a list of appeals that can be made on his behalf to get him on the job. He is a fire patrolman, and for someone who wants to be in the fire service, the fire patrol is the next best thing to the FDNY. But Michael still has a dream. When the list for the New York Fire Department expired, he was just one hundred names away. Although you can be appointed to the department up until you are thirty-five years old, you cannot take the test if you are over 30, and so Michael, now 33, was left behind. It is his life's ambition to be a firefighter, like his father and his brother. I personally can't think of anyone more deserving of the honor of wearing a New York firefighter's badge than a man who lost his father and his brother in the line of duty, serving and protecting the people of New York.

Jimmy Boyle came next, and predictably committed himself to doing everything he could to help Michael. Surely there was a provision somewhere that would grandfather Michael back onto a list, or perhaps a special legislative appointment could be made. Maybe. Maybe. All we can do is hope – and make a few calls.

Lee Ielpi arrives next, with his usual companions, Marty McTigue and Pete Bondy. They have come straight from Ground Zero, and they are upbeat. Several missing firefighters from Squad 288 were found at the site today, and they were all together. They were not where Lee had thought, beneath the Marriott Hotel, but over in the footprint of the south tower.

After dinner, Michael is the first to leave, and I tell the others about how his sister-in-law, Donna, had called and asked him to sit with her when she told her children that their father and grandfather were missing. When I get to the part where Donna says to the children that their father is missing, and the 7-year-old says, 'Oh, Pop will go in and find him,' I notice there are tears in Lee's eyes. He is silent a moment, and then says, 'It makes me think of my little grandson Andrew, only ten. He came into my room every morning for weeks after 9/11, and he would say, "Will you bring my daddy home today, Pop?" And, then suddenly, he just stopped asking, and that made me very sad.'

On December 12, the 93rd Day, I get a call from Pete Bondy. They found Jonathan. I call Lee immediately. 'They found two guys in the south tower, last night,' Lee says.

■ Jonathan was there, and another fellow from 288. Jon was outside of the footprint, though, but right in that line of the stairs where we thought. They called about 11:30 last night, then the phone didn't stop, but they waited until my son Brendan and I got there. You know, I am not a superreligious guy, but listen to this. When I get off the phone, I say I have to get Brendan. He's assigned now to Ladder 157, and I know he is not working tonight. It's about a quarter to midnight. I call him, and I get his girlfriend. She tells me, 'Brendan just went to Great Neck.'

I say, 'Brendan went to Great Neck at 11:30 at night?'

She says, 'Yes, he went to the firehouse. He wasn't tired, and he thought he would just go there.' [All the Ielpis are members of the Vigilant Engine Hook and Ladder Company.]

So I call the firehouse, and they don't know where he is. I tell them if he comes around to call home immediately. With that he pulls up the driveway. Now he has never done that. Never, period. Not at that hour of night, and in these circumstances. He just walks in.

He says, 'Hey, Dad.'

I say, 'Brendan, did somebody call you?'

He says, 'No. What's up?'

So, we go down, speeding all the way, and they kept Jonathan and the other firefighter at the bottom of the hill until we got there. When Brendan and I arrived, they lined everybody up, all the firefighters on each side, helmets off, and we carried him off. ∎

This image of Lee and his son Brendan carrying Jonathan through the midnight mist, over the dirt and mud and buried souls of Ground Zero, and between the lines of saluting firefighters, is for me the most refined definition of familial love that can exist in a fire department.

The funeral is held in Great Neck, in St. Aloysius Church, on a softly cold day. A light rain falls upon the two or three thousand firefighters lined up in the street. Nearly the entire bagpipe band has shown up for this funeral, and the bright red of their uniform jackets sparkles in the wet air. This is Lee Ielpi's son Jonathan. Everyone knows that Lee has worked tirelessly for ninety-three days to bring his son home. In him we have seen the standard of behavior for a bereaved father and a dutiful firefighter. God has answered his prayers.

I think about Ed Schoales still searching for his Tommy, and Rosemary Cain waiting for word about her Georgie, and Marian Fontana waiting for Dave, and David Rosenberg hoping they will find his brother, Pete Vega. I close my eyes in remembrance of their love and their yearning, and I say the prayer I have said each day for these families. I think now of John Vigiano II, James Riches, and young Dennis O'Berg, and their brave and stalwart and relentlessly searching fathers, and I say another prayer. And I remember, finally, the others, the over two hundred firefighters, fifteen police officers, and the more than two thousand victims who have not yet been found, and perhaps who will never be found. 'That will be the saddest day,' Lee has said to me, 'when the site is cleared and the last stone is turned, and we have to tell all those families there is nothing more to find.'

The story of Ground Zero will not be over, though, for over twelve thousand body parts have been recovered, which will take months of DNA analysis before the identifications can be determined. Rudy Giuliani arrives, and with Fire Commissioner Von Essen he follows Lee, Anne, Yesenia, Andrew, Austin, and Brendan into the church. Today's service is different from any memorial or funeral I have attended since 9/11, for no one has been asked to speak.

Lee told me that Jonathan had two dreams. He wanted a house with a large backyard for Yesenia, Andrew, and Austin. And he wanted a transfer into Rescue 2. He would have had both – no one would have doubted that – if he had had more time. He loved being with his family, and he loved those days being with his father in Rescue 2, days that molded that young man into the great firefighter he had become.

Lee Ielpi gets up, the twenty-seven ribbons for heroism prominent on his chest, and he tells us all that he will be able to speak before us about his son Jonathan, and that he will also be able to cry before us. And he does both.

He ends his talk with these words, words that could be spoken on behalf of each man and woman of the 403 emergency workers who went in to help others out, and of all the rest who have perished:

'I know he is in good hands. . . .
'But I wish he was in my hands.'